CLIMATE CHANGE RESEARCH PROGRESS

Climate Change Research Progress

Lawrence N. Peretz
Editor

Nova Science Publishers, Inc.
New York

For permission to use material from this book please contact us:
Telephone 631-231-7269; Fax 631-231-8175
Web Site: http://www.novapublishers.com

Library of Congress Cataloging-in-Publication Data

Climate change research progress / Lawrence N. Peretz (editor).
p. cm.
ISBN-13: 978-1-60021-998-6 (hardcover)
ISBN-10: 1-60021-998-5 (hardcover)
1. Climatic changes--Research. 2. Global warming--Research. I. Peretz, Lawrence N.
QC981.8.C5C51968 2008
363.738'74072--dc22

2007035860

Published by Nova Science Publishers, Inc. ✣ New York

CONTENTS

PREFACE

Climate change refers to the variation in the Earth's global climate or in regional climates over time. It describes changes in the variability or average state of the atmosphere over time scales ranging from decades to millions of years. These changes can be caused by processes internal to the Earth, external forces (e.g. variations in sunlight intensity) or, more recently, human activities. In recent usage, especially in the context of environmental policy, the term "climate change" often refers only to changes in modern climate, including the rise in average surface temperature known as global warming. This new book presents leading-edge research in this fast-moving field.

Chapter 1 - This paper is a synthesis of main recent results the authors obtained jointly with his colleagues from the Limnological Institute (Irkutsk). The authors suggest a conceptual model to reconstruct the Pleistocene history of humidity and environment changes in the watershed of Lake Baikal (East Siberia). The climate events are recorded in high-resolution patterns of more than 25 elements, diatom abundances and compositions of diatom assemblages, sediment physical properties, etc., as well as in the behavior of U and Th isotopes the authors measured in a ~1 Myr long drill core of Baikal bottom sediments and in cores that span the past 300 kyr. There are downcore element profiles that either correlate or anti-correlate with the biogenic silica record. The concentrations of elements in the former group are high in interglacial and low in glacial periods. Spectral analysis of the Baikal records reveals all orbital frequencies of 96, 41, 23, and 19 kyr, and also a distinct 72,-54, 35 and 30 kyr rhythms. The response of the continent to insolation is more prominent than the marine response because power of spectra in the tilt (41 kyr) and precession (23 kyr) bands of some proxies are as strong as or even more the nonlinear 100-kyr cycle.

Direct U-Th dating of all interglacial intervals down to MIS7.1 shows an almost synchronous response of Lake Baikal to insolation forcing which slightly leads the more inertial global ice volume signal.

The authors use the geochemical behavior of U-series isotopes to simulate the paleoflow history in the Baikal watershed for the past 100 kyr, within the limits of the authors conceptual model, proceeding from the mass balance of soluble uranium the tributary rivers carry to the lake. The river flow appears to have been vanishing in dry glacial maximums, which is supported by records of clastic-related elements in sediments.

The core interval spanning the past 150 kyr has been investigated to a decadal (~ 20 yr) resolution by SRXRF scanning at 0.1-1 mm, applied to 8000 horizons. The Baikal watershed shows responses to all Greenland D/O events through the last glacial and short spells of abrupt cooling that intervene all interstadials of MIS 5 (MIS 5.1, 5.3 and 5.5). The clear response of Siberia to the North Atlantic events suggests their global distribution over the Northern Hemisphere.

Chapter 2 - Awareness of the problem of climate change has boosted the implementation of climate policy instruments at the international (i.e., Kyoto Protocol) and national levels in order to tackle greenhouse gas (GHG) emissions mitigation. In turn, this has stimulated research by various social science disciplines, including economics, on the analysis of policies and measures aimed at GHG mitigation. This literature has focused on the functioning and design of the instruments, on the assessment of different instruments according to several criteria (environmental effectiveness and economic efficiency in emissions reductions) and on a comparison of their respective merits and drawbacks. This paper provides a critical overview of the economic literature on climate policy research, indicating the main lines of research and key results from this fruitful literature. It also proposes several avenues for further research, particularly concerning issues to be tackled and methods to be applied. In this regard, the analysis of the interactions and complementarities between different climate and energy policy instruments, the inclusion of technological change in climate change mitigation and the ancillary benefits are particularly worth undertaking.

Chapter 3 - Reconstructing climate over the last millennium is one major target of climate research. The extent of climate variability over recent centuries sets the scale on which ongoing climate changes can be measured. The quality of reconstructions of millennial temperatures in the northern hemisphere has greatly improved over the last decade (*IPCC*, 2007). Nevertheless there is still a lack of precisely dated, high-quality climate reconstructions in the tropics, a lack which is particularly disturbing since this region is a major climate engine for the global water cycle. In recent years, a significant effort has been made to improve the authors knowledge on climatic variability in the tropical Andes by developing new proxy records, new transfer functions and better time resolution of available proxy records. Hence, the aim of this paper is to review the different proxies with multi-decadal to annual resolution available in the tropical Andes (Bolivia, Peru, Ecuador) such as glacial moraines, pollens, isotopic variations in ice core records, and tree rings. For each proxy, the authors focus on recent major methodological progress and on new results revealing climatic shifts over the last millennium.

Chapter 4 - This chapter presents an overview of the application of inorganic geochemistry in peat and lake sediments as a tool for paleoclimate studies. A brief look at the important dynamics occurring in lakes and mires highlights the fact that the authors are examining small-scale processes to deduce large-scale changes. A summary of analytical techniques, including sample preparation and instrumentation are presented and important statistical and interpretive tools are explored. Specific examples from recent publications demonstrate the practical application of these tools and the paleoclimatic interpretations that can then be drawn.

Chapter 5 - Climate changes such as global warming, decadal climatic regime shifts and interannual variability of the ENSO phenomenon can affect ocean conditions, and thus affect the functioning of marine ecosystems and the distribution and abundance of fisheries resources. There has been an increasing trend in sea surface temperatures in the last 100 years which has accelerated in the last decade in the marginal seas of the Northwest Pacific. The recent warming trend in the Korean marine ecosystem is associated with a decline in cold-water species (e.g. walleye pollock) and an increase in warm-water species (e.g. squids, jellyfish, mackerels). There were also some indications of decadal regime shifts in 1976/77 and in 1988/89 in the region. Thus, it is anticipated that climate change will affect the productivity and distribution of species important to Korean fisheries.

There are more than 300 fish species in the marginal seas of the Northwest Pacific of which about 100 are of commercial value. Most commercially important species can be categorized into three fish communities. A demersal community (small yellow croaker, hairtail) exists in the Yellow Sea, and in the East/Japan Sea and the Okhotsk Sea (walleye pollock, Pacific cod). A pelagic community (mackerels, squids, anchovy) exists in the East China Sea. Overfishing has been a serious threat to productivity, resulting in a higher proportion of smaller, very young individuals. Climate related ocean changes also profoundly affect productivity. There has been a decline in the catch of demersal species and an increase in the catch of pelagic species. Chub mackerel, Pacific saury and common squid abundance increased in periods of weak Aleutian Lows and a negative PDO. In contrast, filefish and Jack mackerel productivity declined during these periods but increased when the regimes shifted to strong Aleutian Lows and a positive PDO. Pacific sardine have a well-known climate related alternating cycle of abundance. Warm sea surface temperatures reduce the marine survival of chum salmon fry and restrict the southward migration of walleye pollock. Each ecosystem responded differently to climate impacts under different exploitation systems. It is recognized that future climate change will have important impacts on marine productivity that will require developing a holistic ecosystem-based approach to fisheries management. This requirement will trigger integrated multidisciplinary research in climatology, oceanography and fisheries sciences to understand the impacts of climate changes on marine ecosystems and fisheries productivity.

Chapter 6 - Simulations of the climate--carbon cycle interaction are discussed in comparison with observationally--based estimates for the global carbon cycle characteristics. Since the beginning of the industrial era, the storage of the carbon dioxide in the atmosphere is smaller than the corresponding anthropogenic emissions. This is due to uptake of the atmospheric carbon dioxide to the terrestrial biota and ocean. Moreover, during the the 20th century, the sink of the carbon dioxide from the atmosphere to the terrestrial ecosystems became larger, due to CO_2 fertilisation effect. However, in the 21st century, the global climate models with the carbon cycle project that interactions between climate and carbon cycle basically lead to the stronger growth of the carbon dioxide burden in the atmosphere. These interactions were in the focus of the Coupled Climate--Carbon Cycle Model Intercomparison Project (C^4MIP). One of the basic outcome of this and related activities is positive climate--carbon cycle feedback leading to enhanced buildup of CO_2 in the atmosphere due to response of climate and corresponding changes in the terrestrial and oceanic uptakes of carbon. In particular, the soil respiration growth over compensates the increase of the net primary production and leads to the diminishing sink of the carbon dioxide from the atmosphere top the living biota and soil. In some models and emission scenarios the terrestrial biota even eventually becomes the source of the carbon dioxide. The parameter of the feedback between climate and carbon cycle changes non-monotonically in the 20th and 21st centuries, depicting characteristic periods of the emission growth and the respective climate response. On a century timescale, climate--carbon cycle feedback may saturate. The 20th century observations of the carbon cycle are insufficient to constrain future evolution of the coupled carbon--cycle system. Nevertheless, they are able to narrow the respective uncertainty range. This can be illustrated, in particular, by employing the Bayesian statistical treatment of the respective ensemble. However, for the most probable values of the governing parameters of the coupled system, climate--carbon cycle interactions enhance global warming in the 21st century by about 10% or even more under the SRES marker emission scenarios.

Chapter 7 - Climate change has already affected many aspects of the earth ecological system. Population health should be one of the most concerned. However, the relationship between climate and population health, including both chronic diseases and infectious diseases, is far from clear. A critical review of previous studies on the relationship between climate change and population health has been conducted with focuses on research methodology, including study population, study design and statistical method, and their results. Further research trends on climate change and population health will be addressed based on the critical review.

Chapter 8 - Available evidence suggests that the world is getting warmer and this is expected to have adverse impact on agriculture in the future. Global circulation models predicted that global warming will lead to increased temperatures of up to 8.7^{o}C and cause variability of rainfall by up to 34% in Kenya by the year 2100. From these predictions, the two extreme climate events that may adversely impact on the agricultural sector are drought and flooding in both the arid and semi-arid lands (ASALs) and the higher potential areas. The major challenges to sustainable agricultural production in Kenya are low, poorly distributed and unreliable rainfall, climate and ecological extremes. The impacts of the extreme weather and climate events could have severe socio-economic impacts including water and energy shortages, and also shortage of other essential basic commodities and long term food insecurity. Previous studies show that the country will incur huge losses in farm revenues due to climate change. This makes it important to understand adaptive mechanisms by farmers to circumvent the potential welfare loss due to climate change in Kenya. To address this issue, this chapter examines the impact of climate change on crop selection and also assesses farmers' awareness of climate change. The chapter also examines the main adaptation measures employed by farmers to counter adverse impacts of climate change, upon perceiving climate change. The results suggest that the decision to farm a crop or a portfolio of crops responds to climate change and that the choice of most crops is more sensitive to temperature than to precipitation. Further, choice of most crops exhibit a significant hill shaped relationship with temperature and precipitation.

Chapter 9 - To ensure future world energy supplies and to reduce carbon emissions an urgency is developing to find replacements for fossil fuels. Renewable energies have serious limitations that prevent a significant increase in their contribution to replace fossil fuels. This paper examines the technical reasons why the renewable energies, i.e., solar, biomass, wind, and hydro, currently supply only about 9% of current world energy demand. This percentage dropped from almost 100% three hundred years ago and is still falling. The limitations of renewable energies are such that they are useful in only relatively small niche applications. Hydro is no exception even though it provides electricity when required, is about 75% of total renewable energy, and currently contributes about 7% of world primary energy demand. Research funds must be withdrawn from renewable energies and directed towards areas with potential to replace significant quantities of fossil fuels.

Chapter 10 - Climate models are the main tools used for predicting the effect of global warming. Recently, these models have developed an increasingly sophisticated ability to simulate tropical cyclones, a very demanding task due to the relatively small size and the intense convection of these storms. This paper gives an evaluation of the state of the art of tropical cyclone climate modeling, including an examination of the authors current ability to simulate the correct numbers, structure and intensities of cyclones. The shortcomings of present models are outlined as well as specific recommendations for ways in which they can be improved. Additionally, a summary of current model-based predictions of the effects of global warming on tropical cyclones is given, including an assessment of how the uncertainty of these predictions might be reduced through improved simulation.

SHORT COMMUNICATION

In: Climate Change Research Progress
Editor: Lawrence N. Peretz, pp. 3-9

ISBN: 978-1-60021-998-6

ADVANCES IN METHODS FOR ESTIMATING SOIL ORGANIC CARBON STOCK CHANGE AT REGIONAL SCALES IN DEVELOPING AND EMERGING COUNTRIES

Eleanor Milne
Department of Soil and Crop Science, Colorado State University,
Fort Collins, Colorado, USA

Abstract

The United Nations Framework Convention to Combat Climate Change (UNFCCC) requires all signatory countries to report terrestrial stocks of carbon (C), including C in soils. Such estimates are important for countries wishing to take part in an emerging C market and are particularly relevant to developing or 'non-Annex 1' countries looking to benefit from C enhancing land use changes under clean development mechanism (CDM) projects. Recently, advances have been made in the methods and tools available for use by developing and emerging countries to estimate soil organic carbon (SOC) stocks and stock changes at the regional scale, under different land use and climate change scenarios. Examples include one study that adapted the empirical IPCC (Intergovernmental Panel on Climate Change) method to produce reference C stocks for conditions found in the Indo-Gangetic Plains and Uruguay. Other projects have used dynamic SOC models linked to spatially explicit datasets from developing and emerging countries via GIS. One such project, the Global Environment Facility Soil Organic Carbon (GEFSOC) project, developed a system that makes national SOC stock change estimates using two SOC models (RothC and Century) and one empirical method (the IPCC method). Other sub-national assessments have employed different models such as the Denitrification and Decomposition Model DNDC. These recent advances are discussed in terms of their benefits and constraints and the possible future direction of SOC stock estimates in developing and emerging countries.

1. Introduction

Soils can act as either a sink or a source of carbon (C) under different land management practices, and can therefore lead to the mitigation of green house gases (GHGs) through C sequestration. Batjes (1999) estimated that, with changes in land use and appropriate management of agricultural lands, grasslands and forests, soils have the potential to sequester 14 ± 7 Pg C, globally, over the next 25 years. Methods of quantifying these stocks and changes at a variety of scales are needed to fulfil national reporting obligations under the UNFCCC (United Nations Framework Convention to Combat Climate Change) and to provide information for developing or 'non-Annex 1' countries looking to benefit from C enhancing land use changes under clean development mechanism (CDM) projects.

Until relatively recently, methods for estimating SOC change over time have relied on empirical methods. The Intergovernmental Panel on Climate Change (IPCC) has developed a computational method for estimating SOC stock changes that can be used at the national and sub national scale. An updated version of the method is described in the IPCC Good Practice Guidance (GPG) for Land Use, Land Use Change and Forestry (IPCC, 2004). The method computes projected net stock changes of C over a given period of time (the default period is 20 years), in a one step process. The method can use default information on climate, soil type and land use/management (tillage and productivity) held by the IPCC (a Tier 1 approach) or, if available, country specific data (a Tier 2 approach). One main drawback is that it considers the change in one step (e.g. one stock for year 1 and another for year 20), assuming a linear rate of change over the period. Another drawback is that much of the data available for deriving the empirical factors in the IPCC default approach are from studies in North America and Europe (typically more are available for temperate versus tropical areas and mesic versus arid areas), which may result in bias (IPCC, 2004). Also, as the default method was designed to be as simple as possible and to use limited and highly aggregated data, the method is less well suited for sub-national assessments. Hence, while useful as a default, the IPCC method has some significant limitations. Where data and methods are available, the Good Practice Guidance recommends their use in place of the default (Tier 1) method (IPCC, 2004).

Several methods for estimating SOC stocks and changes at the regional scale have been developed and used in the USA and Europe. Some have used empirical approaches (Kern and Johnson, 1993) and others have used dynamic, process based, soil organic matter models, linked to spatial data through geographical information systems (Paustian *et al.* 1995, 2002; Falloon *et al.* 1998; Ardo and Olson, 2003). Approaches that use process based models have the advantage of being able to account for the complex long-term dynamic changes that determine SOC status. Generally the models employed, such as Century (Parton *et al.* 1988) and RothC (Coleman and Jenkinson, 1996), were originally developed using temperate data sets and therefore require substantial parameterization and evaluation before application to developing country situations where soil types, climate and land management systems can be very different.

In recent years, examples employing both empirical and modelling based approaches to estimate SOC stock change at the regional scale in developing countries have appeared. Some of these are discussed below.

2. Studies in Developing and Emerging Countries

2.1. Empirical Approaches

Several studies have used mapping based approaches to estimate SOC stocks in developing countries. In the case of the Brazilian Amazon Moraes *et al.* (1995) determined stocks of C in the top 100 cm of soils under undisturbed vegetation using 1162 soil profiles from the RadamBrasil survey and a digitized Brazilian soil survey map. Bernoux *et al.* (2002) used the same survey data, enhanced by further measurements and a method for estimating soil vegetation associations, to estimate SOC stocks for the first 30 cm of soil in the Brazilian Amazon. Mapping based approaches are well suited to making estimates that are specific to a single point in time, however, the problem arises of how to use these static stock values to make estimates of SOC stock change. Huang and Sun (2006) used 132 published articles on SOC stock change in different cropland areas of China and a set of statistical relationships to estimate national stock change over the past two decades in croplands of China. Their study concluded that SOC stocks had risen in approximately half of Chinese croplands over this period. The study highlights some of the problems of amalgamating numerous estimates that have been made using different approaches. A standardised approach, such as that developed by the IPCC allows estimates to be produced that are, as far as possible comparable.

An example of a recent a study that used the IPCC method in five different countries is Grace *et al.* 2004. This study adapted the IPCC method to produce reference C stocks for conditions found in South Eastern Australia, Northern Kazakhstan, The Indian Indo-Gangetic Plains, Sweden and Uruguay. The areas chosen varied in terms of soils, climate and land use as well as potential management strategies. The study adapted the IPCC method by using statistically estimated reference C stocks based on detailed in country data. This produced a robust starting point, from which stock change estimates could be made. It also allowed an analysis of the uncertainties associated with stock change estimates. The study used the stock change between years one and 20 to calculate an average annual C stock change rate. The main aim of the study was to assess C sequestration potential in soils and the economic cost associated with that sequestration. The method used was, therefore, developed to meet this aim. Typically 'one step' approaches have been designed to answer questions related to static or fixed conditions or specific land use/economic relationships rather than the dynamic response of SOC to land use change. In order to account for dynamic changes in SOC turnover, approaches incorporating processed based models are needed.

2.2. Processed Based Models Linked to GIS

A recent project funded by The Food and Agriculture Organisation (FAO) considered SOC turnover in tropical ecosystems using SOM models and GIS databases. The study was confined to the watershed and district scale and considered three areas, two in Mexico and one in Cuba. In a preliminary report, the authors (Ponce-Hernandez, 2004) made estimations of above ground biomass in the three watersheds using remote sensing data and ground level transect data. They used these data to estimate C in above ground biomass and below ground biomass using multiplication factors. They then used the above ground estimations of C as inputs for two soil C models, RothC and Century. They devised a graphical user interface

(GUI) for Century, that allows the user to input the minimum of parameters required to run the model and map results to given geographic areas using a GIS. The 'framework methodology' developed in this study, was designed for the watershed scale and involved much collation of new data, something that is rarely possible at the national and sub national scale. In addition, restricting the possible input parameters that the user can enter makes the system less specific to the area it is applied to. This is an important consideration when it is being applied to tropical conditions.

In recent years, interest in using modelling approaches to estimate SOC stock change at the regional scale, has grown in China. This has been precipitated by increased investment in scientific research in China and the global debate on potential quantification methods for CDM projects that sequester C in soils. Li *et al.* (2003), estimated the annual SOC balance for the year 1990 for croplands in China, using the denitrification decomposition model DNDC. County scale model simulations were conducted for all counties with cropland in China for 1990 conditions. At the same time a similar exercise was carried out for cropped areas of the USA. The authors estimated that China's cropland lost 1.6% of total SOC stock (for the first 30cm of soil) in 1990 and that this was due mainly to the removal of crop residues for other uses. In contrast, the authors estimated that the USA lost just 0.1% in the same year. This study modelled SOC change over a one-year period (1990) only. The authors used results from this one-year simulation to make inferences about longer term changes. Modelling longer time periods and extending these into future scenarios has obvious advantages, however such estimations can be restricted by available computing power when a model uses daily climate information (as in this study).

Another recent Chinese study used the CEVSA model, to estimate C sequestration potential of agricultural soils of China (Yan *et al.* 2007). The authors used remote sensing information and a remote sensing based production efficiency model (GLO-PEM) to estimate NPP of different cropping areas in China. The allocation of C in NPP to roots, straw and grain was estimated for four major crops and this information was then used as inputs to a modified version of the CEVSA ecosystem model which accounted for the effects of tillage.

The Global Environment Facility (GEF) co-financed a project between 2002 and 2005 to develop an SOC modelling system built on developing country data (Milne *et al.* 2007; Easter *et al.* 2007). Models developed by research groups working in the USA and the UK (Paustian *et al.*, 1995; Paustian *et al.*, 1997; Falloon *et al.*, 1998) were built on to create a system for estimating SOC stocks and changes at the national and sub-national scale applicable to a wide variety of conditions, including those found in arid, semi-arid, humid and sub-humid areas (the GEFSOC Modelling System). The GEFSOC system links two SOC models (RothC and Century) and the empirical IPCC method to a database of soils, climate and land-use/management information via a GIS. The system was developed using data from four contrasting eco regions; the Brazilian Amazon, Jordan, Kenya and the Indian part of the Indo Gangetic Plains. These regions were chosen for their contrasting climates and soil types and because they provide examples of those areas underrepresented by current soil C models. The Brazilian Amazon is an area dominated by native forest and recent land use change, in contrast, The Indian Indo-Gangetic Plains have been under continuous cultivation for hundreds of years and are subject to changes in land management rather than land use. A first stage of system development was to parameterise and validate the models in the system for conditions found in the four contrasting case study areas. The GEFSOC system includes a graphical user interface (GUI) to allow the user input the soils, climate and land use information needed to run the three methods. A set of program modules runs the three methods to produce SOC estimates under a range of user defined land use and climate change scenarios. Technical details of the system development are given in Easter *et al.* (2007).

The GEFSOC System has been used to estimate current (year 2000) SOC stocks and stock changes between 1990 and 2030 under a business as usual scenario. Current estimates were found to be in the same range as estimates made using mapping based approaches. SOC stocks for the year 2000 for the first 20 cm of soil were 32,000 Tg, 1,400 Tg, 1,300 Tg and 66 Tg for the Brazilian Amazon, Kenya, The Indian Indo-Gangetic Plains and Jordan respectively, according to the Century output of the system (Cerri, *et al.* 2007; Kamoni *et al.* 2007; Bhattacharyya *et al.* 2007; Al-Adamat *et al.* 2007). Losses in SOC stocks were predicted for all for case studies from the year 2000 to 2030 under business as usual scenarios. Losses ranged from 4 to 13%. Differences between the estimates made using the models and the Tier 1 IPCC method, highlighted problems of applying a 'Tier 1' IPCC approach (using default reference SOC stock estimates) to areas with extreme climates. For example when the IPCC method using global default values was applied to Jordan there was a tendency for overestimation. The IPCC default reference C stock levels for the 'warm-temperate dry' region are too high to reflect the situation in the very low productivity hyper-arid rangelands that dominate large areas of Jordan. The use of country-specific data for reference C stocks would improve the method. The GEFSOC System is, to date the only system designed exclusively using developing country data that incorporates both empirical methods and processed based models.

3. Conclusion

The sequestration of C in soils is important both in terms of land rehabilitation and GHG mitigation. Methods of quantifying these stocks and changes at a variety of scales are becoming increasingly important for developing or 'non-Annex 1' countries to allow them to benefit from C enhancing land use changes under clean development mechanism (CDM) projects and to help them demonstrate emissions reductions at the national scale. Many countries and international organisations (such as the GEF and the UNEP) have recognised this and have invested in the development of methods and models to estimate changes in SOC. Systems that incorporate processed based ecosystem models (such as the GEFSOC System) account for the dynamic changes that determine SOC turnover over tens to hundreds of years. This makes them useful tools for land use planners, CDM project planners and policy makers wishing to project SOC change under varying land use change scenarios.

The GEFSOC Modelling System is available for download from:
www.nrel.colostate.edu/projects/gefsoc-uk

References

Ardo, J.; Olson, L. 2003. Assessment of soil organic carbon in semi-arid Sudan using GIS and the CENTURY model. *J. Arid Environ.* **54**, 633-651.

Al-Adamat, R.; Rawajfih, Z.; Easter, M.; Paustian, K.; Coleman, K.; Milne, E.; Falloon, P.; Powlson, D.S.; Batjes, N.H. 2007. Predicted soil organic carbon stocks and changes in Jordan between 2000 and 2030 made using the GEFSOC Modelling System. In: Soil carbon stocks at regional scales. Milne, E., Powlson, D.S., Cerri, C.E.P. (Eds.); *Agric. Ecosyst. Environ.* **122** (1), 35-45.

Bernoux, M.; Carvalho, M.C.S.; Volkoff, B.; Cerri, C.C. 2002. Brazil's Soil Carbon Stocks. *Soil Sci. Soc. Am. J.* **66**, 888-896.

Bhattacharyya, T.; Pal, D.K.; Easter, M.; Batjes, N.H.; Milne; E.; Gajbhiye, K.S.; Chandran, P.; Ray, S.K.; Mandal, C.; Paustian, K.; Williams, S.; Killian, K.; Coleman, K.; Falloon, P.; Powlson, D.S.; 2007. Modelled soil organic carbon stocks and changes in the Indo-Gangetic Plains, India from 1980 to 2030. In: Soil carbon stocks at regional scales. Milne, E., Powlson, D.S., Cerri, C.E.P. (Eds.), *Agric. Ecosyst. Environ.* **122** (1), 84-94.

Batjes, N.H. 1999. Management options for reducing CO_2-concentrations in the atmosphere by increasing carbon sequestration in the soil. Report 410-200-031, Dutch National Research Programme on Global Air Pollution and Climate Change (also published as: Technical Paper 30, International Soil Reference and Information Centre, Wageningen), Bilthoven.

Cerri, C.E.P.; Easter, M.; Paustian, K.; Killian, K.; Coleman, K.; Bernoux, M.; Falloon, P.; Powlson, D.S.; Batjes, N.H.; Milne, E.; Cerri, C. 2007. Predicted soil organic carbon stocks and changes in the Brazilian Amazon between 2000 and 2030. In: Soil carbon stocks at regional scales. Milne, E., Powlson, D.S., Cerri, C.E.P. (Eds.), *Agric. Ecosyst. Environ.* **122** (1) 58-72.

Coleman, K.; Jenkinson, D.S. 1996. RothC-26.3 –A model for the turnover of carbon in soil. In: Evaluation of Soil Organic Matter Models.D. S. Powlson, D.S., Smith, P., Smith, J.U. (Eds.) NATO ASI Series, Vol. 138.

Easter, M.; Paustian, K.; Killian, K.; Williams, S.; Feng, T.; Al Adamat, R.; Batjes, N.H.; Bernoux, M.; Bhattacharyya, T.; Cerri, C.C.; Cerri, C.E.P.; Coleman, K.; Falloon, P.; Feller, C.; Gicheru, P.; Kamoni, P.; Milne, E.; Pal, D.K.; Powlson, D.S.; Rawajfih, Z.; Sessay, M.; Wokabi, S. 2007. The GEFSOC soil carbon modelling system: a tool for conducting regional-scale soil carbon inventories and assessing the impacts of land use change on soil carbon. In: Soil carbon stocks at regional scales. Milne, E., Powlson, D.S., Cerri, C.E.P. (Eds.), *Agric. Ecosyst. Environ.* **122** (1), 13–25.

Falloon, P.D.; Smith, P.; Smith, J.U.; Szabó, J.; Coleman, K.; Marshall, S. 1998. Regional estimates of carbon sequestration potential; linking the Rothamsted carbon model to GIS databases. *Biol. Fert. Soils* **27**, 236-241.

Grace, P.R.; Antle, J.; Aggarwal, P.K.; Andren, O.; Ogle, S.; Conant, R.; Longmire, J.; Ashkalov, K.; Baethgen, Valdivia R.; Paustian, K.: Davison, J., 2004. Assessment of the costs and enhanced global potential of carbon sequestration in soil, International Energy Agency, IEA/CON/03/95, 130 pp.

Huang, Y. and Sun, W.J. 2006. Changes in topsoil organic carbon of croplands in mainland China over the last two decades. *Chinese Science Bulletin* **51** (15), 1785 – 1803.

IPCC 2004. Good practice guidance for land use, land use change and forestry. IGES, Kanagawa, Japan.

Kamoni, P.T.; Gicheru, P.T.; Wokabi, S.M.; Easter, M.; Milne, E.; Coleman, K.; Falloon, P.; Paustian, K. 2007. Predicted soil organic carbon stocks and changes in Kenya between

1990 and 2030. In: Soil carbon stocks at regional scales. Milne, E., Powlson, D.S., Cerri, C.E.P. (Eds.), *Agric. Ecosyst. Environ.* **122** (1) 105 – 113.

Kern, J.S.; Johnson M.G., 1993. Conservation tillage impacts on national soil and atmosphere carbon levels. *Soil Sci. Soc. Am. J.* **57** (1), 200-210.

Li Chansheng; Yahui Zhauang; Frolking, S.; Galloway, J.; Harriss, R.; Moore, B.; Schimel, D.; Wang, X. 2003. Modeling soil organic carbon change in croplands of China. Ecological Applications **13** (2) 327-336.

Milne, E.; Al-Adamat, R.; Batjes, N.H.; Bernoux, M.; Bhattacharyya, T.; Cerri, C.C.; Cerri, C.E.P.; Coleman, K.; Easter, M.; Falloon, P.; Feller, C.; Gicheru, P.; Kamoni, P.; Killian, K.; Pal, D.K.; Paustian, K.; Powlson, D.; Rawajfih, Z.; Sessay, M.; Williams, S.; Wokabi, S. National and sub national assessments of soil organic carbon stocks and changes: the GEFSOC modelling system. In: Soil carbon stocks at regional scales. Milne, E., Powlson, D.S., Cerri, C.E.P. (Eds.), *Agric. Ecosyst. Environ.* **122** (1), 3-12.

Moraes, J.L.; Cerri, C.C.; Melillo, J.M.; Kicklighter, D.; Neill, C.; Skole, D.L.; Steudler, P.A. 1995. Soil carbon stocks of the Brazilian Amazon basin. *Soil Sci. Soc. Am. J.* **59**, 244-247.

Parton, W. J.; Stewart, J. W. B.; C.V. Cole. 1988. Dynamics of C, N, P and S in grassland soils: A model. *Biogeochemisty* **5**: 109-131.

Paustian, K.; Elliot, E.T.; Collins, H.P.; Vernon Cole, C.; Paul. E., 1995. Using a network of long-term experiments for analysis of soil carbon dynamics and global change: the North American model. *Aust. J. Exp. Agr.* **35**, 929-939.

Paustian, K.; Levine, E.; Post, W.M.; Ryzhova, I.M. 1997. The use of models to integrate information and understanding of soil C at the regional scale. *Geoderma* **79**, 227-260.

Paustian, K.; Brenner, J.; Killian, K.; Cipra, J.; Williams, S.; Elliott, E.T.; Eve, M.D.; Kautza, T.; Bluhm, G. 2002. State-level analyses of C sequestration in agricultural soils. In: Agriculture Practices and Policies for Carbon Sequestration in Soil. Kimble, J.M, Lal, R. and Follett, R.F. (Eds.). Lewis Publishers, CRC Press, Boca Raton, Fl., USA. pp. 193-204.

Ponce-Hernandez, R., 2004. Assessing carbon stocks and modelling win-win scenarios of carbon sequestration through land use change. FAO Report, ISBN 92-5-105158-5, Rome.

Yan, H.; Cao, M.; Liu, J.; Tao, B. 2007. Potential and sustainability for carbon sequestration with improved soil management in agricultural soils of China. *Agric. Ecosys. Environ.* **121**, 325-335.

Research and Review Studies

In: Climate Change Research Progress
Editor: Lawrence N. Peretz, pp. 13-85
ISBN: 978-1-60021-998-6

Chapter 1

TRACER ELEMENTS AND URANIUM-SERIES ISOTOPES IN THE BOTTOM SEDIMENTS OF LAKE BAIKAL AS GEOCHEMICAL CLIMATE PROXIES FOR HIGH-RESOLUTION RECONSTRUCTIONS OF REGIONAL CHANGE

E. L. Goldberg
Limnological Institute of Siberian Branch of Russian Academy of Science,
Laboratory of Paleoclimatology, Irkutsk, Russia
Archaeology Institute of Siberian Branch of Russian Academy of Science,
AMS – center, Novosibirsk, Russia

Abstract

This paper is a synthesis of main recent results I obtained jointly with my colleagues from the Limnological Institute (Irkutsk). We suggest a conceptual model to reconstruct the Pleistocene history of humidity and environment changes in the watershed of Lake Baikal (East Siberia). The climate events are recorded in high-resolution patterns of more than 25 elements, diatom abundances and compositions of diatom assemblages, sediment physical properties, etc., as well as in the behavior of U and Th isotopes we measured in a ~1 Myr long drill core of Baikal bottom sediments and in cores that span the past 300 kyr. There are downcore element profiles that either correlate or anti-correlate with the biogenic silica record. The concentrations of elements in the former group are high in interglacial and low in glacial periods. Spectral analysis of the Baikal records reveals all orbital frequencies of 96, 41, 23, and 19 kyr, and also a distinct 72,-54, 35 and 30 kyr rhythms. The response of the continent to insolation is more prominent than the marine response because power of spectra in the tilt (41 kyr) and precession (23 kyr) bands of some proxies are as strong as or even more the nonlinear 100-kyr cycle.

Direct U-Th dating of all interglacial intervals down to MIS7.1 shows an almost synchronous response of Lake Baikal to insolation forcing which slightly leads the more inertial global ice volume signal.

We use the geochemical behavior of U-series isotopes to simulate the paleoflow history in the Baikal watershed for the past 100 kyr, within the limits of our conceptual model, proceeding from the mass balance of soluble uranium the tributary rivers carry to the lake. The river flow appears to have been vanishing in dry glacial maximums, which is supported by records of clastic-related elements in sediments.

The core interval spanning the past 150 kyr has been investigated to a decadal (~ 20 yr) resolution by SRXRF scanning at 0.1-1 mm, applied to 8000 horizons. The Baikal watershed shows responses to all Greenland D/O events through the last glacial and short spells of abrupt cooling that intervene all interstadials of MIS 5 (MIS 5.1, 5.3 and 5.5). The clear response of Siberia to the North Atlantic events suggests their global distribution over the Northern Hemisphere.

Key words: Lake Baikal, Bottom sediments, Climate change, Orbital forcing, Dansgaard-Oeschges and Heinrich events, Millennial scale climate variability, U-Th isotopes, Elemental time series.

1. Introduction

In recent years significant concern has risen about the global warming problem. The IPCC-team predicts that further increase in atmospheric CO_2 will amplify the current trends in regional climates, landscapes, productivity and water supply [1-2]. Siberia has experienced an especially large warming increase over the past 140 years. The Siberian region occupies an enormous area of Eurasia but its climate history remains far less explored than elsewhere in the world. The knowledge of past changes in Siberian climates gained through simulations with realistic climate models or paleoclimate reconstructions from climate proxies in natural archives is still scanty. The climate change signals from the Atlantic and Pacific regions or China cannot be directly extrapolated to Siberia which is sited in the middle of the Asian continent far away from the polar ice and oceans and has specific patterns of heat and moisture input. The Siberian climate obviously changed under the intricate interaction of the Siberian High, Westerlies and, perhaps, East Asian Summer Monsoons.

Lake Baikal is located in southern East Siberia near the Russian/Mongolian frontier (fig 1). It is one of the world oldest lakes with an age of around 20-30 million years. Baikal occupies a deep and narrow rift basin produced by tectonic subsidence along major border faults. Being the world's largest and deepest (1637 m) lake, it stores 23,000 km^3 of fresh surface water. Its immense watershed covers an area of 570,000 km^2 [3, 4]. The lake water budget is maintained by several large inflowing rivers, which carry more than 80% of water and about 200 minor rivers and streams, and the only river (the Angara) flowing out. The major tributaries are the Upper Angara (14%) and Kichera (3%) from the north, the Barguzin from the northeast (7%), and the biggest river Selenga (51%) from the southeast (fig 1). The Selenga forms a large delta area in the southeastern part of Lake Baikal. The Selenga drains a large catchment area in the trans-Baikal region in southernmost East Siberia (37% in Russia) and in semi-arid regions of Northern Mongolia (63%) (fig 1).

Unlike many other deep lakes, the extremely deep lake of Baikal has its entire water column constantly oxygenated due to regular mixing of deep water with oxygen-rich surface waters and deep-water renewal [4-8]; the oxygen content in the bottom water is never less than 230 mmol/l [4, 7, 8]. The current water residence time is about 400 years and water mixing occurs every 12-16 years [4-7]. The lake water column consists of two main zones that differ in temperature profiles and mixing intensity. The upper zone down to 200 m is subject to temperature gradients and the water below 200 m is stratified and has almost constant temperatures of ca. 3– 4^{0}C. Water in Lake Baikal has ultra-fresh calcium bicarbonate chemistry of 7.6 pH, with less than 100 mg/l of dissolved salts [4, 7-9]. The concentrations of most elements in the water column are controlled by weathering of surrounding rocks, and the riverine input is consequently the major source comprising up to >99% of load (97% on average) [7]. The sediment surface is orange-brown and enriched in Fe and Mn oxyhydroxides. The oxidized layer is from 1 to 20–30 cm thick, depending on the oxygen penetration depth and sedimentation rate [9].

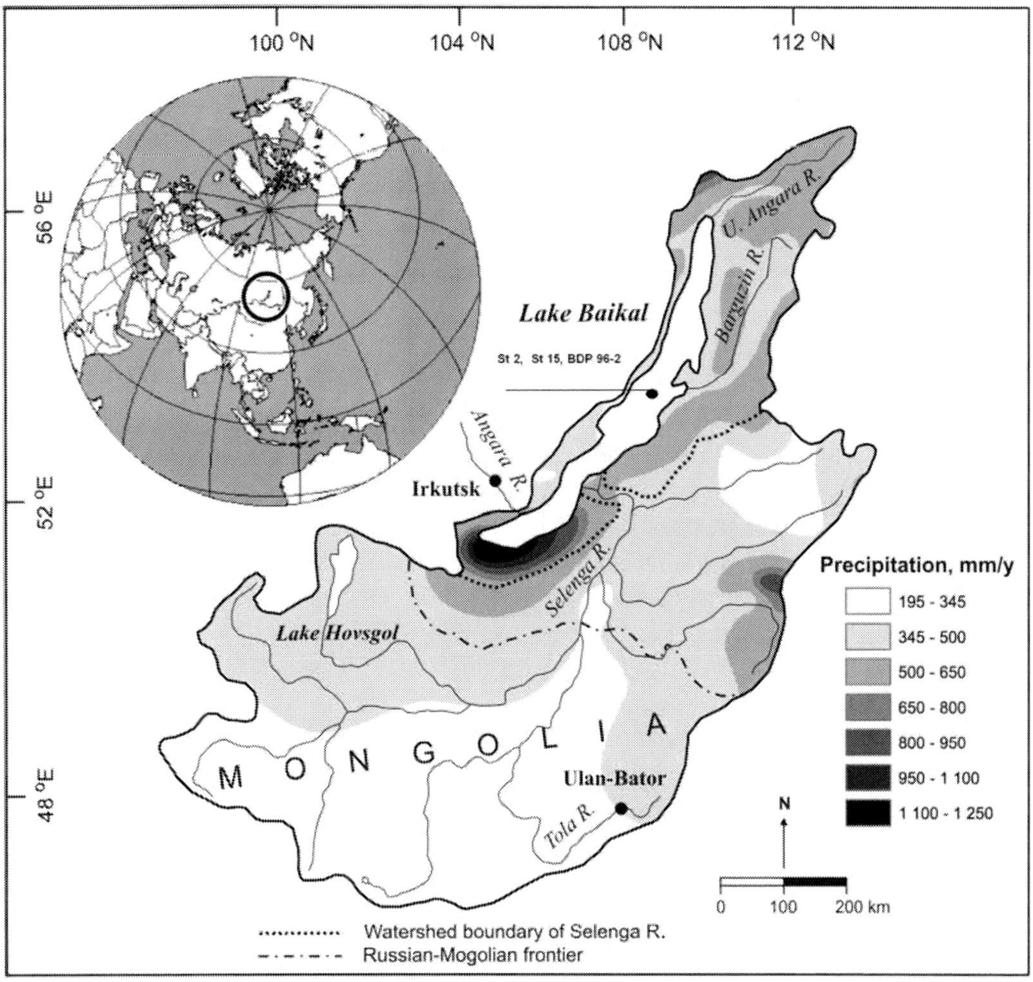

Figure 1. Map of the Lake Baikal area. Also plotted are the location of the investigated cores st 2, st 15 and BDP96-2, on the Academchesky Ridge in the central part of the lake; the main rivers and their watersheds; regional distribution of atmosphere precipitations.

Lake Baikal lies in a highly rugged terrain with its surface topography built by active tectonic processes in the Baikal Rift system [10, 11]. The lake is flanked by mountains of the Primorsky and Baikalsky Ranges that stretch along its western coast, the Barguzin Range in the northeast, and the Khamar-Daban Range in the south. The submerged topographic highs of the Akademichesky Ridge and the Buguldeika saddle divide the lake basin into the northern, central and southern subbasins.

The lake basin has accumulated extremely thick sediments (over 8000 m locally) through many millions of years of continuous and steady deposition [11-13]. The lake sediments can be expected to store a faithful record of direct insolation forcing and nonlinear ocean/ice sheet interaction [14] because Baikal lies in the high-latitude (52-56 N^0) Asian continental interior, as far from the influence of oceans and past ice sheets as no other place on Earth [14]. Thus, due to its continental interior position and long continuous sedimentation, Lake Baikal is the exceptional paleoclimate site, a perfect undisturbed archive of the continental climate history.

Most of the well documented paleoclimate and paleolimnological records for Lake Baikal have been received from the Academichesky Ridge in the lake center, which is isolated from turbidite flows and river influence and receives only a slow rain of hemipelagic sediment [14]. Mean sedimentation rates on the ridge appear to have been almost constant (ca.4-5 cm/kyr) [14-24], and the changes in its sedimentary record can be thus correlated with the global climate archives of the marine oxygen isotope and insolation 60^0N [14-30], and Greenland ice records [31-33].

A number of sensitive climate proxies have been revealed in the Baikal sediments which record changes in the lake productivity [14-17, 19, 20, 22-30], vegetation history [13,15,17, 34-36] and variability of the sources of terrigenous input [17, 18, 21, 31-33, 38-46]. The content of biogenic silica in the sediments was one of the first proxies recognized as responding to glacial/interglacial cycles [14-17, 19-23, 26-30, 37]. Biogenic silica primarily reflects diatom productivity in Lake Baikal, which, in turn, is subject to limnological controls from water temperature, duration of the ice-free season, turbidity, and availability of nutrients [14]. The records of tracer chemical elements in the Baikal sediments may shed more light to the history of terrigenous input sources [18, 21, 31-33, 37, 38-46].

Uranium isotopes are of particular importance in this respect. Their concentrations presumably depend on the volume of river flow [42-45], the assumption that makes a basis for quantitative reconstruction of the humidity history of East Siberia [45] reported below.

In this paper I synthesize the most important results obtained through the recent four or five years by my research group at the Limnological Institute (Irkutsk) and suggest new interpretation of data and revise some of earlier assumptions.

2. Methods and Samples

Cores. We have studied cores taken on top of the underwater Akademichesky Ridge of the Lake Baikal (see fig 1). Stratigraphy and lithology of the sediments of Station 15 ($53^0$33'19"N, $108^0$00'43"E, depth 430 m) obtained in 1994 is described in Ref. [17, 40]. The sediments of Station 2 ($53^0$33'04"N, $107^0$54'53"E, depth 400 m) were obtained by a gravity core (st2-GC) in 1998 and PC corer in 1999 and 2001(st2- PC-1999; st2-PC-2001) [32, 44]. St2–GC covered 0-100 kyr period and both PC cores covered period from 50 up to 160ky. The composite cores from st2 were obtained by the correlations of the records of diatom complexes, biogenic silica (BiSi) and water contents (WC) for the sediments of st2-GC, st2-PC-1999 and st2-PC-2001. Sampling rate for core st2 was each 1 cm for ICP-MS and U-Th isotope investigations [44] and 1mm for scanning SRXRF analysis [32, 33, 46] .

The BDP-96-2 drilling core was obtained in 1996 due to the international Baikal Drilling Project [15, 16, 19]. Results obtained with the first 3365 cm of the aligned geological section that spans the Brunhes magnetic epoch up to 780 kyr BP) [15,16 19] are described in the present paper. The uppermost part of the sediment (> 2 m) in BDP-96-2 including Holocene and uppermost Pleistocene was lost at drilling. Samples for elemental analysis were taken from cores at ca. 10 cm intervals (ca.2 kyr). The same sampling rate was applied to the sediments of st15. Samples for determination of BiSi, WC and dry bulk density (DBD) were taken at 1-5-cm intervals (ca. 0.2-1 kyr). The data obtained were linearly interpolated onto even intervals (length 10 cm), and subjected to three-point Gaussian smoothening. Pure diatom frustules do not contain majority of the elements studied [47]. For this reason, we used two sets of element content data, one expressed as micrograms of element per gram of dry bulk sediment, and the other as micrograms of element per gram of terrigenous part of sediment. The later records, unlike the initial ones, are "cleaned" of the effect of mere dilution of clastic matter with BiSi and directly reflect changes in the concentrations of elements in the clastic sediments.

Elemental SRXRF analysis. X-ray fluorescent analysis with synchrotron radiation beam (SRXFA) was performed at the station of elemental analysis in the Novosibirsk's Institute of Nuclear Physics. Synchrotron radiation beam have been received from the VEPP-3 storage ring of the electron-positron collisions accelerator. Samples of sediment having a mass of 35 mg were pressed into tablets with the diameter of 6 mm. The tablets were irradiated with a monochromatized and polarized SR beam. Two incident beam energies were used. The 26 keV band was used for the determination of elements with an atomic mass less than that of Mo. The 45 keV band was used to determine light lantanoids, Sb, Sn, I, Ba. The method is described in detail in [48-50]. Neutron activation analysis (NAA) was performed using the nuclear reactor of the Tomsk Institute of Nuclear Physics. We determined the concentrations of the following elements: K, Ti, Mn, Fe, Cu, Zn, Mo, Pb, Rb, Ba, Sr, La, Ce, Y, Yb, Eu, Pr, Nd, Tb, Sn, Sb, U, Th, As, Sc, Nb and several other. Accuracy of the methods was evaluated by intercomparison of data obtained with the control samples by means of NAA, SRXRF and ICP-MS [48] and it is evaluated as ca.10-30%. BiSi was determined according to [17]. BiSi records in the BDP-96-2 core [15] were provided in the digital form by M.I. Kuzmin.

Scanning SRXRF was performed at the same station using the high-precision positioning scanner "Baikal" [32, 33, 46, 51, 52]. The scanner permits the pulling of a sample with a length of up to 300 mm across a monochromatic SR beam and a reception of XRF spectra. The minimal spatial step of scanning is equal to 0.1 mm. Special software completely controls the spectra acquisition during the scanning of a sample. Besides that, the software tests the presence of the SR beam from the storage ring and performs repeats of the spectra acquisition if the electron beam changes in the storage ring [51]. We cut out blocks of raw wet sediment from core along its axis using aluminum containers. The sediment slabs were about 27 cm long, 0.5-1cm of thick and ca. 1 cm width. To keep the samples wet we coated the aluminum containers by thin polymer film. The slabs were scanned continuously under monochromatic synchrotron (SR) beam, with both beam aperture and scanning step of 1 mm. Thus, sampling rate of scanning SRXRF was ca.1 mm. Temporary resolution of the records obtained was about 20 years. A number of horizons in core investigated by this way were ca. 8000. For details of the sampling and treatment procedures see [46, 52].

ICP-MS isotopes analysis. Subsamples of the sediments after taking for diatom analysis and determination of BiSi, DBD and WC were ground in a agate mortar. 0.7 g sediment specimens were mixed with 2.4 g $LiBO_2*2H_2O$ in graphite-glass crucibles and ignited in a muffle furnace for 20 min at 1000 °C. The crucibles were placed into a ceramic box and argon was supplied inside the box through a hole to prevent the crucibles from burning. The pearl obtained was left for a week for dissolution in a 15 ml polypropylene tube with 10 ml 8M HNO_3. Then the extract was centrifuged and the supernatant was carefully separated from the silica precipitate. The residue was rinsed twice with 5 ml 8M HNO_3, the supernatants were united and supplemented with a 100 mkL spikes solution of ^{229}Th and ^{236}U, 10 ppb each, evaporated to dryness and redissolved in 4 ml 8M HNO_3. 3 ml of the solutions received was loaded onto an ion-exchange column. Extracts (3 ml) were applied on a polypropylene column with 2 ml of Biorad AG1-X8 anionite pre-washed with 10 ml 8M HNO_3. Elution was performed with 4 ml 8M HNO_3 followed by 30 ml 0.4M HCl. The united eluate was evaporated to dryness and the residue was dissolved in 4 ml 3% HNO_3.

Measurements of U-Th isotopes were performed with a VG Plasma Quad II PQ instrument (VG Elemental, UK) in repeated cycles "standard-washing-7 samples-washing-standard", 3 min for a sample, at an uptake rate of 1 ml/min. U and Th isotopes were determined in Baikal water taken from a 500 m depth 1.7 km offshore in Listvenichny Bay. 4.5 L of this water was filtered through 0.1 mkm capon filters; 100 mg $FeSO_4*7H_2O$ were dissolved in every 1.5 L bottle, and 2 ml of concentrated NH_4OH added. After 16 hours, the $Fe(OH)_3$ precipitate was collected by decantation, followed by filtration and dissolution in 5 ml 8M HNO3. The solution was evaporated to dryness and redissolved in 1 ml 8M HNO_3. U and Th were separated by anion-exchange chromatography (see above) and the isotopes were determined by ICP-MS. Accuracy of measurement of absolute concentrations of ^{238}U and ^{232}Th were 1-3% , but their ratios ~ 0.3 - 0.8%. Typical values for ^{230}Th and ^{234}U were ca. 2-8% and from 2 to 6% for ^{230}Th / ^{232}Th and ^{234}U/^{238}U. Analytical-grade HNO_3, $LiBO_2*2H_2O$, HCl, and $FeSO_4*7H_2O$ were tested prior to analysis and found free from U and Th. Details of the ICP-MS procedure of U-Th measurements are published in [44, 53].

3. Geochemical Climate Proxies in the Milankovitch Scale

3.1. Overview

It has become a universal premise that the tens- to hundreds-kyr fluctuations of global climate are the response to orbital forcing associated with variations in the amount of solar radiation (insolation) in the Northern Hemisphere [54-63]. Insolation changes with Earth's orbit shape (eccentricity) and with the tilt of the Earth's rotation axis relative to its orbital plane (obliquity). Eccentricity varies at periods of 413, 125, and 96 kyr and obliquity fluctuates in 41- and 54-kyr rhythms. There are also precession cycles of 23 and 19 kyr due to changes in the direction of the Earth's axis and in the Earth's position on the orbital ellipse, which control especially the seasonal insolation contrasts. Periodicities of the same lengths as the orbital cycles of 100, 41, 23, and 19 kyr [59-60] are found in many climate records stored in marine sediments, Antarctic (Vostok, EPICA) ice cores, and loess sequences of China and in Lake Baikal [14-16, 18-20, 38-40, 54-58, 61-68].

The exact driving mechanism of the 20-100-kyr climate changes is not fully understood but the available models that are worthy of note simulate it as a complex nonlinear ocean-atmosphere-land interaction in response to changes in boundary conditions (insolation is only one of them) complicated with internal feedbacks or threshold effects [69-75]. Moreover, the mechanisms of climatic response based on laws of stochastic processes have been recently published [71-75]. There are the excellent reviews of climate change mechanisms in [56, 57, 70, 73]. So, the climate processes at the 100-kyr scale are more or less explained only in the sense that at least one their cause is known quantitatively .

Global cooling during glacial periods apparently decreased evaporation from the ocean surface and thus reduced the moisture transported to the continents, to make the climate more arid in the high and middle latitudes of the Northern Hemisphere (NH). Inverse processes took place at climate warming, what is similar to the present day situation [76]. The latent energy of evaporation currently accounts for about 70-80% of heat carried by air flows to the middle latitudes, and the heat budget of Siberia is thus controlled by both insolation and Atlantic atmospheric transport [see review in 77]. As a result, air humidity in Siberia apparently correlates with temperature at large time scales [78-80]. The wet/dry climate contrasts must be especially well defined in middle latitudes of continental Asia where moisture arrives with constant west Atlantic winds (Westerlies transport). Thus, one can expect a prominent climate response from mid-latitude Siberia, as far as Yakutia, the easternmost area reached by the Atlantic influence.

During long and strong glaciations, the mountains around Lake Baikal became thickly covered with ice [81, 82 and review for Siberia 83]. The growth of glaciers and cycles of freezing-thawing of bedrocks were attendant with degradation of rocks and influx of "glacial milk" into the lake. On the other hand, the aridization of global – and also Siberian – climate apparently decreased the river input and the respective riverine fluxes of solids to Lake Baikal. Since suspended load in rivers and "glacial milk" particles have different chemical compositions, the downcore patterns of sediment chemistry should reflect the redistribution of fluxes from different sources and thus record the climate change. Then, the geochemical records can be used to identify and investigate glacial/interglacial cycles in the drainage basin of Lake Baikal.

3.2. Elemental Records and Integral Geochemical Indicators

The presence of orbital frequencies in the spectra of Baikal sedimentary records in ~250 kyr cores was first noted by Peck [18] in the magnetic susceptibility and by Colman et al. [14] in biogenic silica (BiSi) profiles. However, their spectral analysis spanned only two 100-kyr cycles. Orbital cycles are also embedded in the spectra of water content (WC) records for the past 800 kyr in the BDP-96-2 core [20]. For the same core we obtained profiles of tracer elements by SRXRF and applied spectral analysis to the element concentrations and their ratios [40]. Figure 2 shows several representative geochemical profiles for sediments from station 15 (Akademichesky Ridge [40]) compared to the BiSi record. The element concentrations used in fig. 2 are only those related to the terrigenous component of sediments (see method), which was necessary to avoid the effect from dilution of sediments with BiSi. Earlier we proved [47] that biogenic silica is virtually transparent to almost all tracers we studied. From fig. 2 it can see the records of some elements are in positive correlation (panel in fig.2 a,c) and others are in anti-correlation (fig.2 b,d) with the BiSi record. The former are hereafter called *warm* and the latter *cold* tracers. BiSi is high over warm periods of MIS1, 5 and 7 and low during glacial periods (cf. ocean's isotope index with BiSi records on fig. 2 e,f.)

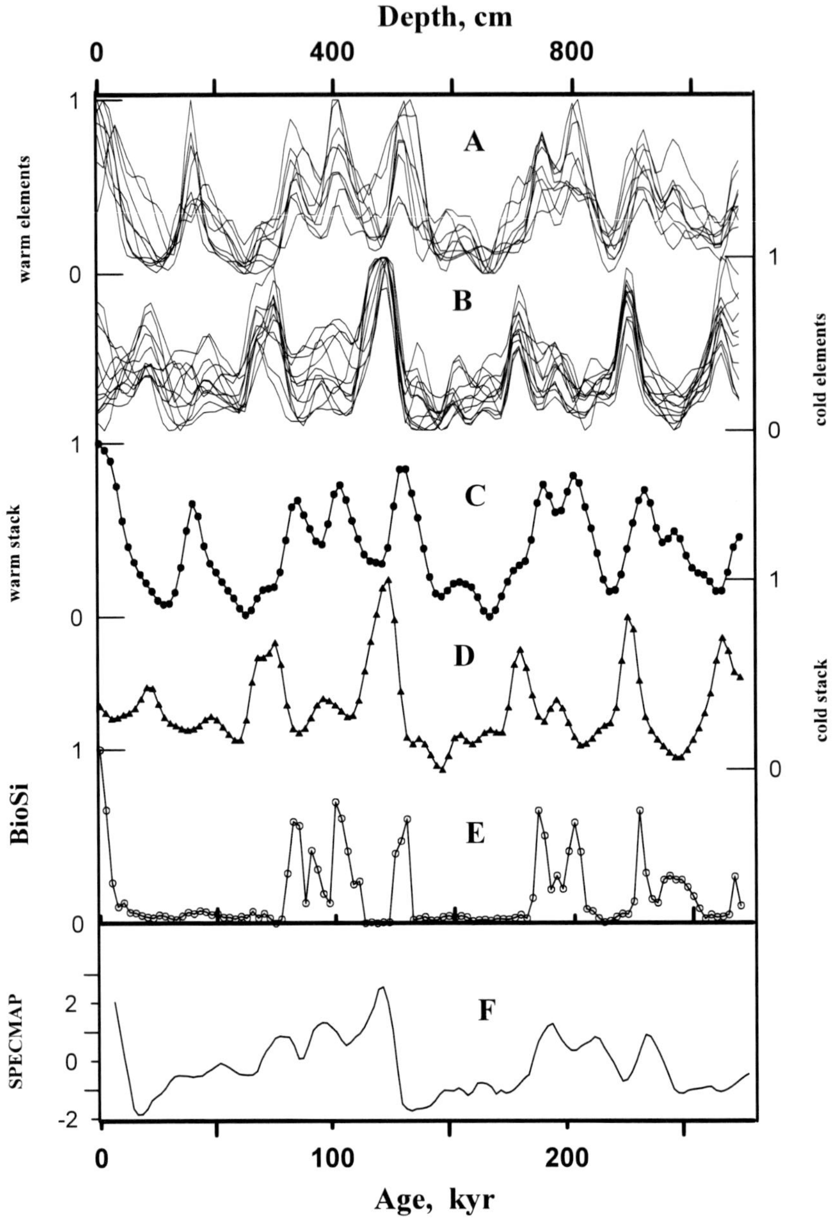

Figure 2. Dimensionless records of elements in st2 core. a – overlapping records of "warm" elements and their ratios that correlate with BiSi (Mo, U, HREE – Eu, Tb, Yb, Sr/Rb (Ba,Cs), U/ Th. b- "cold" elements and their ratios that anti-correlate with BiSi (Th, Ba, Rb, La, Ce, Cs, La (Ce)/Yb, Ba(Ce,La)/Zr), c – warm (interglacial stack), d – cold (glacial) stack, e- BiSi record, f- marine oxygen isotope curve [62] as a global ice volume proxy. All records are normalized (eq.1) to variation range {0-1} and are thus drawn in the same scale.

Stacks. In processing core data we use the *stacks* of elements obtained as follows. First we estimate average concentrations of tracer elements and their STD and then normalize the profiles by subtracting the means from each current value and dividing the difference by the STD, which gives the records of relative variations. A stack is an averaged sum of normalized records for few elements re-normalized to the variation interval {0-1} according to

$$Y_{new} = (Y - Y_{min}) / (Y_{max} - Y_{min}) \quad (1)$$

where Y, Y_{max} and Y_{min} are the current, maximum and minimum values of the variables, respectively. Unlike the usual practice of stacking *uniform* profiles, such as foraminiferal δO^{18} or δC^{13} in marine sediments, we used the averaging procedure to obtain stacks of different but correlated profiles and thus cleaned the signal from random noise and highlighted the common regularities shared by the analyzed records [33, 40]. Lines in fig. 2c,d show a *warm stack* obtained by averaging the U/Th, Sr/Ba, Sr/Cs, Sr/Rb, Eu/Sc, U, Mo, Yb and Tb records and a *cold stack* as a normalized averaged sum of the Th, Ba, Rb, Cs, La, Ce, La/Yb, Ce(La)/Y, Ce/Yb, Ce(La, Ba)/Zr and Th/Sc profiles. Note again that we avoided the BiSi dilution effect in the stacks by using either element ratios or the concentrations in the terrigenous portion of sediment. The warm stack generally agrees with the BiSi profile as its peaks are in phase with the BiSi peaks, but there are some essential differences. For instance,

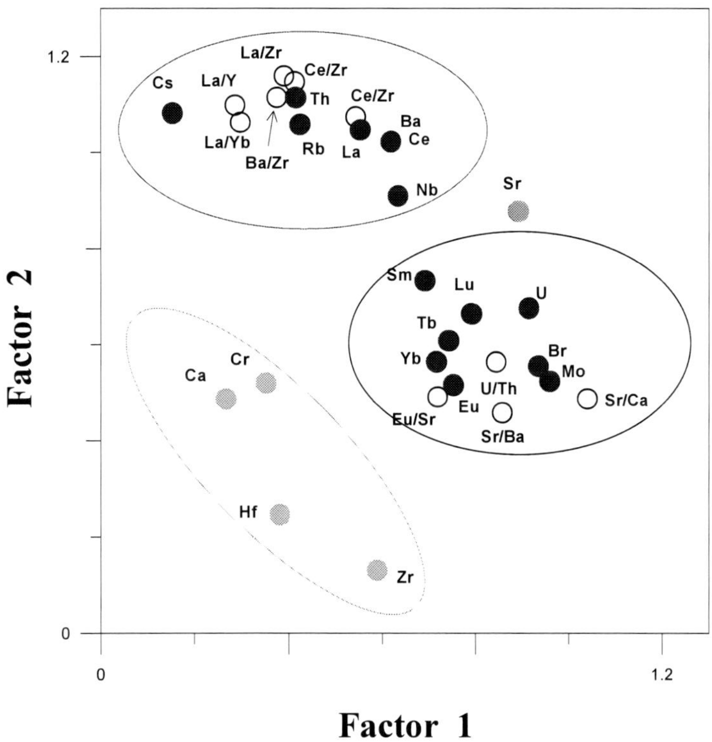

Figure 3. Factor diagram of elements in st15 core. Ellipses contour separate clusters of tracer elements. Elements fall into two distinct groups: “cold” : La, Ce, Ba, Nb, Cs, Rb, Th and La (Ce)/Yb (Y,Zr) ratios and “warm”: U, Br, Mo, Yb, Tb, Lu, Sm and Sr/Rb (Ba) and Eu/Sc ratios. Note the particular response of Cr, Hf, Zr, Ca.

the period corresponding to the core depth of 120-200 cm (MIS 3) appears rather warm in the warm stack, but the respective peak is relatively small in the BiSi record. Diatom numbers, a more sensitive climate proxy (though not expressed in mass units), gives evidence that diatoms did live in open Baikal at that time but had solely a low abundance [14-17, 20] (see also section 4 and 5).

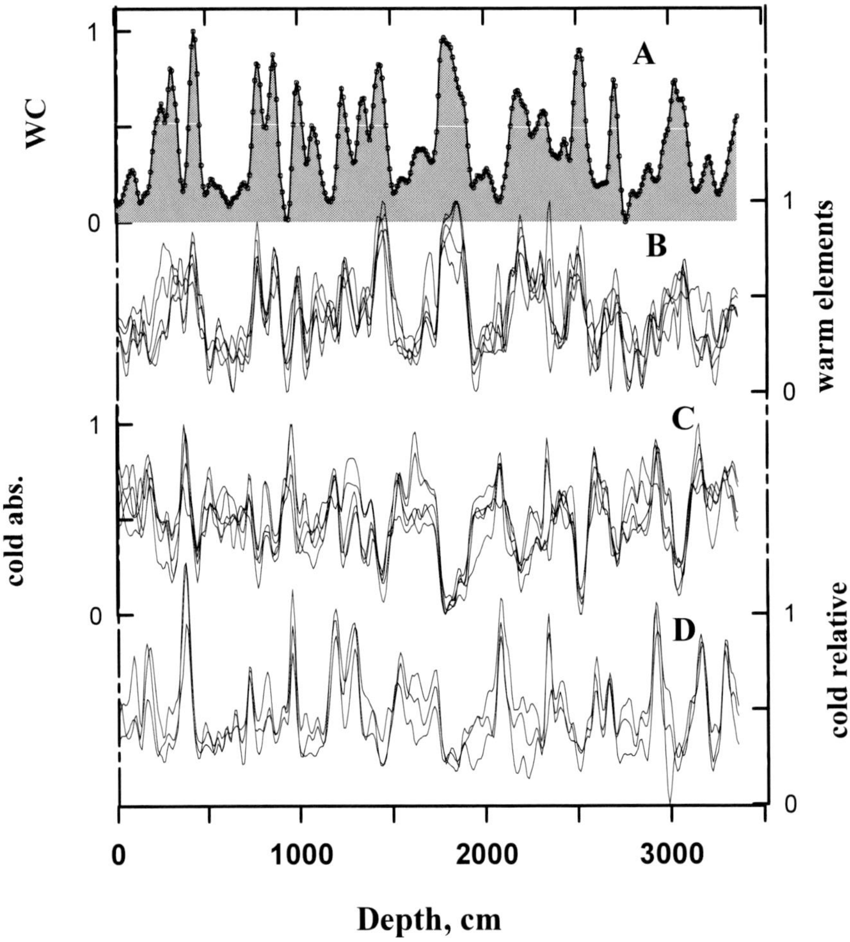

Figure 4. Dimensionless records of elements in BDP-96-2 core. a – water content (WC), b – overlapping records of "warm" elements and their ratios that correlate with BiSi (Zn/Nb, Sr/Rb(Ba,Ti) and Br, c- "cold" elements (Ba, La, Ce, Rb, Ti) and d "cold" ratios that anti-correlate with BiSi (La/Y, Ce/Zr and Ba/Zr). Note more prominent signals of the element ratios relative to the elements themselves.

Comparison of the cold and warm stacks shows that the peaks in the cold stack rather mark warm-to-cold transitions than the glacial peaks. This is especially well evident in the core interval between 550 and 700 cm spanning a long interglacial, where both stacks show low values. We formalized and extended the intuitive inferences from data in fig. 2 due to factor analysis of the geochemical tracers (see the results for st15 in fig. 3). The set of element concentrations and ratios falls into two clusters (fig. 3) of which one correlates with BiSi. The best representative tracers in the "*warm*" cluster are U, Mo, Br, Yb, Lu, Eu, Tb,

U/Th, Sr/Cs, Sr/Ba, and Sr/Fe, and those of the "*cold*" cluster are La/Yb, Ba/Zr, La/Zr, Ce/Zr, Th, Ba, Rb, Cs, La, Ce, and Nd. The cold cluster, in turn, consists of two subgroups: The subgroup of La/Yb, La/Y, Ba/Zr, La/Zr, and Ce/Zr ratios is slightly farther away from the warm cluster than the subgroup of Th, Ba, Rb, Nd, La, and Ce concentrations. Therefore, the cold stack is not quite uniform, and its part based on La/Yb, La/Y, Ba/Zr, La/Zr, and Ce/Zr better records the contrasts between climate cycles (figs. 3,4).

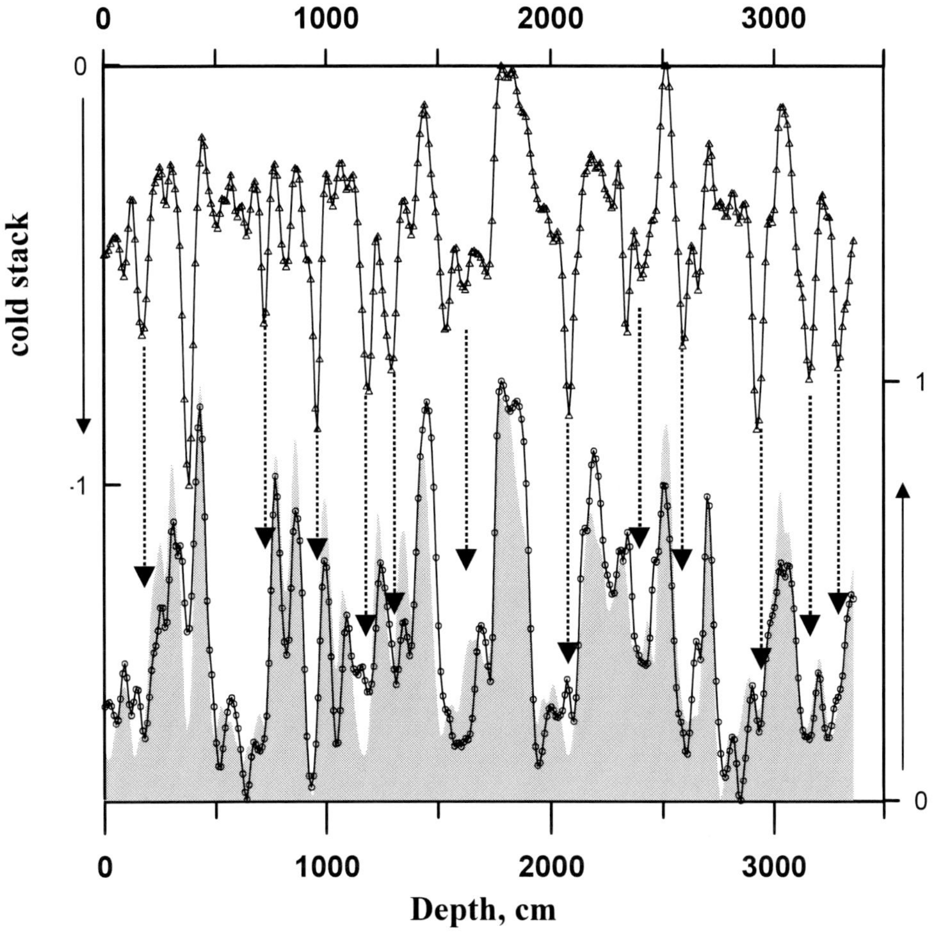

Figure 5. Geochemical stacks in BDP-96-2 core for Brunhes chron (B/M reversal is found at 3365 cm [16, 86]). Cold stack is above (inverted axis), warm stack (line) and water content – WC are below. The WC content profile is in gray. Arrows show peaks of the cold stack (cooling).

The element patterns in the BDP-96-2 core cluster in a similar way (see few overlapping records in fig. 4): The profiles of Br, Sr/Ba, Sr/Rb, Sr/Fe, Sr/Ti, Mo, and Zn/Nb are the warm tracers (panel fig. 4, b) and Ba, Rb, La, Ce, Nb, Ti, Pb, Ce/Y, La/Y, Ce/Zr, La/Zr, and Ba/Zr are the cold tracers (fig 4c,d). As in the st15 core, the latter ratios make a subgroup of prominent cold tracers. The U and U/Th ratios mark warm climates and Th concentrations track cold cycles throughout the Brunhes chron, as it was first noted in [50]. The stacks for BDP-96-2 are shown in fig.5. The warm stack for both cores (cf. with the st15 data in fig. 2) matches perfectly the BiSi and WC profiles. Peaks in the cold stack as a rule mark the onset rather than the maximum of glacials. The element ratios are likewise more climate-sensitive than their concentrations as in the core of st 15.

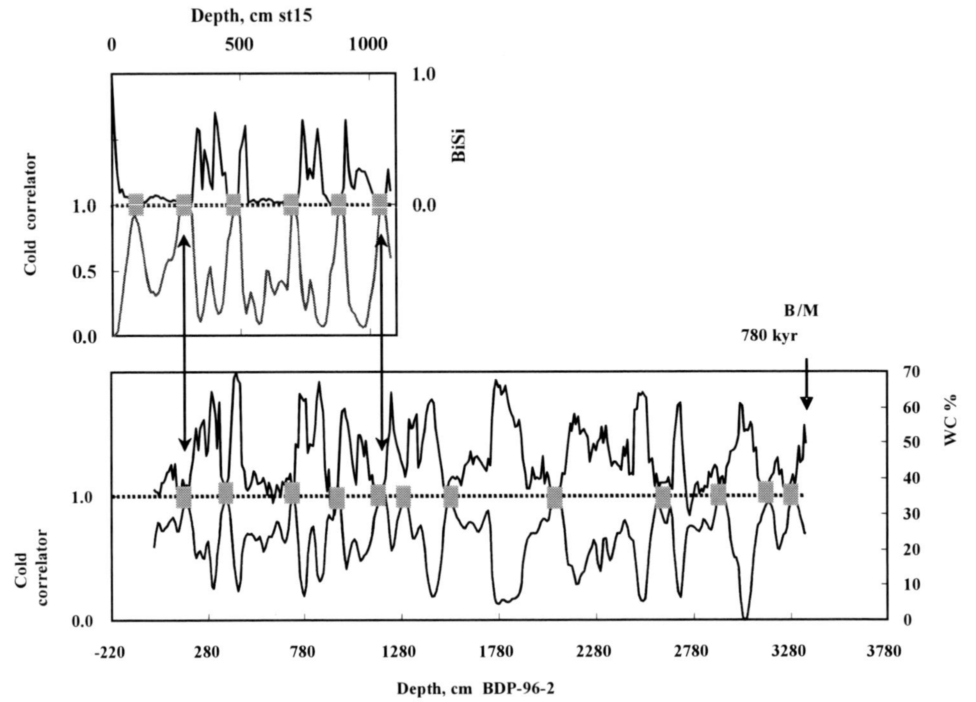

Figure 6. Glacial correlators for both (st15 and BDP 96-2) cores. Upper panel shows BiSi (or WC) records for each station. The glacial correlator shows weight percentages of terrigenous component in glaciers input. Peaks in the correlator indicate horizons with maximum glacial contents (gray rectangles). Dips in the correlator correspond to periods of riverine sediment input.

Correlator. The best way to show how the stacking results can be used is to evaluate weight percentages of inputs from a glacial (or interglacial) source in the terrigenous component of sediments. For this we select only the narrow core intervals that contain peaks of cold stacks (or warm) and calculate the interval-average element compositions of a standard "cold" (or "warm") sample by averaging the element concentrations in a series of these typical intervals. Then the sediment composition at some core horizon is presented as a vector in an n-dimensional space, with its coordinates as n dimensionless concentrations. The semblance of a sample from any horizon to the cold (or warm) standard is measured by the correlation ratio of each vector (horizon) to the standard "cold" (or "warm") vector. Note that the correlation coefficient remains unchangeable against dilution of sediment with BiSi or any other "element-transparent" admixture (f.e., quartz concentration of which can vary from 15 to 35%) [84]. Large correlation coefficients mean high semblance of the sample to the standard, and at small coefficients the two can be expected to have more or less different spectra of element ratios (i.e., element composition). Because dimensionless concentrations are used, all measured elements contribute *comparable shares* to the correlation coefficient, irrespective of their absolute contents, i.e., the contributions of elements with concentrations from 1 to 10 ppm are comparable to those whose concentrations vary from 1 to 10 %. The set of correlation coefficients of the sample vectors against the standard vector is hereafter called *correlator*. See the correlators for the st15 and BDP 96-2 cores in fig.6. The coefficients of correlation of vector compositions from different core horizons to the standard cold vector (glacial correlator) and warm vector (warm correlator) were obtained as described above and renormalized to the variation ranges {0-1}.

It is easy show that the value of thus obtained glacial correlator for a given horizon is proportional (or is equal) to the weight percentage of "typically cold" material in the terrigenous sediment component, assuming sediment supply from only two terrigenous sources (a warm-riverine and a "glacial milk") at each moment of time. The same statement is valid for the warm correlator. If the record of warm correlator is negative mirror with the record of glacial correlator (it is certainly so, as we see on fig11a,b), we may assume that sediment composition at any time is presented a linear combination of compositions from a glacial and an interglacial sources. Thus, fig. 6 shows cold correlators - indexes of percentages of a glacial source in the terrigenous component of the st15 (upper panel) and BDP 96-2 (lower panel) cores, and the BiSi and WC profiles for comparison. The same result (fig11a,b) appears in correlation to the warm vector which in a first approximation corresponds to the composition of modern surface sediments. We stress again that the element concentrations in the vectors were calculated for the terrigenous component and are free from the BiSi dilution effect. Thus, all data in fig. 6 represent the variability of sources responsible for the terrigenous input to Lake Baikal from its watershed during climate changes.

The cold correlator for st 15, especially for the past 250 kyr, show prominent 40-kyr periodicity with a contribution from the 100-kyr cycle (see fig. 6). Therefore, glacial material came from one constant source presented by some weighted average flux from all watershed provinces during all stronger or weaker glaciations for the past 250 kyr. Thus, the integral geochemical climate proxies, the cold and warm correlators, faithfully track the glacial-interglacial rhythms (fig. 6) being very sensitive to climate change even during cold interstadials. The use of correlators allows gaining more information from the Baikal climate archives. According to [14-17,37], the interval of 120-200 cm for the st15 core corresponds to MIS 3, the 520 cm horizon to MIS 5.5, and the warm interval at 720-820 cm to MIS7. Brief warm and cold excursions (fig. 6) appear within the MIS 3 and MIS 6 interstadials (530-720 cm). The similar behavior we can see and for deeper intervals of BDP 96-2 (see, fig 6.)

3.3. Orbital Forcing and Fourier Dating of the Baikal Records

Paleomagnetic data for the BDP-96-2 core [15,16, 19, 86] indicate an almost linear depth dependence of age for the past 2 Myr. Then, the depth of a horizon (Z) is presumably related to its age (t) as

$$Z = \alpha\, t + \varepsilon\,(t)\,,\ \varepsilon\,(t) << \alpha\, t \qquad (2)$$

$$dZ\,/\,dt = \alpha + d\,\varepsilon\,(t)\,/\,dt$$

where $\varepsilon(t)$ is a small alternating function, a residual. The residual $\varepsilon(t)$ being small ($\varepsilon\,(t) << \alpha t$) does not mean that local sedimentation rate (dZ/dt) is always constant. The residual may change rapidly and its derivative ($d\varepsilon(t)/dt$) may be much greater or much less than α. The small magnitude of $\varepsilon(t)$ only indicates that faster sedimentation within a short interval is cancelled by slower sedimentation in the next short interval.

Proceeding from the astronomical theory of climate [54-63], we expected that the signals we obtained are modulated by orbital frequencies. A signal that bears a component with a period T_i years should, by (2), have a nearly periodic "spatial" component with the period $Z_i = \alpha T_i$. Therefore, if the signals bear depth-periodical modes with Z_1, Z_2, Z_3, etc., the spatial period ratio for any two modes with the periods Z_i, Z_k will equal the ratio of the respective orbital periods T_i, T_k. The ratio of the spatial periods to the respective orbital periods should remain constant for all modes and equal the mean sedimentation rate (over the interval): $\alpha = Z_i/T_i$. Using this algorithm, one can find mean sedimentation rates (α) in quite long core intervals.

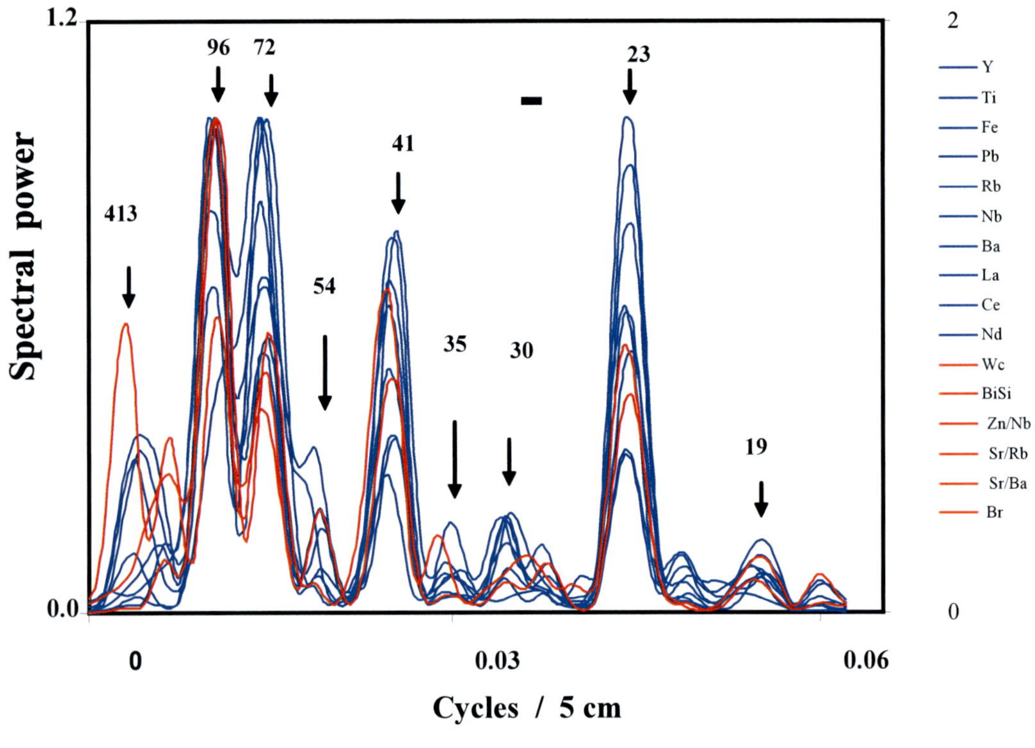

Figure 7. Fourier spectra of some geochemical records (Sr/Ba, Sr/Rb, Zn/Nb, BiSi, WC, Br in red and Y, Ti, Fe, Rb, Pb, Nb, Ba, La, Ce, Nd in blue) in BDP-96-2 core. Sedimentation rate in this core is about 5cm/ky (see table 1). Numerals are astronomic orbital periods (in kyr) of eccentricity (415, 123, and 96), obliquity (41), and precession (23 and 19). See clear signature of combined cycles at 72, 54, 35 and 30 kyr bands [40]. The responses of the Baikal watershed to precession and tilt variations are in some cases comparable with (or higher) the response to the nonlinear 100-kyr cycle

Figure 7 shows the Fourier spectra of some element profiles for the upper 22 m of the BDP 96-2 core. Both biogenic and terrigenous signals are clearly periodic and have the same set of spatial frequencies (but different phase angles). Numbers on fig. 7 are the orbital periods, which correspond to each peak , known from the Milankovitch theory and observed in marine records. Fourier analysis of data from fig. 7 is summarized in Table 1. The first line lists spatial periods (in cm) found in the Baikal records and averaged over five profiles of different signals (BiSi, Sr/Rb, Ba, Nb, La, Ce). Asterisk marks the "reference period of the 100-kyr cycle" prominent in global oceanic records [54-75]. The third line gives the astronomic orbital periods we correlate to the spatial periods in the records. Lines 2 and 4 are the ratios of the spatial and orbital periods, respectively, to the reference period. The two sets of period ratios agree well for all identified peaks (fig.7, Table 1). The resulting mean sedimentation rate over the uppermost 22 m of BDP 96-2 averaged over five profiles and six peaks (20 values) is 4.85 ± 0.05 cm/kyr.

The spectra of the Baikal records clearly resolve the eccentricity band (96-kyr cycle) and distinguish the 96-kyr and 125-kyr cycles, unlike the marine records where the two show weak or no distinction. Another strong cycles are the 54-kyr rhythm, which has a low amplitude in the insolation spectrum and is rather weak in marine spectra, and the 72-kyr cycle likewise weak in the latter. The clear 72-kyr periodicity, as well as the 54, 30, and 35 kyr cycles corresponding to the combined orbital bands (Fig. 7) may be due to strong nonlinearity of the climate response to insolation. Note in this respect the dominance of the 100-kyr eccentricity cycle which shows up in global change for the past 1 Myr though the eccentricity contribution to insolation variations is almost two orders of magnitude smaller than that from precession (23 kyr) and obliquity (41 kyr) [58, 59]. The Baikal records are apparently very sensitive to insolation, as well as to other orbitally forced climate controls such as the North – South air (or surface Atlantic) temperature gradients and/or precipitation. Perhaps, these additional frequencies may be caused by nonlinearity associated with the "memory effect". For instance, a mathematical model simulating the ice sheets waxing and waning [85] as a nonlinear system with "memory" reveals strong additional lines in the spectra corresponding to the periods of 72, 54, and 30-35 kyr. The 72 and 54-kyr modes in this model are comparable in amplitude to the 96- and 41-kyr modes, and the predicted spectrum of the climate response [85] is similar, in all parameters, to our spectra of the Baikal geochemical profiles (fig. 7). The reason why these lines are absent from the marine spectra may be either a low time resolution or a larger climate noise in the ocean relative to that on the continent [69-75]. The first the existence of the 72, 54, and 30-35 kyr combinational astronomical cycles in geochemical records from Lake the Baikal were found out in our investigations [20,40]. Recently the existence of these combinational astronomical cycles have been supported by spectral analysis of other proxies in the Baikal sediments, including magnetic properties [86], SiO_2 [87] and BiSi [88] profiles, and grain size composition [39].

The intensity of the 72-kyr rhythm and the appearance of the combined 30-35- and 54-kyr bands may be specific to local changes of continental climate in East Siberia. On the other hand, the combined frequencies can be quite trivially caused by orbital modulation of sedimentation rate which is probably also a climate-sensitive parameter. If sedimentation rate oscillates about its average at orbital frequencies, rather than stochastically as it is implicitly assumed, the spectra of linear-scale records will inevitably contain additional bands, exactly the same as those appearing in our records.

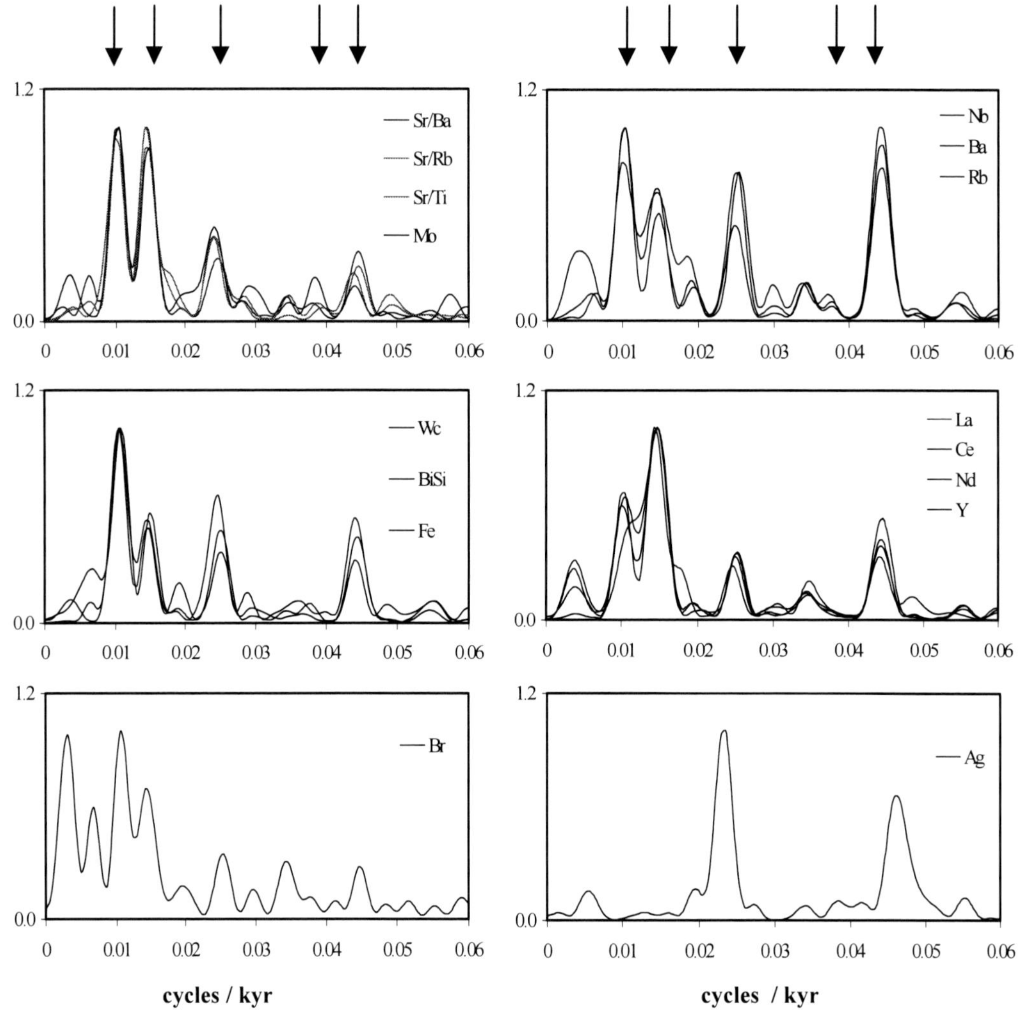

Figure 8. Fourier spectra of some groups of geochemical records in BDP-96-2 core. Note the peculiar responses of Ag and Br. Ag responds only to precession and obliquity; only Br responds to the long eccentricity cycle of 415 kyr. Slightly different responses of different element groups to orbital forcing is evidence that the processes are more complex than mere alternation of glacial and riverine sources. Arrows above show main orbital periods of 96, 72, 41, 23, 19 kyr.

Figure 8 shows the spectra of some element profiles within the same depth interval grouped according to their responses to orbital forcing, which are slightly different. The 100-kyr cycle is less clear than the 72-kyr cycle in the profiles of La, Ce, Nd, Y, and the purely insolation peaks of 41 and 23 kyr are especially prominent in the Nb, Ba, Rb group. The response of "warm" and biogenic tracers is between these extremes. Silver has a very strange response: it shows no long-period bands but bears only the 41- and 23-kyr peaks.

This notable discrepancy in the responses of different element groups indicates that the fluxes to the lake actually followed more complex patterns than in the simplified model with two sources. The average composition of terrigenous material during glacials was likewise non-uniform and the terrigenous input was alternating between few sources. These may have been a glacial and/or an aeolian sources, where the glacial one was from the region and the other was, possibly, dust carried from geographically distant areas. Or, it may have been glacial input from the western and eastern lake sides, which appears to be a more plausible scenario.

Table 1. Spectral analysis of uppermost 22 m of BDP96-2 core

"Spatial" period Z_i, cm	2005	471.8*	348.7	258.7	205. 7	111.4	91. 2
Z_i / Z*	4.3	1	0.74	0.55	0. 44	0. 24	0. 19
Orbital period T_i, kyr	413	96*	72	53.6	41	23	18.9
T_i / T*	4.3	1	0.75	0.56	0.43	0.24	0.19

Z* - "100-kyr cycle "

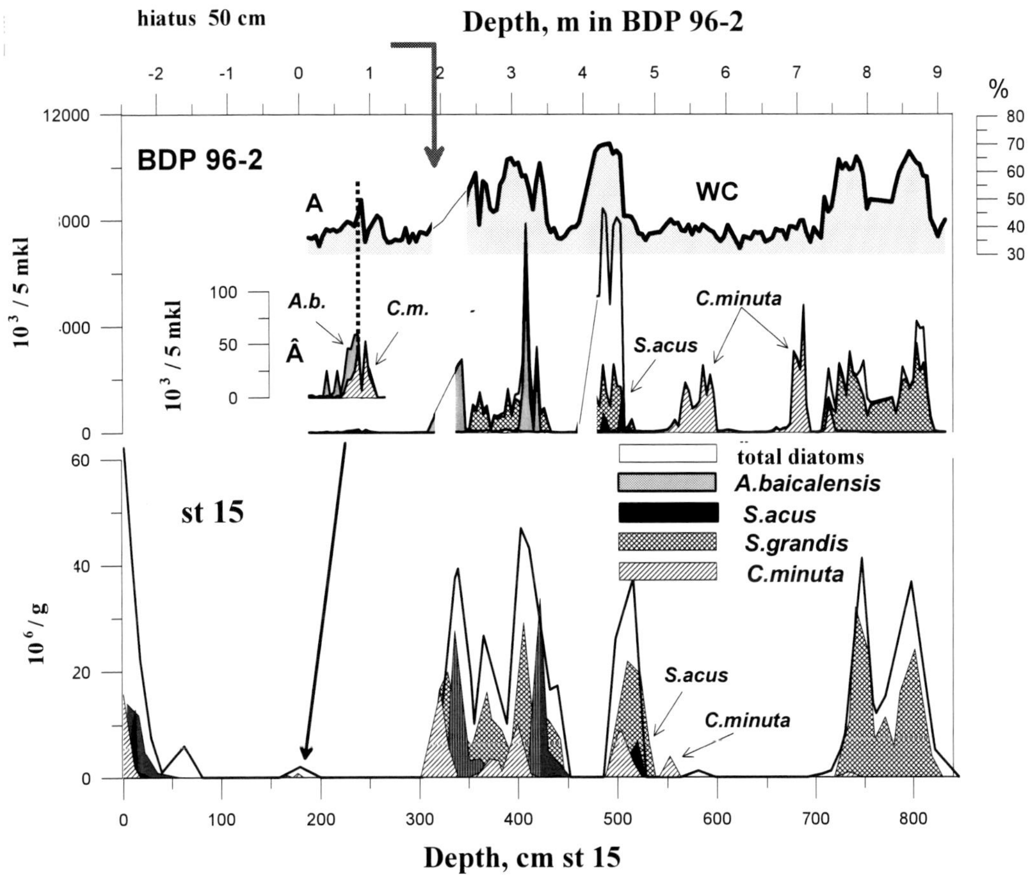

Figure 9. Correlation of diatom assemblages in st15 and BDP-96-2 cores [17, 37, 20]. Top panel: A - WC profile and diatom ensembles at BDP-96-2 (B – inset enlarges the Karga warming interval equated to MIS3); bottom panel: diatom ensembles at st15.

Table 2. Mean sedimentation rates at selected core intervals

Core interval BDP-96-2, cm	0 - 1000	0-435	435-775	500-2330	2380-3360
Mean sedimentation rate. cm/kyr	4.85	5.04	4.6	4.82	4.2

With the above algorithm, spectral analysis can be applied to estimate sedimentation rates which are further used to construct orbitally-tuned chronologies. In this study we use the simplest method of sliding window. The BDP-96-2 core is divided into three intervals of 0–1000 cm, 500–2330 cm, and 2380–3370 cm. The gap between 2330 and 2380 cm includes a "Mn anomaly" with Mn reaching ~ 2%, or an order of magnitude higher than its rather uniform contents throughout the core. This gap may separate zones of different sedimentation rates [89] and here may be a break in the age-depth model. Sedimentation rates in each of the three core intervals were estimated as shown above, and the final results are listed in Table 2. First we date the top horizon of the BDP-96-2 core, which misses the uppermost 1-2 meters, and then construct an orbitally tuned chronology.

The chronology can be obtained using several ties. Namely, the 0–1000 cm interval includes diatom peaks that correlate in species composition with peaks in the st15 core (Fig.9) where two interglacials have measured U-Th ages (see next section). The two control points are 129±4 kyr at 435 cm and 203±6 kyr at 775 cm core depth. Another tie is the paleomagnetic age (780 kyr, B/M) of the deepest horizon we measured in the BDP-96-2 core (3365 cm) [13-16, 19, 86, 88]. Let the mean sedimentation rates within the intervals 0–435 cm, 435–775 cm, and 0–775 cm be $\alpha 1$, $\alpha 2$, and $\alpha 3$, respectively. We reasonably assume that the mean sedimentation rate estimated for the 0–1000 cm interval (Table 2) holds at the slightly shorter interval of 0 to 780 cm. Since $\alpha 3$ is a weighted mean of $\alpha 1$ and $\alpha 2$ with the weights proportional to the respective interval lengths,

$$\alpha 3 = \alpha 1 \, (435 / 775) + \alpha 2 \, (775 - 435) / 775. \tag{3}$$

Using the obvious relationships

$$\alpha 2 \equiv (775 - 435) / (203 - 129) \tag{4}$$

and

$$\alpha 1 \equiv 435 / (129 - t0) \tag{5}$$

where t0 is the unknown age of the core top, and substituting (4) and (5) into (3), we obtain t0 = 42.7 kyr. According to (5), the sedimentation rate $\alpha 1$ at 0–435 cm is 5.04 cm/kyr. The equation like (5) for the whole interval from 0 to 780 cm

$$\alpha 3 = 775/(230 - t0), \tag{6}$$

gives t0 = 43.1 kyr.

With the found core top age and sedimentation rates at five overlapping core intervals (Table 2) we can obtain the time scale for BDP-96-2 by estimating the ages of horizons by simple integration over the five intervals:

$$t(z) = t0 + \int_0^z dz \,/\, \alpha(z) \qquad \text{at } 0 < z < 2330 \text{ cm} \qquad (7)$$

$$t(z) = 780 - \int_z^{3360} dz \,/\, \alpha(z) \qquad \text{at } 3360 > z > 2380 \text{ cm} \qquad (8)$$

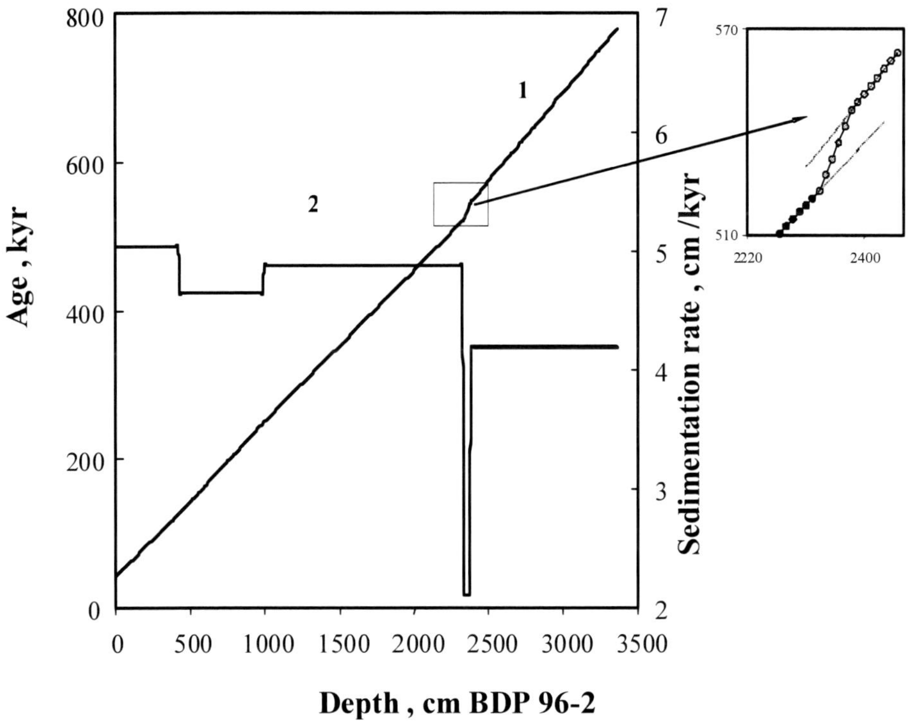

Figure 10. Orbitally tuned time scale for BDP-96-2. Inset shows conjugation of the age-depth (T-Z) model in the region of the Mn anomaly (at 2300 cm).

The 2330–2380 cm interval with the Mn anomaly was timed by linear interpolation from (7) and (8), for no *a priori* data is available on the behavior of the system in this interval. Thus, (7) and (8) give the sought chronology (fig. 10). Figure 11a,b shows the record of glacial (index of glacial material in sediment) and interglacial (index of riverine material in sediment) correlators for st15 and BDP 96-2, with the obtained orbitally-tuned time scale compared to June insolation at 60 N and ocean oxygen isotope ^{18}O. We can see almost synchronous response of Lake Baikal to the insolation variations. Data in figs. 2, 4-9, and 11a,b indicate that the response of the Baikal watershed correlates with the global signals but is, however, less inertial. The Baikal systems respond more strongly than the global ice volume to the glacial/interglacial cycles at the obliquity (41 kyr) and precession (23 kyr) bands, and these responses are as prominent as to the 100 kyr-cycle (fig 7,8,11). It means that the systems of the Baikal watershed are highly sensitive to insolation, what appears reasonable in view of its continental interior position and sharply continental climate.

3.4. Causes of the Orbital Response of the Baikal Watershed

The air temperature and humidity obviously had different patterns during glacials/interglacials in different parts of the huge Baikal watershed. The strongest climate response can be expected in the southern part of the area, in the Selenga catchment (Northern Mongolia), where the today's continental climate is very dry (see fig1), with high seasonal temperature contrasts. The change of climate periods inevitably influenced the terrigenous input to the lake. Lower river flow in a generally drier climate of glacials may have been attendant with greater glacial input. Furthermore, even with reduced riverine input, its relative percentages from the south and the north may have been redistributed because of different aridity intensities in the north region of Lake Baikal watershed and Mongolia (see fig 1). We expect that these changes are recorded in the behavior of tracer elements.

The reported data give reliable evidence of at least two different sources of terrigenous input which increase in different climates: one associated with sediment load carried by rivers, especially the Selenga, in warm periods and mainly a glacial source in cold and dry periods of lower river flow (see section 4). The glacial source supplied material enriched in Ba, Rb, Cs, La, Ce, Th, Nd, and Ti with the La(Ce,Ba)/Yb(Y,Zr) ratios higher than in times of warming (figs. 2, 4-6). Particles with this composition can be disseminated monazite from granites in the flanking mountains. Glacial input of terrigenous material grew due to faster mechanic erosion of rocks by glaciers, slope erosion, and glacial milk fluxes to the lake. It was the greatest at the onset of glaciations when glaciers expanded rapidly. It is the motion of glaciers rather than their mere existence that is the key agent of erosion. Peaks of glacial input during sudden warming right after long glacial periods (terminations) are produced by a burst discharge of terrigenous material deposited by quickly thawing glaciers in the watershed. Accelerated sedimentation at that time is an obvious and essential consequence of increased erosion during glaciations. Below we support this inference with direct U-Th dating.

Terrigenous material carried by rivers during warm periods has high U, Mo, Br, Sr/Ba(Rb,Cs,Ti), Zn/Nb and HREE (Yb, Tb, Lu, Eu, Y) and low La(Ce)/Yb(Y,Zr) ratios. We begin with discussing the component responsible for high U, Mo, Br and relatively high Sr/Rb(Cs,Ti) ratios. Bobrov et al [90] discovered a horizon in the BDP-93 core with very high Sr/Ba(Rb,La,Ce) ratios and explained it by high carbonate content in the sediment. Authors assumed that carbonates proceeded from erosion of carbonate complexes in Olkhon Island and its surroundings which have Sr/Ba ratios about ten times the crustal abundance [91]. Sr in these carbonates reaches 2000-8000 ppm (0.2-0.8%) while Ba and LREE are low. Of course, the particles of calcium carbonate, as well as dolomite, virtually fail to deposit in the lake sediments as they inevitably dissolve in the Ca-deficient Baikal water [3, 7, 92, 93]. Nevertheless, greater erosion of carbonates in the watershed during interglacials appears very probable. The more rapid leaching in warm periods increased the concentrations of Ca and Sr ions in the Baikal water and thus facilitated authigenic precipitation of calcium phosphates. The product of ion concentrations Ca2+, PO43-, and OH- in today's Baikal exceeds the solubility product for hydroxyl apatite [94] and, hence, its authigenic precipitation is compatible with thermodynamics. Experimental measurements by Zmodik et al [95-96] revealed Fe-rich apatite-like phosphate at several depths. Deike et al [97] reported that iron-manganese crusts in Lake Baikal are high enrichment by Ca, P and some other elements. Our SRXRF scanning with 100 mkm resolution of P-bearing layers [95] in a radiographically measured sample indicated very high contents of U, As, Sr, Mo, Br, Y and some other

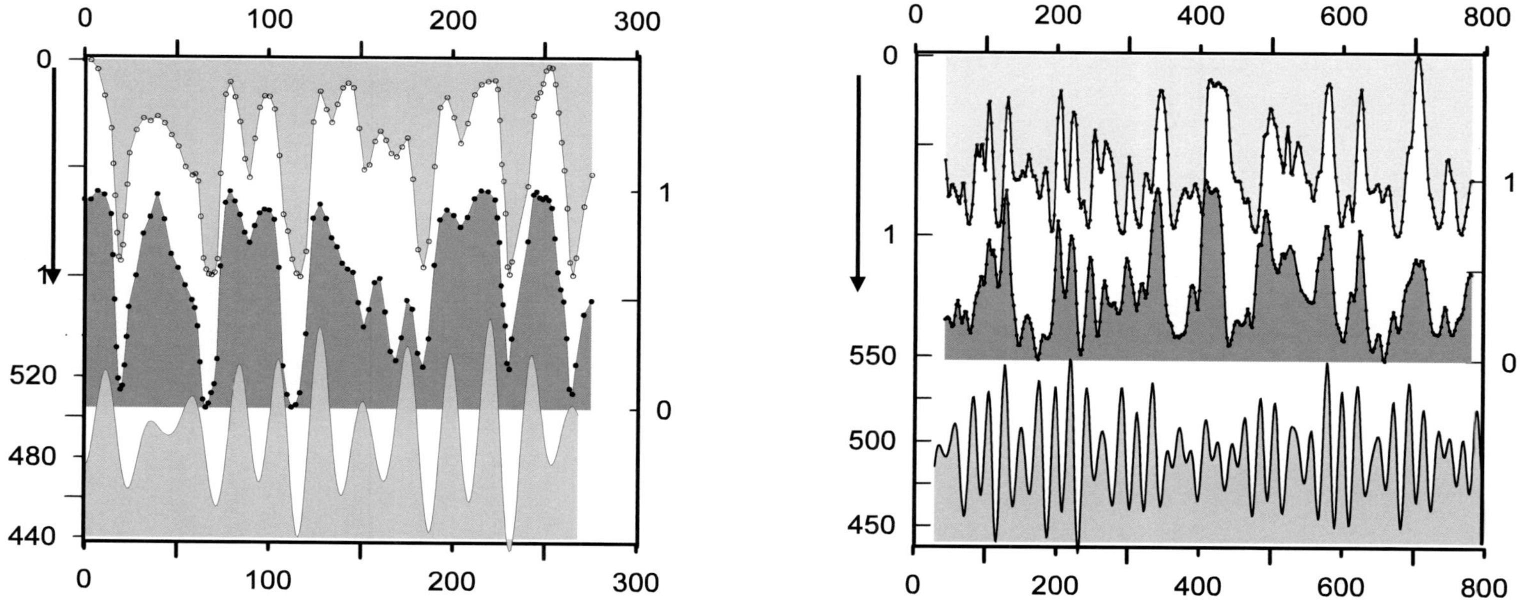

Figure 11a. Glacial and interglacial correlators for st15 and BDP-96-2 cores at the obtained orbitally tuned scales. Note almost synchronous Baikal responds to June insolation in both cases. Both correlators track almost all precession maximums (and minimums) of insolation. Both cores show mirror symmetry of the glacial and interglacial correlators (see text for explanation).

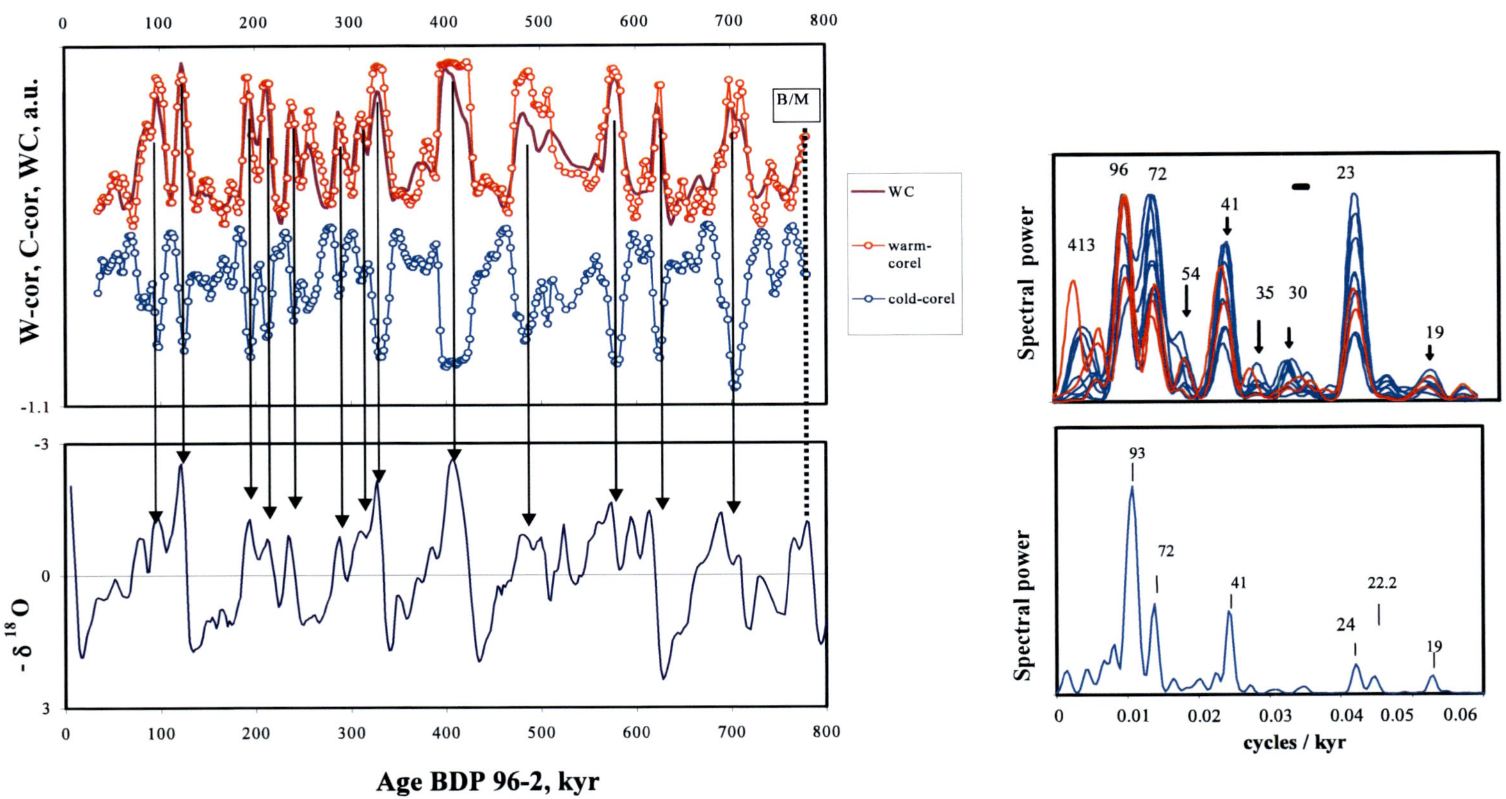

Figure 11b. Glacial and interglacial correlators for BDP-96-2 core at the obtained orbitally tuned scale together with ocean oxygen isotopes (left). Power spectra for Baikal region (continental Asia) and ocean are shown (right). Well-marked high frequency (41, 23 kyr) periodicity on continent is clear.

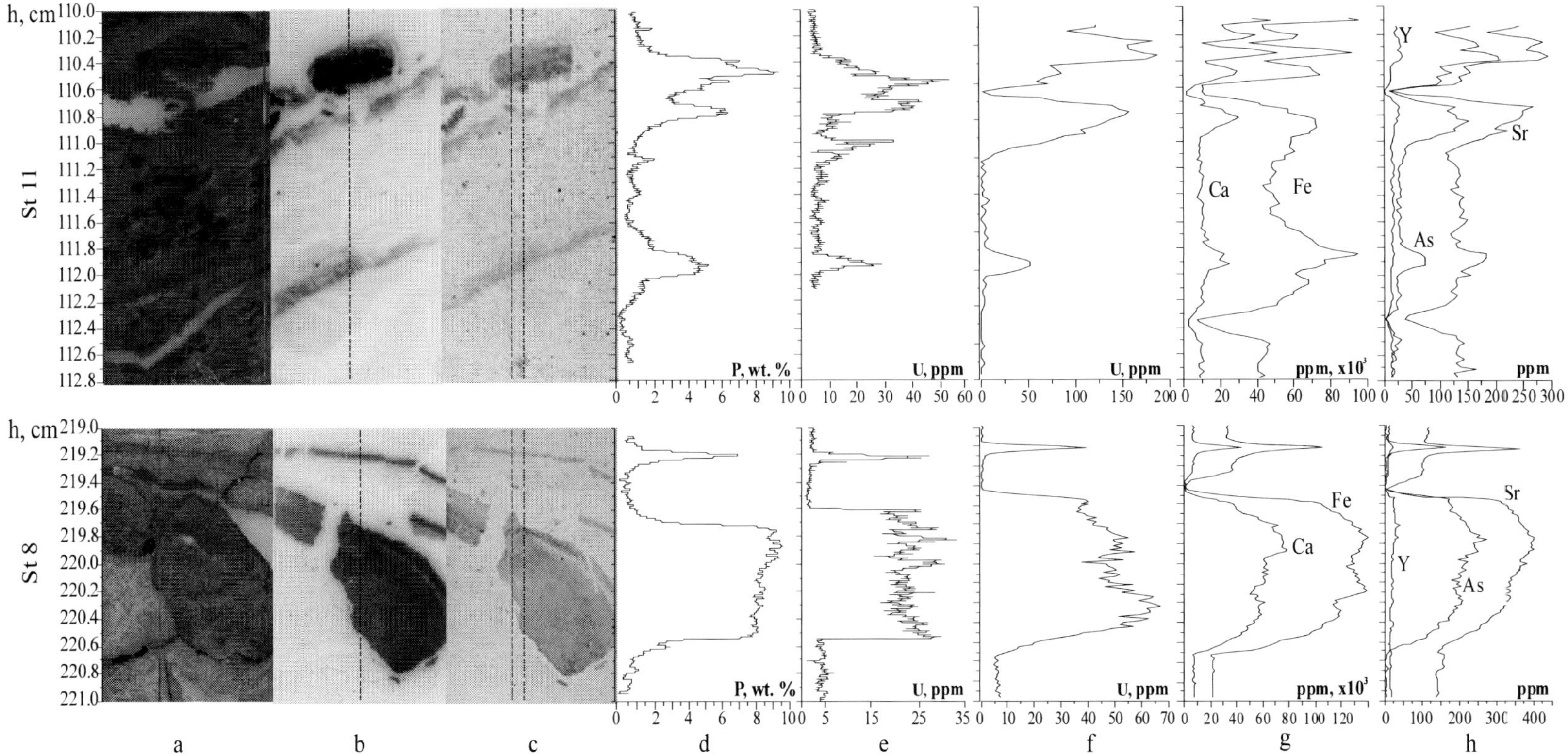

Figure 12. Autoradiography and element patterns in thin sections of phosphate layers. Cores st 8 and st11 are from Akademichesky Ridge, Lake Baikal.
a - photographs, d–autoradiography of phosphorous profile, e – autoradiography of uranium profile, f, g,h – element profiles measured by scanning SRXRF. Vertical lines are photometry lines in b, and SRXRF scanning band at a collimated 3 mm and 100 mkm beam in c. Data are from [95, 96]. The phosphate layers are highly enriched in elements (U, Sr/Rb, Ca/K, As, Y) typical of interglacial periods in lake sediments.

elements of the warm stack at those depths (Fig. 12) [96]. Almost the same elements are accumulated by iron-manganese crusts [97, 98]. If phosphates or remnant crusts Fe-Mn crusts are quite frequently encountered in interglacial periods, the presence of some warm-stack elements which come to the lake in the dissolved form can be accounted for by their irreversible adsorption and/or co-precipitation with iron hydroxides and/or phosphates.

Thus, the geochemical response recorded in the bottom sediments of Lake Baikal is the climate-induced periodicity of increase/decrease in terrigenous input from different sources which became dominant in different climates. The transport of solids being closely linked to water flow, the element signal bears contributions from the water input itself controlled by climate humidity and from the composition of the water that carries the particles; it also bears imprint from different compositions of waters flowing from thawing glaciers (in cold times) and from plainland inlets (in warm times).

4. U-Th Ages

U-Th dating of Baikal sediments stems from decay of ^{238}U according to (simplified scheme without short-living isotopes)

$$^{238}\text{U} \underset{4{,}5\cdot 10^{9}\text{ y}}{\longrightarrow} {}^{234}\text{Th} \underset{24\text{ days}}{\longrightarrow} {}^{234}\text{U} \underset{2.48\cdot 10^{5}\text{ y}}{\longrightarrow} {}^{230}\text{Th} \underset{7.5\cdot 10^{4}\text{ y}}{\longrightarrow} {}^{226}\text{Ra}$$

In old bedrocks the slow decaying parent ^{238}U is in secular equilibrium with its progeny ^{234}U and ^{230}Th, and their activities (A_8, A_4 and A_0, respectively) are equal. Chalov and Cherdyntsev, Russian geochemists, found out in the 1950s that uranium dissolved in natural water always has its activity ratio $A_4/A_8 > 1$ [99-103]. Up to present day the causes of the parent-daughter disequilibrium remain not quite clear but most models invoke recoil-effects as a reason of preferential mobilization of ^{234}U atoms from minerals to the surrounding microenvironment. By alpha-decay of ^{238}U atoms, the emitted daughter short-living ($\tau_{1/2} \sim 24$ days) ^{234}Th atom can either fall directly into the water phase from the near surface layer (~200Å) of particle or move from the crystalline domain to defect sites (fig. 13 [101, 103]).

There it becomes more easily washed out, as ^{234}U, by water percolating the defects if the characteristic leaching time exceeds the half-life of ^{234}Th (24 days).

Edgington et al. [42, 43] have found that the Baikal water is not an exception but the initial ^{234}U-^{238}U disequilibrium of the soluble uranium in the lake is high enough: $A_4/A_8 = 2$. Note for comparison that this ratio in natural waters reaches 8 in some areas of Kyrghizstan and Kazakhstan and is locally as high as 10 in Georgia [99-102]. Diatomaceous sediments in the Baikal cores from the Akademichesky Ridge are enriched in ^{238}U (U anomalies first discovered by Gavshin et al. [104]). Furthermore, the A_4/A_8 ratio is much above unity in diatom-rich layers and approaches unity in diatom-barren layers (fig.14, from data [42]). Deep diatomaceous sediments also have ^{230}Th above the secular equilibrium ($A_0/A_8>1$).

These facts were interpreted using a geochemical model [42, 43] in which soluble uranium is adsorbed by iron-hydroxide films which cover the surface of terrigenous particles and thus becomes buried in lake sediments. The adsorbed uranium inherits the A_4/A_8 ratio of the soluble uranium. The A_4/A_8 ratio in terrigenous particles produced by erosion of bedrock along the lake shore is, on the contrary, about unity.

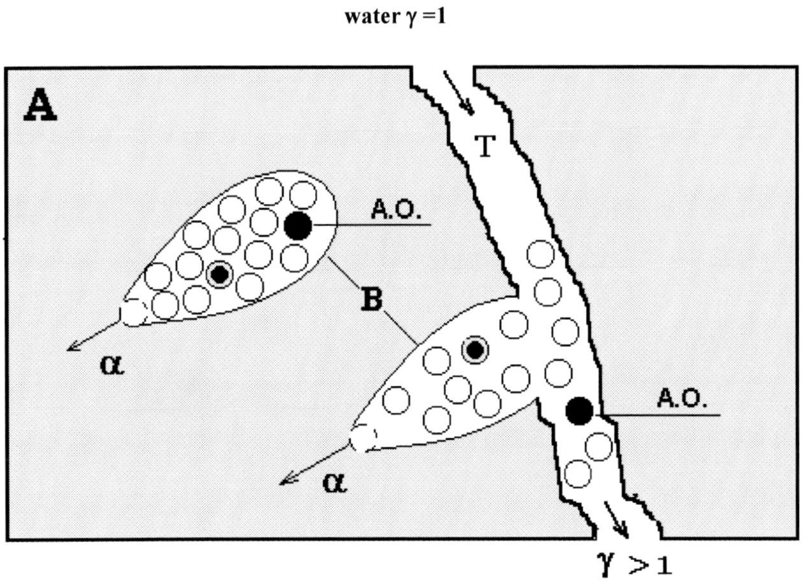

Figure 13. Sketch of the Chalov-Cherdyntsev effect, after Chalov [100, 101]. Formation of tracks (lattice disordering sites) by recoil atoms with $\gamma = {}^{234}U/{}^{238}U > 1$. In the case of ^{238}U decay, ^{234}Th (τ1/2~24 days) is the nuclear recoil atom. ○ - ^{238}U atom, ◉ - ^{234}U atom, ●- recoil atom (R.A.) ◌- position of ^{238}U before decay, arrow points to direction of α emission, A-undisturbed lattice domain. B-lattice domain disordered by recoil atom (shift peak, track). T– defect site in crystal (microcrack, grain boundary, dislocation wall, dislocation nuclei, etc.). After α decay, the recoil atom may, to certain probability, fall into the domain T where it disturbs the U-U equilibrium and increases the $^{234}U/^{238}U$ ratio. Both isotopes are leached from the domain T at the same rate but the leachates have $^{234}U/^{238}U > 1$ because of high concentrations of recoil atoms at defect sites. The shift peak (track), which is located inside the crystal and does not contact the surface defects, eludes the effect of weak solutions that leach uranium from defects; the disturbed domain relaxes to the original state in some time (~3-10 kyr on average) [101].

Excess of adsorbed (authigenic) ^{234}U buried in the Baikal sediments gradually decays into disequilibrium ^{230}Th which, in turn, decays according to the above scheme and, eventually, the activities of all ^{238}U series isotopes become identical. This is exactly the prerequisite for U-Th dating of sediments. The time dependence of the activity ratio $R(\beta, t) = A_0/A_4$ in the authigenic component is [53]

$$R(\beta, t) = [1-\exp(-\lambda_0 t) + (\beta-1) f] / [1 + (\beta-1) \exp(-\lambda_4 t)] \quad (9)$$

where $f = [\exp(-\lambda_4 t) - \exp(-\lambda_0 t)]\ \lambda_0 / (\lambda_0 - \lambda_4)$

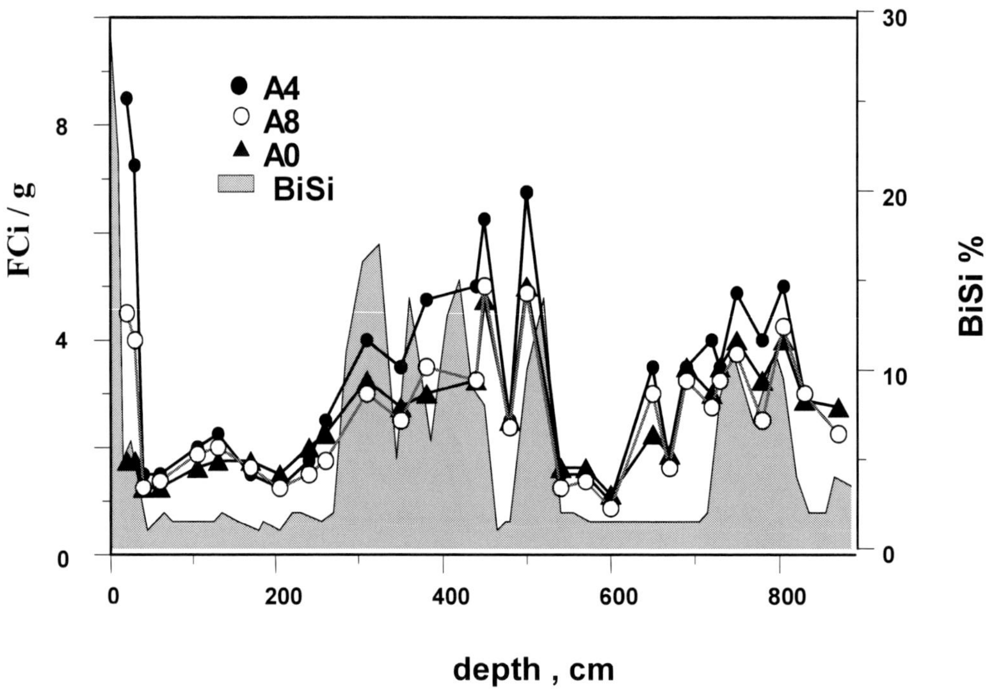

Figure 14. Isotope data borrowed from [42] for st18 (Akademichesky Ridge, Lake Baikal). Warm periods are characterized by both high activities of all U-isotopes and $^{234}U/^{238}U$ ratio [42].

There λ_0 and λ_4 are the ^{230}Th and ^{234}U decay constants of $9.193*10^{-6}$ and $2,83*10^{-6}$ years^{-1}, respectively, t is time (years) elapsed since the sediment burial, and β is the activity ratio A_0/A_4 of soluble uranium to authigenic uranium at the time of burial. The dependence R(β, t) for β = 2 is shown in fig.15. Note that R(t) weakly depends on β (fig. 15): the age uncertainty even for a 200 kyr sample is no more than 10 kyr at β variations as large as from 1.5 to 2.5.

Had uranium in the Baikal sediments been only authigenic, dating would reduce to measuring the A_0/A_4 ratios at each horizon and estimating its age from R(β, t). However, uranium is of authigenic and terrigenous origin, and the measured activities are the sums of the respective components:

$$A_8 = A_{8aut} + A_{8ter} \tag{10}$$

$$A_4 = A_{4aut} + A_{4ter} \tag{11}$$

$$A_0 = A_{0aut} + A_{0ter} \tag{12}$$

The ^{238}U decay being very slow, the activities A_8, A_{8aut}, and A_{8ter} are independent of sediment age. The geochemical model implies that the ^{238}U-series isotopes in the terrigenous fraction are in secular equilibrium, i.e., $A_{8ter} = A_{4ter} = A_{0ter}$ = Ab at each horizon. Ab for each horizon is independent of isotope half-life and of time since the burial but depends on the sediment composition. It can vary with mineralogy and is thus, in the general case, a function of depth and, hence, of age. Thus, the ratio A_0/A_4, being a function of time *t* due to radioactive decay and a function of depth Z due to mineralogy variations, cannot be used for dating without additional assumptions.

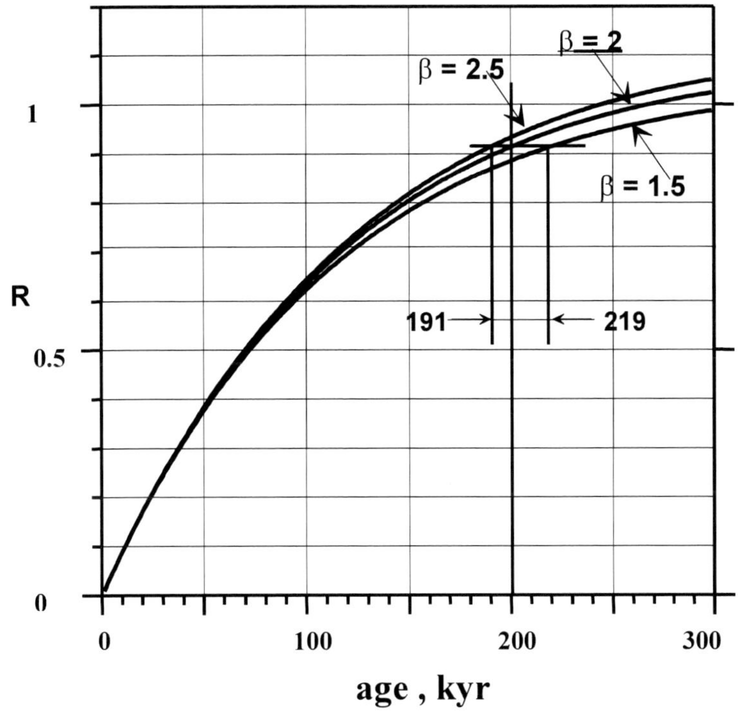

Figure 15. Time dependence of R(β, t). Age uncertainty is within 1.5 kyr at β variations in a broad range in sediments with ages about 100 kyr [53], but the uncertainty caused by β departure (±0.5) from 2 is as high as 10 kyr for ~ 200 kyr sediments.

Edginton et al. [42] estimated mean sedimentation rate in the Akademichesky Ridge proceeding from correlation of U profiles to the SPECMAP record. At station 18 it was ~3.5 cm/kyr for the past 230 kyr. Direct isotope dating were not applied because of its quite low precision [42] and the isotope ratios were used only as a proof that the obtained time scale is reasonable. They measured isotope activities by classical alpha spectrometry [42] which is time consuming and low efficient. We developed an alternative analytical technique based on inductively coupled plasma mass spectrometry (ICP-MS) [44, 53]. See the cited papers for samples treatment and analytical procedures and [53 and references therein] for the processing algorithms used for dating. Figure 16 shows data from [53] where we applied ICP-MS to investigate the patterns of U and Th isotopes in the st15 core (Akademichesky Ridge). We estimated the U-Th ages of two BiSi-rich intervals of the st15 core corresponding to MIS5.5 (Kazantsevo interglacial) and MIS7.1 using the methd of "narrow interval" developed in [53].

As follows from (9)-(12),

$$A_0 = A_4 \cdot R(\beta, t) + Ab \cdot [1 - R(\beta, t)] \qquad (13).$$

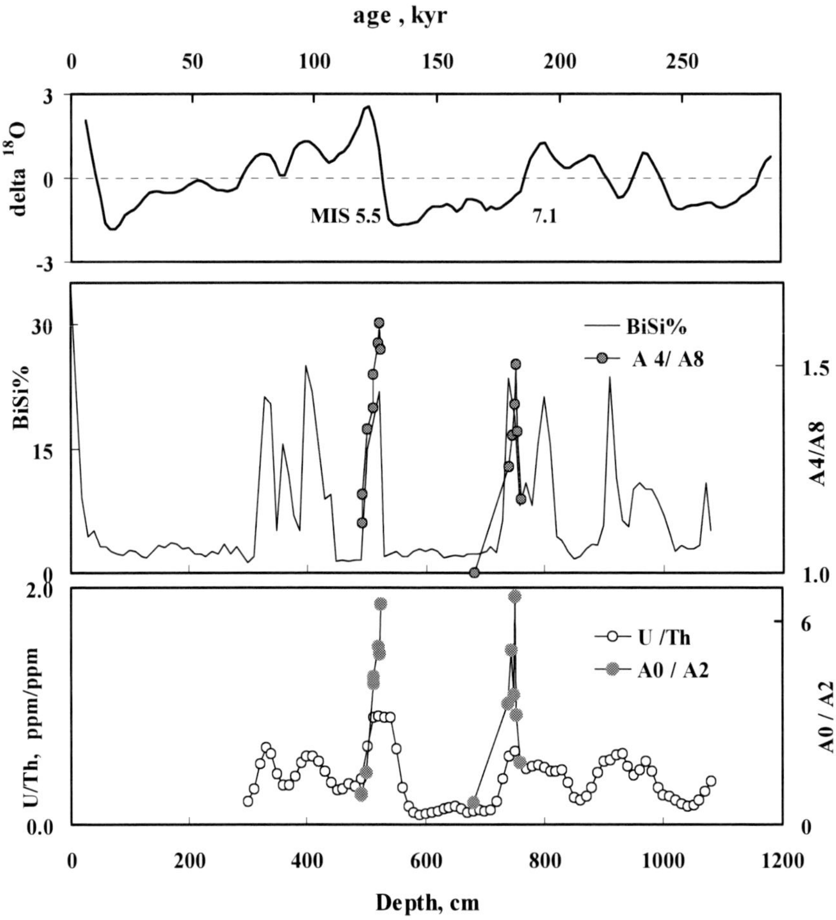

Figure 16. A4/A8 and A0/A2 ratios in st15 core against the BiSi profile. The quasi-isochrone method is applied to selected narrow intervals with high gradients of isotope ratios (see text, [53]).

Besides the terrigenous matter, dry Baikal sediments contain biogenic silica which influences Ab(z) through dilution because biogenic silica almost lacks uranium [47]. Therefore, we used the activities A_0 and A_4 divided by the activity A_2 of ^{232}Th, which does not belong to the ^{238}U series, rather than A_0 and A_4 proper:

$$A_0/A_2 = R(\beta, t) \cdot A_4/A_2 + (Ab/A_2) \cdot [1 - R(\beta, t)] \tag{14}$$

The A_0/A_2 and A_4/A_2 ratios should bear a weak or no imprint of dilution with BiSi.

Method of "narrow intervals" (quasi-isochrones). We may assume that within a narrow interval the age increment (Δt) is very small ($\Delta t \ll t$) and that $R(\beta, t)$ is constant and measures the average age of the interval. Then, assuming an invariable terrigenous source within this short interval, we may infer that the Ab/A_2 ratio is also constant (though unknown). The A0/A2 vs. A4/A2 dependence can be found for a number of samples from the short "quasi-isochrone" interval. This dependence is linear according to (14): ($y = a + bx$), where the slope

(b) equals the average R(β, t) in the selected narrow interval and the intercept is a = (Ab/A_2}·[1- R(β, t)]. The slope and the intercept of the quasi-isochrone are determined to the best accuracy in the core intervals where A_8/A_2 gradients are high (i.e., correspond to the rise and/or dip of ^{238}U peaks, see fig. 16) and the variation ranges of A_0/A_2 and A_4/A_2 are the largest. See the isochrones for two intervals of the st15 core in fig. 17 where the data fit well equation (14). Having measured the slope of the isochrone R(β,t) and the intercept (Ab/A_2)·[1- R(β, t)], the $(^{238}U/^{232}Th)_{ter}$ ratio in the terrigenous sediment component for the given interval is found as Ab/A_2.

The activity ratio A4/A8 we measured by ICP-MS for soluble uranium in today's Baikal turned to be 2.09, which is compatible with the values of 2.0 to 2.3 in [42]. We found R(β,t), assuming that β = 2 as the isochrones were obtained for warm intervals of the st15 core (fig. 17), and then estimated the average ages of MIS5.5 and MIS7.1 at 131 ± 4 kyr for the middle of MIS 5.5 (R= 0.75 ±0.013) and 202 ± 9.7 kyr for MIS 7.1 (R= 0.916 ± 0.017).

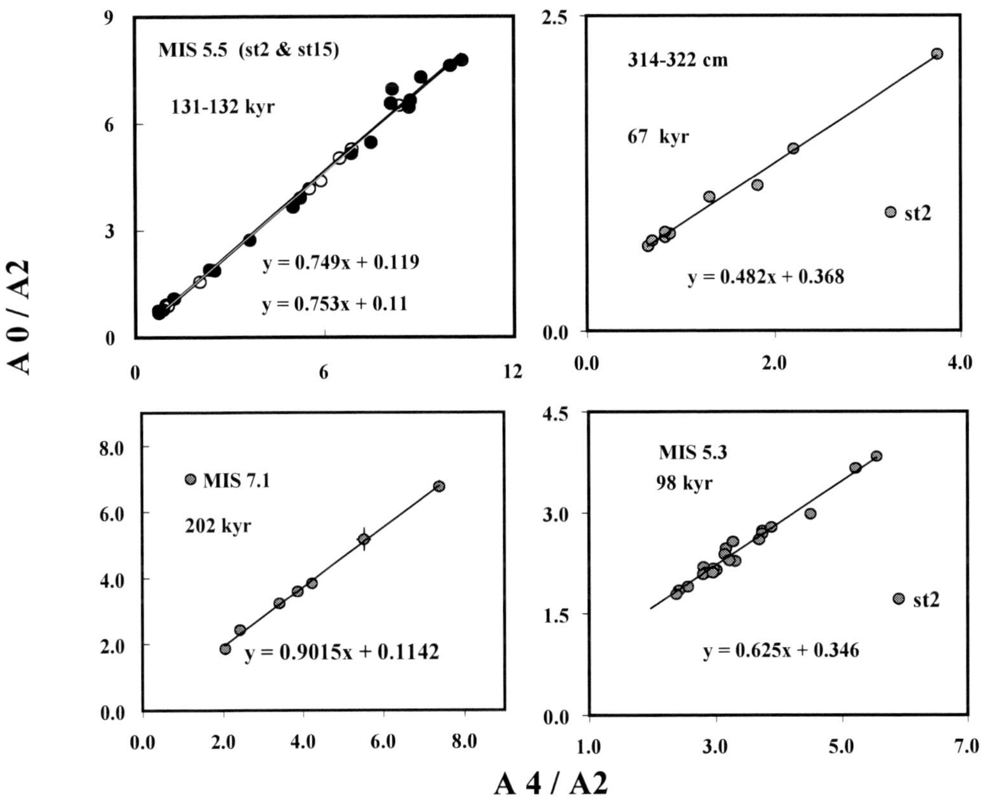

Figure 17. Quasi-isochrones for st2 and st15 cores. Intervals and ages are shown in the figures. Data for MIS5.5 in the two cores lie on the same line. The average age of the Kazantsevo interglacial in Siberia (MIS5.5) is ~ 130-132 kyr for both cores.

These ages generally agree with the age-depth model of Colman et al. [14] though there are small lags (~3-4 kyr) relative to SPECMAP data (see below). Thus, our dating method allowed direct identification of MIS5 and MIS7 in core st 15.

The younger core intervals were dated through further isotope studies. Chebykin et al [44] determined the concentrations of ^{238}U, ^{234}U, ^{230}Th, and ^{232}Th at 520 horizons of the st2 core recovered at station 2 in the vicinity of station 15 in the Akademichesky Ridge (see fig.18 for the results [44]).

With such a large collection of data, we expected to reconstruct the continuous age-depth profile. The problem has not been completed so far because of high natural noise and imperfection of the geochemical model. Our key assumption of the equilibrium of the terrigenous component, i.e., the equality of A_0, A_4 and A_8 and the lack of ^{230}Th excess (see above), failed because the activity A0 at cold and transitional intervals turned to be much above the equilibrium (fig.18). The simplest explanation for the disequilibrium is that meteoric water washes out highly soluble uranium from terrigenous sediments staying long exposed on the surface while less soluble thorium, including excess ^{230}Th produced by the decay of excess ^{234}U, remains in the watershed. Excess ^{230}Th gets into bottom sediments during rapid terrigenous deposition, e.g., associated with formation of glaciers. Thus we failed to obtain reliable U-Th dates for MIS3 and determined the ages only for the intervals where the A4/A8 ratio was notably above 1.

The peak of A_4/A_8 profile at st2 at 540-570 cm on fig. 18 is clearly identified as MIS5.5 by correlation of diatom assemblages in the st2 and st15 cores. Furthermore, the quasi-isochrones for the peaks MIS5.5 at both stations are perfectly aligned (fig.17). This match is more so surprising that the analyses were made with different instruments (one in Brussels and the other in Irkutsk), preconditioning procedures, and analytical details, and, finally, the very cores were different.

Other diatom peaks of st. 2 were dated in several ways. The first, we applied the above quasi-isochrone method in the vicinity of the horizons where A_4/A_8 was above 1.4 (fig.17) and obtained an age of 98±4 kyr for the peak at 420-460 cm (fig.17), almost coinciding with the age of MIS5.3, and 67±4 kyr for a single peak at 314-322 cm. The quasi-isochrones for MIS 5.1 (~350 cm) and MIS3 (the Karga) peak (about 180 cm) are very noisy and the respective ages are poorly reliable without additional assumptions. The second, we calculated $R(\beta,t)$ by averaging data from several narrow intervals corresponding to warm stages (for the intervals see fig. 19). For each interval it was found as a ratio of values averaged over 15-20 neighbor horizons: $R(\beta,t) = (A_0/A_2 - Ab/A_2)/(A_4/A_2 - Ab/A_2)$. This became possible because sampling rate was as high as to highlight the background Ab/A_2 (hatched area in fig. 19). The background ratio of 0.4-0.6 agrees well with data from st15 (see below). The age determined after calculating $R(\beta,t)$ was attributed to the middle of the interval.

The third, we calculated $R(\beta,t)$ at each horizon, where $A_4/A_8>1.45$, i.e., for the warmest horizons, assuming the same ratio Ab/A_2 as in the second step. The rigorous selection criterion (the warmest horizons with $A_4/A_8>1.45$) provides reasonable depth-dependent aging at a mean sedimentation rate of 4.4 cm/kyr (see fig. 19 for the dating results). The ages obtained in the three ways, though slightly not identical, are quite proximal. The warm interval centered on 350 cm with the age ~80-90 kyr corresponds to MIS5.1 (fig.19), right as we expected. Both the quasi-isochrones and the calculation of average $R(\beta,t)$ for the 150-250 cm core interval gave an age of 40-50 kyr. Although the dates for this interval are of lower quality (large scatter) than for other peaks, it undoubtedly corresponds to MIS3 (Karga warming in Siberia) which has an average age of 40-50 kyr.

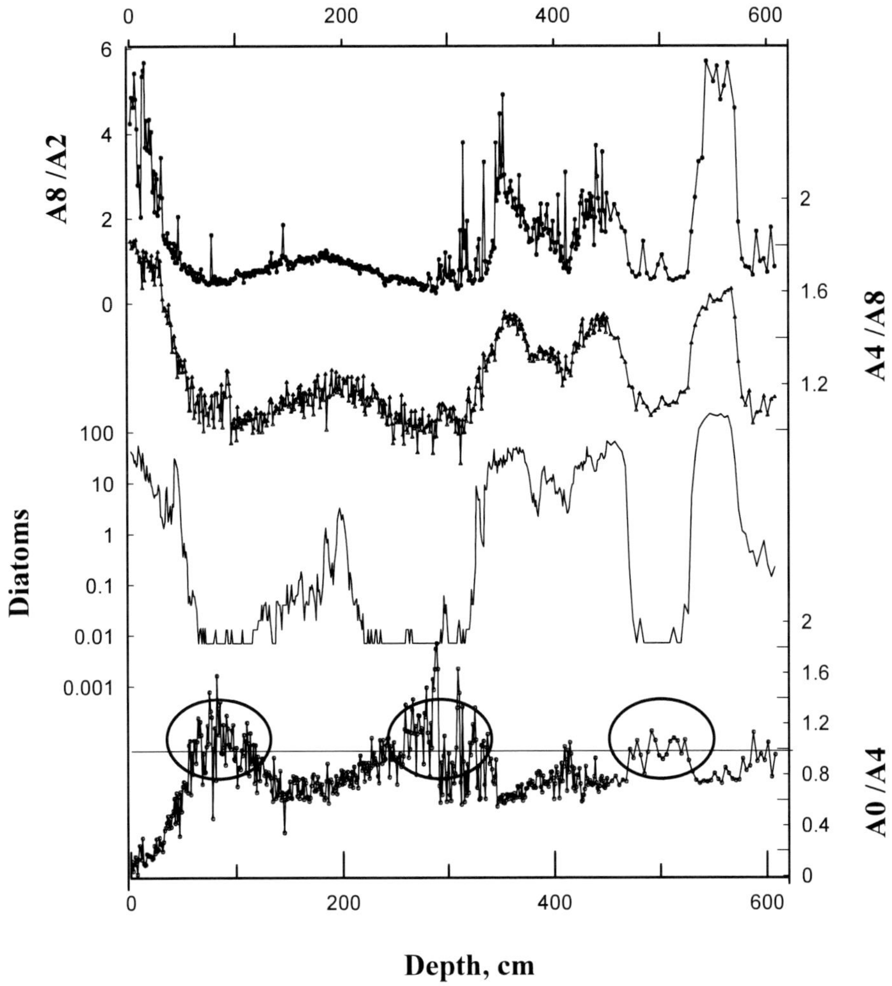

Figure 18. Records of diatom abundances and uranium-series isotopes in st2 core [44]. Ellipses contour areas of uranium equilibrium or/and excess ^{230}Th. Excess ^{230}Th ($A_0/A_4>1$) is found during the reliably identified glaciations of MIS2, MIS4, MIS5.4.

Thus we obtained several direct age determinations beyond the 30 kyr limit of AMS ^{14}C age methods and clearly identified all warm intervals for the past 140 kyr. This is direct evidence that the Baikal ecosystem responded to climate events almost synchronously with global change. Below we discuss how accurate this synchronicity is.

The concentrations of ^{234}U, ^{238}U, and ^{230}Th are the highest, much above the terrigenous background, and A8/A2 are the highest in MIS 5.5 (figs.16, 18), and we expected to obtain the most accurate ages for that interval. For this interval we used more sophisticated ways of dating, remaining within the limit of our geochemical model [53]. As we wrote above, the isochrones for st2 and st15 lie exactly on the same straight line (fig. 17) which indicates good quality of the data. We obtained the same average age of MIS5.5 130-132 kyr±4 kyr for both cores, with a quasi-isochrone covering the whole stage (see fig.17). To improve our dating procedure we used quasi-isochrones with sliding window at several narrow intervals within MIS 5.5 on st15, followed by averaging, yielded an age of 129±4 kyr for the BiSi peak corresponding to MIS 5.5 (R = 0.74 ±0.014) and the average intercept $Ab/A_2 = 0.58 \pm 0.08$ over MIS5.5 [53]. The same procedure for the BiSi peak corresponding to MIS7.1 in st15 gave 198±4 kyr (R=0.909±0.07) and $Ab/A_2 = 0.65 \pm 0.41$. [53]

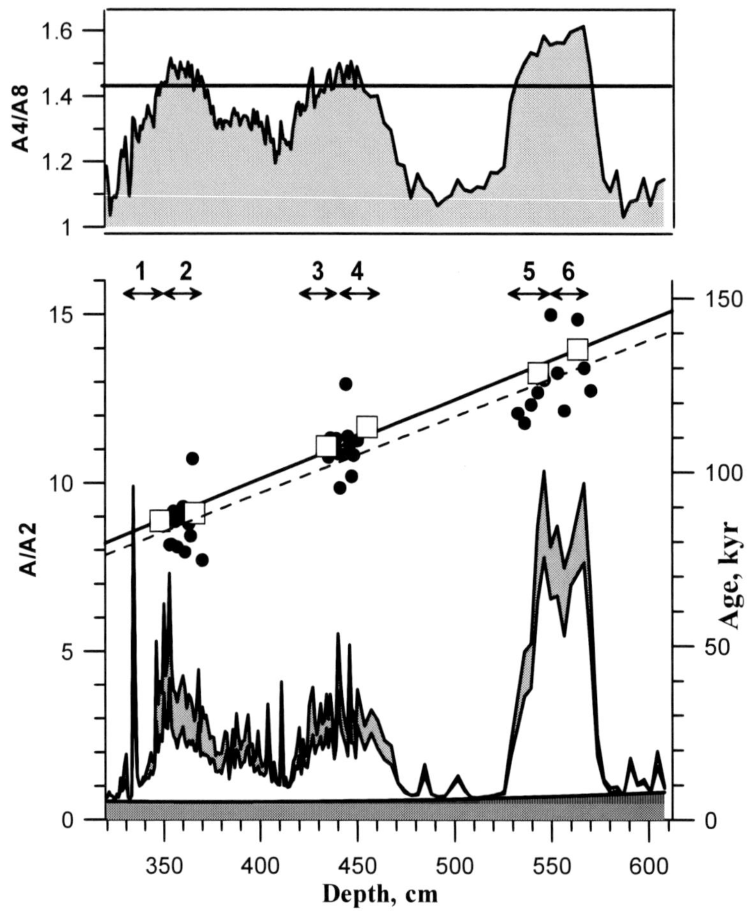

Figure 19. Ages of horizons in st2 core determined by different methods. Arrows with numerals show data averaging intervals (their ages are shown by squares). Black dots are ages of horizons with A4/A8 >1.45 (points of peaks in top panel). Dashed line Z-T drawn through the black dots gives a core-average sedimentation rate of 4.3 cm/kyr over the 300-560 cm interval (MIS5.1–MIS5.5).

The uncertainty of β varying as $\beta = 2 \pm 0.2$ gives only a minor age uncertainty (see fig 15). If we obtained the average terrigenous background for MIS5.5 ($Ab/A_2 = 0.58$) which comes after the quasi-isochrone procedure, we may evaluate ages of each horizon in the core with a uncertainty ca 1-2 ky after substitution of $Ab/A_2 = 0.58$ in equation (14) with β=2.

Moreover, since Ab/A2 = 0.58 comes from averaging quasi-isochrones within MIS5.5, we are able to perform dating in each horizon in the core by using the ^{234}U-^{238}U geochronometer (15) additionally to the ^{238}U-^{230}Th one (16). The substitution of Ab/A_2=0.58 into (15)-(16) gives the complete system of initial equations of ^{238}U-^{234}U-^{230}Th unequilibrium, of which (9) was obtained as a specific case and solution of the system gives us both the age *t* and the initial disequilibrium β in water at each horizon, thus obtaining a set of ages within MIS5.5.

$$A_4/A_2 = A_8/A_2 + q \cdot (A_8/A_2 - Ab/A_2) \cdot \exp(-\lambda_4 t) \quad (15)$$

$$A_0/A_2 = Ab/A_2 + (A_8/A_2 - Ab/A_2) \cdot [1 - \exp(-\lambda_0 t) + q f] \quad (16)$$

where $q = \beta - 1$

Figure 20 shows the age-depth profile for the BiSi peak corresponding to MIS 5.5 at station 15 after solution of (15), (16) with Ab/A2 = 0.58. The resulting ages are ~129 kyr for the peak middle, 130-131.5 kyr for its maximum, 134-135 kyr for the transition from MIS6 to MIS5.5 (525 cm), 113-115 kyr for 5.5/5.4 transition and the duration of the MIS5.5 interglacial in Siberia is about 20 kyr. These estimates correlate with the ages of the respective transitions in the SPECMAP curve, in the Vostok ice-core record [65], and in the Devils Hole records from Nevada [105, 106]. Nevertheless, there is some systematic discrepancy (fig. 21). The events we dated slightly (4-5 kyr) lead the respective events in the marine $\delta^{18}O$ curve, but Transition II in Lake Baikl took place almost simultaneously with insolation increase (cf with fig11). Moreover, duration of MIS 5.5 in Lake Baikal (~20 kyr) is almost the same as in the Antarctic (~18-20 kyr) and Nevada (~23 kyr). This is quite an expected result because the marine $\delta^{18}O$ is a global ice volume proxy and the polar ice sheets obviously have a great inertia and their thawing may postdate global temperature rises or regional insolation and climate events. U-Th-coral timing of sea level fluctuations reported recently by [107-110] gives solid evidence that the sea stood only 10-20 m below the today's level 132-134 kyr ago, when the regional climates approached the optimum, while the total amplitude of sea level rise at the glacial-to-interglacial transition reached 126 m.

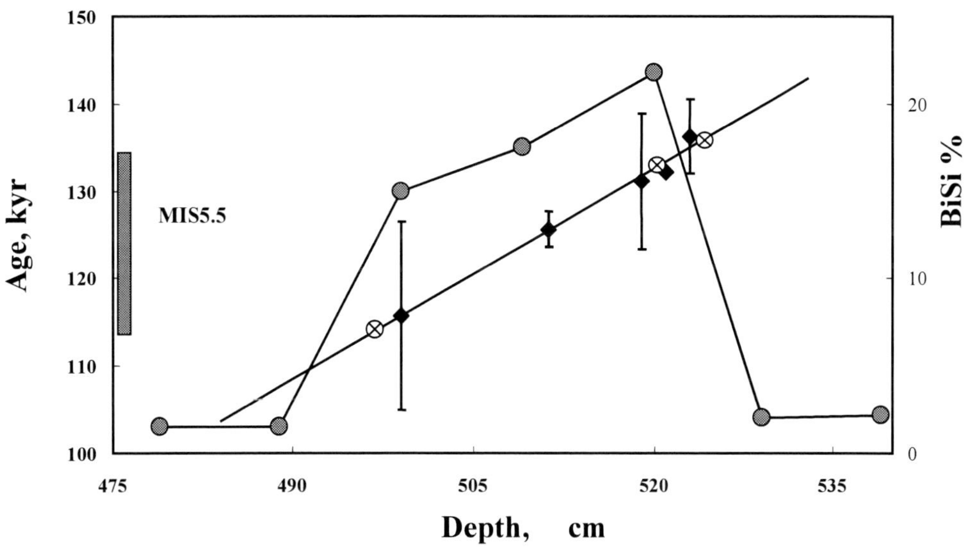

Figure 20. Isotope dating of MIS5.5 at st15. The MIS5.5 beginning, end, and maximum are marked by circled crosses (in BiSi record).

Thus, the new data are at odds with the uniformly accepted SPECMAP profile and the respective age of termination II (129-130 kyr BP [62, 63]). The onset of warming was almost synchronous in different regions and followed insolation increase (fig. 21), but inside the warm stages the regional climates responded to insolation events with lags of several millennia. Insolation fell to the minimum 118 kyr ago but both the ice sheets and the continents "missed" it, and the interglacial continued as least 4-5 kyr longer. Something similar appears to happen today: insolation peaked 11 kyr ago and is now declining to the minimum. Does it mean that a new glacial is not far off?

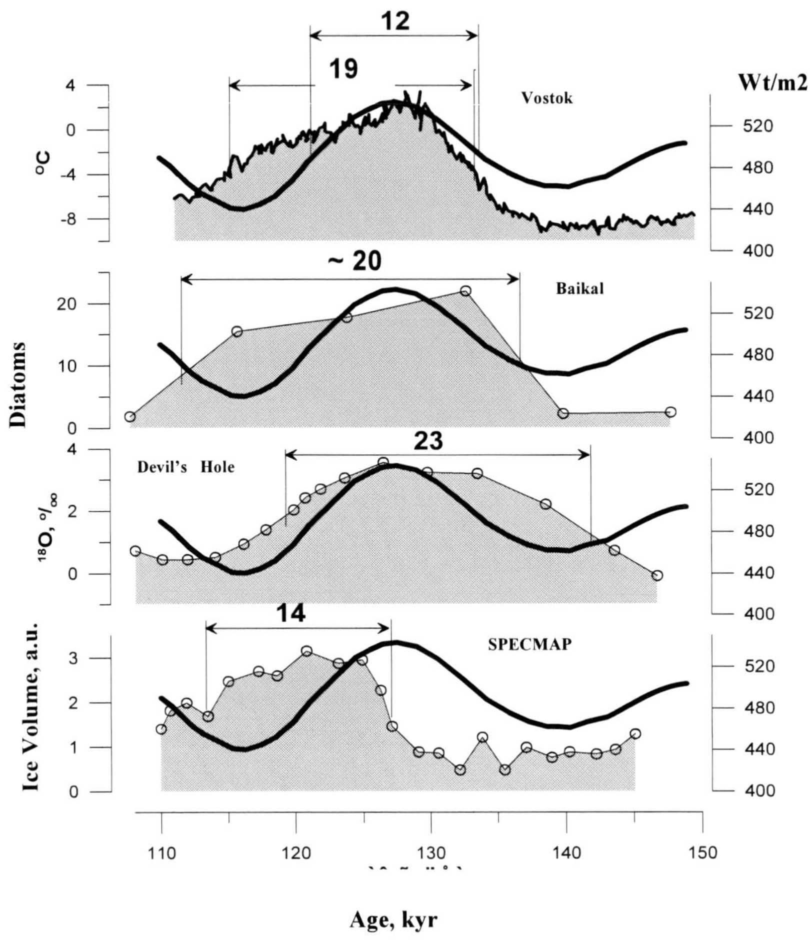

Figure 21. Baikal and global Vostok [65], Nevada [105, 106], and marine $\delta^{18}O$ (SPECMAP) [62] records at the respective time scales, compared with July insolation at 65N (black line) [60].

Turning back to the Baikal sediment, we note that its U-Th age at MIS 5.5 increases monotonously with depth (fig. 20). Using the dates within MIS 5.5 we can estimate the sedimentation rate. It turns out to be ~ 1.25 cm/kyr (fig. 20), or about four times less than the core average for st15 (525 cm/128 kyr = 4.1 cm/kyr) over a time span including glacials and interglacials. Of course, the numerical derivative in fig. 20 may bear a large error, but slower sedimentation during MIS5.5 was proven independently by estimating sedimentation rate with (15)-(16). It is easy to show that as follows from (15), (16)

$$\lambda_3 t = \ln (f_0 + 1/q) - \ln [f0 - (A_0 / A_8 - 1) / (A_4 / A_8 - 1)] \quad (17)$$

where $\lambda_3 = \lambda_0 - \lambda_4$, $f_0 = \lambda_0 / (\lambda_0 - \lambda_4)$.

With the mean sedimentation rate V introduced for the studied interval, (17) gives the relation of the depth *Z* and the function $Y1 = -\ln [f0 - (A_0/A_8 - 1)/(A_4/A_8 - 1)]$:

$$Y1 = \lambda_3 Z / V + Y0, \tag{18}$$

where Y0 is some constant including the term ln (f0 + 1/q). At small variations of *q*, the function (18) is a straight line in the coordinates Y1 – Z with a slope angle λ_3/V. In an arbitrary case, equation (18) can be used for approximate estimates of sedimentation rates. This function is shown in fig. 22 for MIS5.5 in st15. The slope angle corresponds to a sedimentation rate of ~0.8 cm/kyr, which is comparable with the above rate of 1.25 cm/kyr and supports the slower sedimentation during the warm period of MIS 5.5. This rate is less than that in the Holocene (about 3 cm/kyr at station 2), even with the correction for sediment compaction (1.3 times) in MIS 5.5. Sedimentation during other warm periods was apparently also slower than during glacials. Here we note that had the sedimentation rate within MIS 5.5 been the same as the core-average for st15 (~ 4 cm/kyr), the 25 cm wide MIS5.5 interval (fig. 16, 19) would be as short as 25/4 ~ 6 kyr (10 kyr according to the interval base). This duration appears improbable being times shorter than at least 18-20 kyr in the Antarctic records [64-66] and 22-23 kyr in the Nevada record [105, 106] (see fig 21), and even less than the minimum of 11-13 kyr estimated by correlation of BiSi records with SPECMAP data [87]

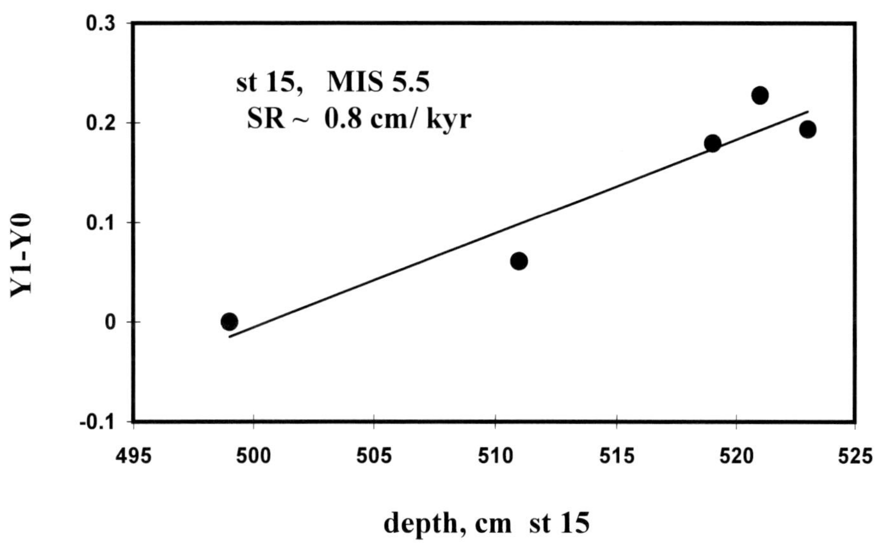

Figure 22. Depth (Z) dependence of $Y1 = -\ln [f0 - (A_0 / A_8 - 1) / (A_4 / A_8 - 1)]$ for MIS5.5 at st15. Slope is proportional to mean sedimentation rate in the interval (see eq. 18).

5. Millennial-Scale Climate Oscillations in Lake Baikal

5.1. Overview

Abrupt millennial-scale climate oscillations has been a key discovery of today's paleoclimate science [111-113]. First millenial-scale events were discovered in Greenland ice-core records of the last glacial period [114-123]. Dansgaard et al., [114] reported evidence for instability of past climate through the last glaciation and found numerous events of rapid (within a few decades) and high (7-15°C) increase in mean annual temperature [118] and greater snow accumulation rates [123] (Dansgaard-Oeschger events, D/O cycles). D/O events lasted 300 to 400 years in which temperature in the Northern Atlantic region remained as high as half the present values or higher and then fell, as rapidly as it was rising, to the typically glacial level. Few spells of warmth within Greenland MIS3 were as long as 800-1200 years or longer. Trains of several short warming/cooling cycles (Bond cycles [119]) ended with rafting of ice from the Canada and Greenland ice sheets into the Atlantic ocean and were followed by ~ 2 kyr long very cold and dry periods (Heinrich, or H-events [115]). H-events are marked by IRD (ice-rafted debris) horizons in North Atlantic sediments between 45 and 60° N composed of large terrigenous particles icebergs carried to the ocean from the continent [119-122, 124-126]. The ice-rafting events occurred every ~ 6000-9000 years and the D/O cycles every ~ 1500-3500 years, in average. The climate instability was not confined to the last glaciation but appears also in the records that span at least 500 kyr of the past climate history [124,126], with excursions of abrupt warming or cooling in glacials as well as in interglacials.

All D/O events through the last glacial were anyhow attendant with abrupt climate wetting, expansion of peatlands and the ensuing increase in atmospheric methane, the marsh gas [128-131]. The events began with temperature rise that was followed, in 20 to 70 years, by methane increase [127]. During the cold phases of each D/O event, North Atlantic thermohaline circulation (THC) slowed down, less heat was transported in the global oceanic conveyer belt [131-138], and the North Atlantic deepwater formation zone (NADW) moved south to 50-55°N. Over H-events the oceanic conveyer transported heat to the Southern rather than Northern hemisphere. Such THC change caused extreme cooling and drying in the Northern Hemisphere and relative warming in the Southern Hemisphere [111, 112, 132, 138].

Less prominent Holocene climate oscillations left imprint in Atlantic sediments, in mountain glaciers of the tropic zone, in pollen assemblages and in lake sediments of mid-latitude America, Africa, and Europe, in records of thermohaline circulation, etc [139-149, 150]. Therefore, even the relatively stable Holocene climate was repeatedly disturbed by abrupt events, which were weaker than in the Pleistocene but strong enough to induce shoaling of large lakes, displacement of vegetation belts, droughts and desertification (e.g., Sahara desert appeared about 6000 years ago [145]). The climate instability may have been responsible for bloom and collapse of civilizations and historic migrations [146-147].

Millennial cycles cannot be due to orbital forcing. Their rapidity and great amplitude prompts that they must be driven by complex nonlinear mechanisms [111-114]. The millennial-scale events bear no clear periodicity but the distribution function of intervals between them has two peaks at ~ 1.5 and ~ 5 kyr [121-123]. Furthermore, the high-frequency component of the Fourier spectra of the Greenland ice-core record has a peak corresponding to a periodicity of ~ 1.5 kyr (see [121-123] and references therein). Thus, there must be some pacemaker that switches on every 1-2 kyr and triggers warming/cooling and wetting/drying of global climate [121-123].

Interest in millennial-scale climate instability has been quickened recently in view of evident rapid global warming increased through the 20th century (0.6°C for 100 years on average), especially since 1975. The economical and political consequences of sudden warming are obvious. Some densely populated areas risk to be flooded because of polar ice melting and sealevel rise while long intense droughts will inevitably occur in other areas [1, 2]. Note that scientists are not unanimous about the true causes of warming in the past 100-150 years. Many paleoclimate experts doubt about man-induced forcing and report evidence for natural mechanisms [141, 142, 149]. For instance, Broeker interprets the 20th century global warming as a common D/O event that immediately followed the Little Ice Age associated with THC slowdown (LIA in Europe between 1350 and 1850 was one of coldest spells over the past 10 kyr [141,142])

Therefore, investigation into Siberian millennial-scale climate events and their pacemaker appears to be a top-priority objective.

5.2. Uranium and Diatom Signals as Humidity Proxies in Lake Baikal. A Conceptual Model of Uranium Response and Century Resolved Reconstruction of Humidity History in East Siberia for the Past 100 kyr

The annual water budget of today's Baikal consists of riverine input from the watershed (Q_1, 58.8 km^3/yr), output through the Angara, the only river that flows out of Lake Baikal (Q_2, 60.4 km^3/yr), precipitation to the lake surface and condensation (Q_3, 10.1 km^3/yr), and evaporation from the lake surface (Q_4, 10.3 km^3/yr) [3-6]. Mean annual precipitation (P_{modern}) in the Baikal watershed is currently 400 mm/yr [3-6]. With the watershed surface area of 557,000 km^2, this corresponds to a moisture input of 223 km^3/yr. Thus, 164 km^3 or 75% of incoming moisture evaporates yearly from the watershed (E_{modern} = 295 mm/yr). The past precipitation and evaporation patterns obviously changed through time. The climate was dry during glacials when both evaporation and precipitation decreased. If, however, evaporation in the watershed remained in excess of moisture input, Baikal might have become a closed lake.

We modeled the humidity history in East Siberia for the past 100 kyr from the profile of authigenic uranium in bottom sediments of Lake Baikal [45]. Authigenic uranium was estimated from excess ^{234}U relative to the secular equilibrium with its parent ^{238}U, because the input of excess ^{234}U can have been only in the dissolved form according to the mechanism of the Chalov-Cherdyntsev effect (see section 4). The ^{234}U and ^{238}U concentrations in the 450 cm long U-Th dated core (see section 4) were used to estimate the current activities A_4 and A_8. The activity of ^{234}U at the time of burial (A_{40}) was found from the U-U geochronometer equation as

$$(A_{40}-A_8) = (A_4 - A_8) \exp (\lambda_4 t) \quad (19)$$

Proceeding from the geochemical model, we assume that uranium came to the lake (i) with terrigenous particles (U_{ter}) where it is in secular equilibrium (A_4=A_8) and (ii) as disequilibrium soluble uranium carried by rivers (U_{diss}) where $A_4/A_8>1$. Uranium was adsorbed in the lake by sedimenting terrigenous particles and buried on the bottom (U_{ads}).

Diatoms are not active uranium absorbers: the greatest portion of uranium (10-15 ppm) concentrates in fine sediments with grain sizes less than 2 mkm while the diatom fraction contains as little as <0.4 ppm U [47]. This proportion was corroborated by recent careful ICP-MS measurements of uranium abundances and ratios in modern diatoms which yielded, respectively, 0.1-0.2 ppm or even less than reported in [47], and $A_4/A_8=1.97$, or equal to that in the lake.

Uranium becomes adsorbed from water most likely by colloids or iron-hydroxide films that deposit on fine clay particles or is bound by colloidal phosphates which either form in water or are adsorbed onto iron hydroxides (see section 3). Such processes are well-known [151-159]. So, uranium sorption by ferric hydroxides and phosphates is an appreciated tool in industrial water cleaning [156-158]. Tight sorption of phosphates onto iron-hydroxide particles, likewise well known, is responsible for 10 to 40% uranium buried in marine sediments [153]. Furthermore, purposeful removal of iron from sediment in our laboratory experiments gave almost ten times lower uranium sorption from Baikal water.

Whichever be the further history of uranium after its burial, it remains immobile in chemically reducing sediments and stable against any post depositional processes. Not going far into ways and mechanisms of uranium adsorption onto sedimenting particles, we only postulate that this uptake – suggested and grounded in [42, 43] – is the key agent of authigenic uranium supply to Lake Baikal.

Uranium adsorption is characterized by the distribution coefficient $K_D = [U_{aut}]/[U_{diss}]$ (concentration on particles/concentration in water) [42, 42, 151, 155]. Authigenic uranium in this ratio refers only to the terrigenous sediment component because of no sorption by diatoms. Thus found K_D is an apparent rather than a true constant which is inversely proportional to the specific area of sediment-water interface and grows as the adsorbing terrigenous particles become of smaller linear sizes.

Thus, in our geochemical model, the variability of sedimentary A_4/A_8 (fig. 18,19) documents the alternating contributions from the U_{aut} and U_{ter} components. The former increased with riverine input of soluble uranium in warm periods (U_{riv}) and the latter was associated with terrigenous input from glaciers that thickly covered the mountains around Baikal in cold periods. The qualitative model with two uranium sources is compatible with the above model of geochemical response to orbital forcing which implied large terrigenous fluxes to the lake with tributary rivers in warm periods and from glaciers in cold periods. Thus, increased river flow and ice waning (warm cycles) are expected to be attendant with increase in authigenic uranium in sediments whereas climate drying (cold cycles) is assumed to cause reduced river flow together with growth of terrigenous uranium influx and relative decrease in its authigenic component.

In the conceptual quantitative model of river flow history, we extended the initial assumptions of our geochemical scenario. First, we consider a one-box model and, hence, an average watershed without division into provinces [45]. It has a time-dependent paleoflow and rock lithology averaged over all provinces and the respective averaged uranium isotope concentrations and ratios in tributary rivers. The model can be further developed by giving up the rigorous initial premises. First we obtained the concentrations of authigenic uranium (adsorbed from water) at the moment of burial from the measured bulk A_4, A_8 using eq. (19) and assuming validity of Edginton's model $\beta = A_{40}/A_8=\text{const}= 2$ [42].

Then we surmise, as in any isotope model, that the sediment becomes a closed isotope system after uranium burial. Another assumption is that the sorption properties of sediment and the K_D coefficient have been constant through the past 150 kyr. Here we proceed from the idea that the sizes of terrigenous particles reaching the Akademichesky Ridge are controlled by offshore distance to their deposition site and are thus invariable. Later we may cancel this assumption by measuring the specific surfaces of sedimentary material at each horizon and applying the respective corrections to local K_D coefficients. We use the age-depth model (U-Th dating, section 4) and here we only use a few dated points with linearly interpolated ages.

The mass balance of dissolved uranium in Lake Baikal is given by

$$V \cdot dU_{dis}/dt = Q_1 \cdot U_{riv} - Q_2 \cdot U_{dis} - S \cdot \hat{J}_{ads} \qquad (20)$$

where V is the lake volume (23,000 km^3), U_{dis} is the current concentration of dissolved uranium in Baikal, U_{riv} is the weighted mean concentration of soluble uranium in tributary rivers, S is the lake surface area (31,400 km^2), and $\hat{J}_{ads}$ is the mean flux of adsorbed uranium U_{ads} to the lake bottom. The terms Q_3 and Q_4 that represent precipitation to and evaporation from the lake are excluded from the water budget as not being the uranium carriers. Furthermore, Q_1 and Q_2 differ very little and the evaporation-precipitation difference returns to the lake as a flux of water condensed in the flanking mountains [5, 6]. Thus, in the first approximation we assume $Q_1 \approx Q_2 \approx Q$, as it currently is. According to the definition of K_D, the flux of U_{ads} is

$$\hat{J}_{ads} = \hat{J}_{ter} \cdot U_{ads} = K_D \cdot U_{diss} \cdot \hat{J}_{ter} \qquad (21)$$

where $\hat{J}_{ter}$ is the mass flux of terrigenous particles to the bottom and U_{ads} is the concentration of authigenic uranium adsorbed onto terrigenous particles estimated with regard to BiSi dilution of sediment. Substituting (21) into (20) gives

$$V \cdot dU_{dis}/dt = Q \cdot \{U_{riv} - U_{dis}\} - S \cdot K_D \cdot U_{dis} \cdot \hat{J}_{ter} \qquad (22)$$

Knowing the current values of U_{ads} and $\hat{J}_{ter}$ and using (22), the momentary value of the paleoflow Q can be found as

$$Q = (V \cdot dU_{ads}/dt) / (K_D \cdot U_{riv} - U_{ads}) + S \cdot K_D \cdot U_{ads} \cdot \hat{J}_{ter} / (K_D \cdot U_{riv} - U_{ads}) \qquad (23)$$

The second term in the right-hand side of (23) is a stationary solution and the first term is a nonstationary addition. The uranium concentrations in present Lake Baikal are 0.42 ppb U_{dis} and 1.05 ppb U_{riv} [92, 93], which are only slightly different from those we measured and those reported by Edgington [42]. Then we calculated K_D in present Baikal for terrigenous particles only, proceeding from the mean concentration of adsorbed uranium U_{ads} in five uppermost horizons of a 5-cm sediment layer, a 27% BiSi content, and the known concentration U_{dis}. The resulting ratio is $K_D \equiv U_{ads} / U_{dis} = 37400 \pm 300$. Note that our laboratory experiments with uranium sorption by the BIL1 standard sample of Baikal sediment gave almost the same value.

The profile of the terrigenous flux (j_{ter}) to the Akademichesky Ridge (fig. 23) was obtained using the current water content in sediments, measured dry sediment bulk density (DBD), sedimentation rate (SR), and known biogenic silica contents (BiSi).

The flux j_{ter} was then used to calculate the modern flux of adsorbed uranium $j_{ads} = U_{aut} \cdot j_{ter}$, which in turn was compared with the mean lake flux of buried adsorbed uranium per Baikal unit area found from the difference of input (with rivers) and output (through the Angara) fluxes of soluble uranium. The total modern uranium input and output were estimated from measured U concentrations in rivers and in the lake (the latter is same as in the Angara) reported in [92, 42]. Thus, the modern budget of adsorbed uranium was found as

$$\hat{J}_{ads}^{mod} = Q^{mod} \{U_{riv}^{mod} - U_{dis}^{mod}\} / S \qquad (24)$$

The lake mean uranium flux ($\hat{J}_{ads}^{mod}$) relates to that at the Akademichesky Ridge (j_{ads}^{mod}) as $\varepsilon = \hat{J}_{ads}^{mod} / j_{ads}^{mod} = 12.3$, which obviously corresponds to the proportion of terrigenous accumulation rates in the lake and at the Akademichesky Ridge ($\hat{J}_{ter}/j_{ter}$). Slower terrigenous deposition at the Akademichesky Ridge lying far from the lake sides, the sources of terrigenous particles, is not surprising. Indeed, mean sedimentation rate at the Akademichesky Ridge is times as low as in the South Baikal basin (4-5cm/kyr against ~ 40 -70 cm/kyr) where the greatest amount of terrigenous particles comes with the Selenga, the largest Baikal tributary [183]. Therefore, the expected ratio of the lake mean terrigenous flux to that at the Akademichesky Ridge is ~10-16 and is compatible with the above estimate from the uranium fluxes ($\varepsilon \sim 12.3$). This ratio is a direct consequence of the fact that only the finest particles can reach the Akademichesky Ridge. Since the lake dimensions did not change much for the past 100 kyr, we may assume $\varepsilon = \hat{J}_{ter}/j_{ter}$ invariable through that time span.

The residence time of soluble uranium in the modern Baikal water is about 130-160 years [92, 42], which is much less than 330-400 years, the time of water residence (renewal). The high rate of uranium adsorption onto terrigenous particles makes us expect that the nonstationary term in (23) due to paleoflow variability is much less than the stationary term, i.e.,

$$(V \cdot dU_{dis} / dt) / S \cdot K_D \cdot U_{dis} \cdot \hat{J}_{ter} << 1 \qquad (25)$$

Then we may use the quasi-stationary solution to (23) and assume that

$$V \cdot dU_{aut} / dt \approx 0$$

in (22). This drives us to a simple solution:

$$Q = \varepsilon \cdot S \cdot K_D \cdot j_{ter} \cdot U_{ads} / (K_D \cdot U_{riv} - U_{ads}) \qquad (26)$$

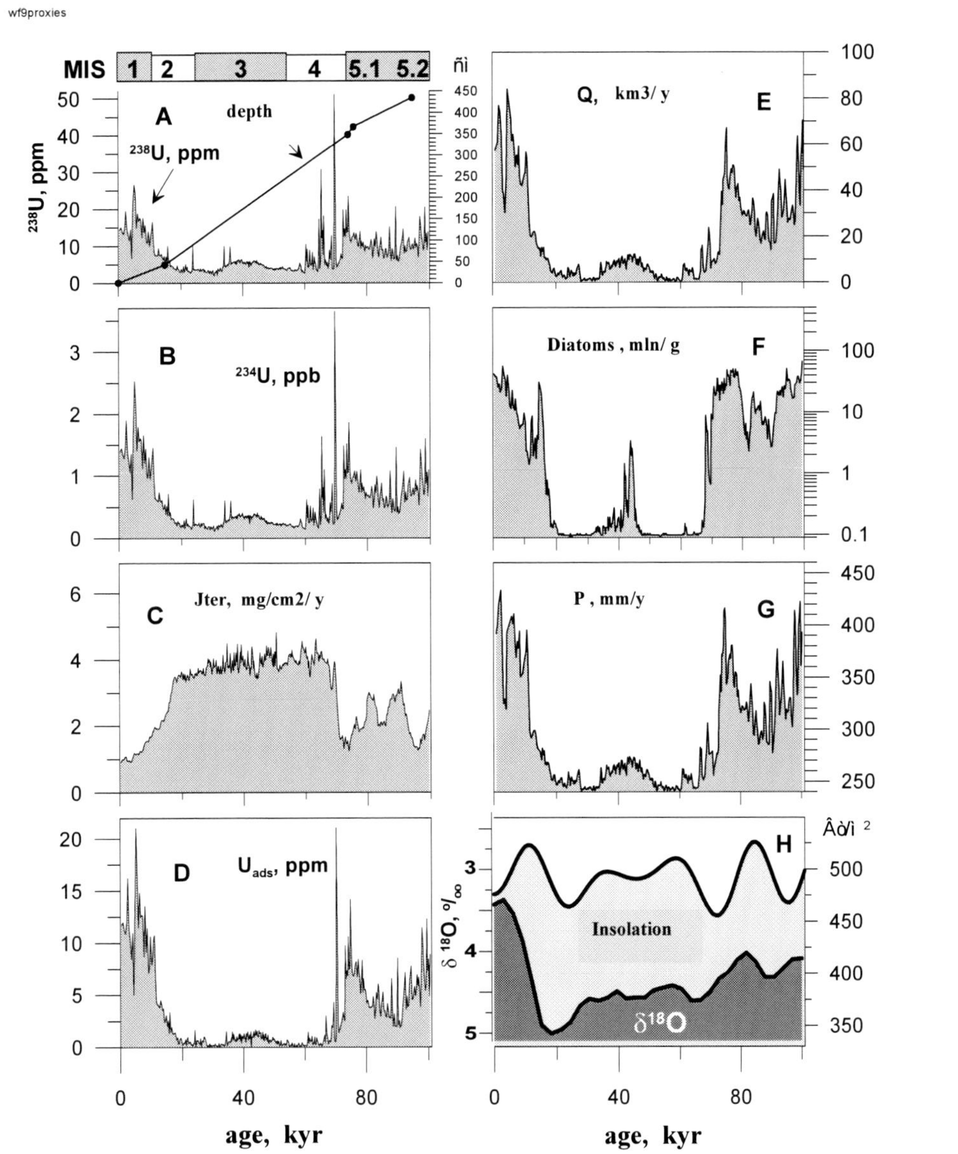

Figure 23. Initial isotope data for st2 (A-D), diatoms (F), and reconstructed parameters: paleoflow (Q) and precipitation (P) for the past 100 kyr. The simplest model with the ages of *C.Flabellatus* (14.5 kyr [23, 24] the onset of B/A) and linear fitting for U-Th ages obtained are used for the time scale [45].

All data for (26) are available: K_D and ε were found from the today's patterns, U_{ads} is the authigenic uranium profile we calculated according (19) with β=2 from bulk U-measurements (fig. 23). The coefficient of uranium distribution (K_D) on suspended oxide and clay particles is pH-dependent [159]. This dependence is, however, quite weak at pH within 7-8 and we assume hereafter that K_D remained invariable and equal to the experimental value for the modern Baikal conditions. To solve (26), one has to know uranium concentration in rivers U_{riv}. Its dynamics is poorly constrained as it can vary with river flow as well as with leaching rate of rocks in different climate periods. There are two possible extreme cases of U_{riv}

changes: $Q*U_{riv} \approx$ const at very wet climate conditions and $U_{riv} \approx$ const at very arid conditions. Uranium supply is limited by its leaching rate from rocks in the former case and furthermore by the time of water-rock contact in the latter case. Concentrations of soluble uranium in rivers (U_{riv}) ranging between these two extremes, closer to the latter case, were measured in the study of uranium fluxes to Lake Issyk Kul (Tien Shan mountains, Kirghizia) and the Aral Sea (western Central Asia) and years-long monitoring of U isotopes in the tributaries of the two lakes [101]. Note that seasonal air temperature contrasts in the lake surroundings in the two areas reached ~ 45^0C and river flow changed for over 15 times (it approached zero in winter) [101]. Studies of the same kind in Lake Baikal, though in few campaigns, gave similar results, U_{riv} varied ~100% at greater seasonal river flow (Q varied in 10 times) and air temperatures contrasts. Simplifying the model, we assumed $U_{riv} \approx$ const in (26) because the main uranium flux (about 80% [3-6]) is carried by the Selenga from highly arid Mongolia (see fig 1). This assumption is of course too rigorous and is to be updated later on when we proceed from the conceptual to the quantitative model.

The concentrations of riverine uranium U_{riv} were borrowed from [92, 42]. The profile of past river flow Q (including the Selenga) with disequilibrium soluble uranium (fig. 23) estimated by Eq. (26) shows that it was almost zero during the cold periods of MIS 4 and 2. The flow of Baikal tributary rivers vanished during dry glacials because moisture coming to the watershed evaporated rapidly. During the last glaciation, Siberia received two times less moisture than now [79, 80], i.e., 200 mm/yr in the Baikal watershed. With this low precipitation, the Baikal outflow must have stopped when evaporation was 68 % the today's level (200/295= 0.68). According to available estimates [160], cooling-induced evaporation during glacial maximums was about 20 % lower than now (by a factor of 0.74 on average). Then, flow can have been zero (at P=E) at a precipitation of 236 mm/yr (295*0.8=236). The mean of these two estimates of precipitation at the last glacial maximum is 218 mm/yr. Thus, we have two extreme points in the 'precipitation – evaporation' diagram: one for the present and the other for the last glacial maximum (fig. 24).

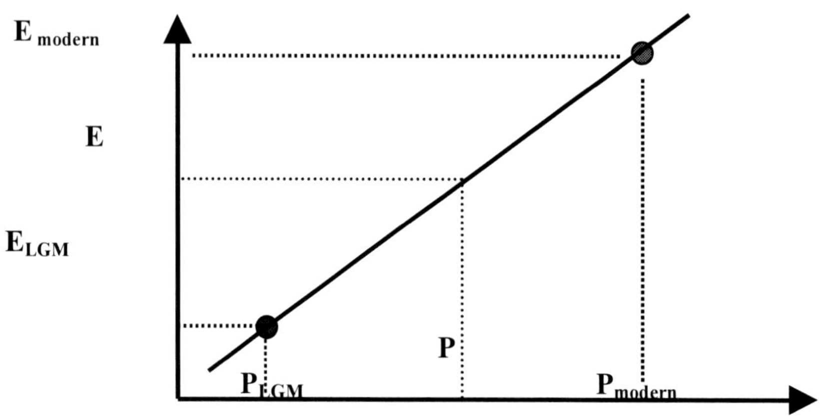

Figure 24. P(E) linear parameterization, implying linear dependence of precipitation (P) on evaporation (E) between two extremes (end members) corresponding to the modern conditions and last glacial maximum (LGM)

The water budget of East Siberia is governed by transport of moisture that evaporates from the Atlantic ocean and distillates on its way eastward to the Asian continental interior. The lower the global air temperatures, the less the moisture reaching East Siberia and the less the precipitation. On the other hand, cooling decreases evaporation from the watershed. The simplest form of the increasing function in the *P – E* diagram from the modern point to the last glacial maximum (LGM) is the straight line. Then, the current *P* and *E* values are related as

$$(E - E_{LGM}) / (E_{modern} - E_{LGM}) = (P - P_{LGM}) / (P_{modern} - P_{LGM}) \quad (27),$$

Together with the water budget in the watershed (28) which becomes

$$Q = (P – E) \cdot S_{watershed} \quad (28)$$

Using (27), (28) we calculated the current values of precipitation and evaporation in the watershed after Q was found by (26). Figure 24 shows the precipitation profile estimated by (26) – (28) with the modern and LGM values (see above) of $P_{modern} \sim 400$ mm/yr; $E_{modern} \sim 295$ mm/yr; $P_{LGM} = 218$ mm/yr; $E_{LGM} = 0.74 * E_{modern}$; $S_{watershed} = 557{,}000$ km^2.

It is reasonable to presume that precipitation to and evaporation from the Baikal surface during the last glacial differed from their modern counterparts to the same degree as those in the watershed. Then, the moisture imbalance, which is now almost zero ($Q_4–Q_3$ = 0.2-0.3 km^3/yr), must have been much greater during the last glacial with a moisture deficit at least ~ 2.5 km^3/yr or more ($Q_4*0.74 - 0.55*Q_3 \sim 2{,}5$ km^3/yr). Given this moisture deficit, Lake Baikal would have dried out in about 8,000 years, but the lack of carbonates in glacial sediments indicates that this never happened. According to estimates with data from [3-7, 92, 93, 42], carbonate deposition would have started even in a half-evaporated lake, i.e., in ~ 4,000 years.

Recent synthesis of bathymetry, high-resolution seismic profiling, and analysis of sediment lithology in the Selenga delta [161] provide convincing evidence that the level of Baikal during the last glacial was only ~30 m below the present level. Sediments of the respective age contain no sand layers, i.e., there was no river flow. As indicated by seismic discontinuities, the lake stood 30 m lower than now for quite a long time. This prediction is compatible with our paleoflow reconstruction. Seismic profiling in the Posolsk Bank [162] gives more evidence that the Baikal level never fell below a few tens of meters during glacials as the bank top at a lake depth of 35 m is covered with only a thin layer of recent sediments.

There arises an apparent paradox. On the one hand, almost zero river flow during glacial maximums (Fig. 23) and a moisture deficit of > 2,5 km^3/yr would break the outflow from the lake and cause its relatively rapid evaporation and, on the other hand, the available data indicate only a minor lake level fall [161]. Furthermore, the absence of river flow during glacials obviously contradicts rapid increase in sedimentation rate (section 4) with the terrigenous flux four times that during interglacials (fig. 23), i.e. terrigenous input increases though the rivers almost do not flow. However, we can cancel the contradiction by assuming that the moisture deficit of 2.5 km^3/yr returned to the lake. It may have been melt water released from glaciers in summer seasons simply because the highland around Baikal, which was much colder than now, provided greater moisture condensation. Moisture that evaporated from the lake thus maintained ice growth in the flanking mountains while melt water that returned to the lake carried large amounts of mechanically weathered terrigenous material.

The question is whether that melt-water flux was able to provide terrigenous transport to such a high sedimentation rate. Yes, it was. During the last glacial, the terrigenous input at the Akademichesky Ridge was 4.5 $mg/cm^2/yr$ against 1 $mg/cm^2/yr$ at present. Therefore, a glacial water flux of 3-5 km^3/yr would correspond to a concentration of fine particles in glacial milk 4.5*(60/5) ~ 60 times greater than the present mean content of suspended load. The characteristic content of suspended solids in the modern lake tributaries is 6-10 g/m^3 and it becomes an order of magnitude greater (up to ~100 g/m^3) during short spring floods [3-7]. Then, our estimate gives ~ 400-600 g/m^3 of terrigenous particles in glacial milk, which is a reasonable amount comparable to that of suspended load in rivers during modern seasonal floods.

If the suggested model is correct, the decrease (or disappearance) of terrigenous material carried by rivers and the proportional increase in glacial input during glacial periods, as well as the respective changes in warm cycles, through the past 100 kyr should be recorded in the behavior of chemical elements in sediments. We measured concentrations of some elements at every 1 mm (resolution ~ 20 years) of the st 2 core (see the next section for experimental details). The sediment compositions were correlated to a standard "warm source" found in the same way as for the st15 and BDP-96-2 cores (see section 2) by averaging the sediment compositions from the upper 5 cm of the core investigated (past 1 kyr). The sediments of the warm standard were brought to the lake uniquely by rivers as there were no glaciers around Baikal in the Holocene. The element composition of a standard cold source was assumed corresponding to that of the MIS 4 glacial. Then we determined the records of both interglacial and glacial correlators. Their records are similar and "warm record" is shown on fig. 25. As it was noted in section 2, the interglacial correlator is the measure of the contribution of the riverine sediment source and the glacial correlator represents the contribution from the glacial (or any other non-riverine) source. The riverine contribution is indeed small during glacials when the main input is from a glacial source (fig.25). The profiles of paleoflow predicted by the two methods are consistent despite the basic difference in the respective approaches. The discrepancy in the reconstructions is obviously due to the adopted approximation. However, the preliminary results being available, the convergence of the two models can be improved through refusing some of the above rigorous assumptions. Thus, the suggested reconstruction of the precipitation and flow histories in the Baikal watershed (fig. 23) appears realistic.

The discussion below details the models in figs.23 for the period of last glacial-Holocene transition. The almost zero flow typical of glacial maximums begins to grow about 17-18 kyr ago, in the end of the glacial and increases rapidly at the onset of the Boreal about 9.7-9.9 kyr BP with a ~ 1.5 kyr lag relative to the beginning of the Holocene (11.5 kyr BP). The first humidity peak is abrupt and occurs at about 9.5 kyr. The Holocene humidity has several peaks at ~ 2, 5, 7 and 9.5 kyr intervened by short arid excursions. Especially prominent are two minimums at 2.5 and 8 kyr. The former coincides with the arid phase of the Subboreal (SB3) and the latter with a global cold spell at 8.2 kyr BP [112-117, 140]. The stable Holocene humidity optimum spans a period between 4 and 7 kyr BP during the Atlantic with a peak about 5 kyr. The climate has becoming ever drier for the past 2.5 kyr: At present it is drier than 1.5-2 kyr ago and than in the Holocene optimum when the flow reached 80 km^3/yr against 58 km^3/yr today.

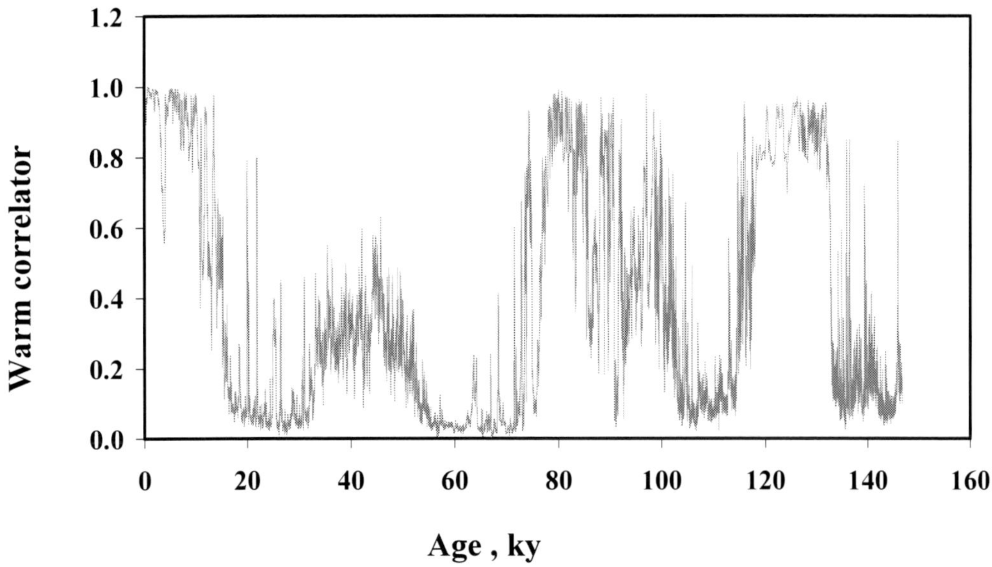

Figure 25. The records of warm correlator for st2 (the resolution is 1 mm, ~ 20 years). During cold MIS2, MIS 4, MIS5.4 and MIS6 riverine runoff was almost zero. Over termination I riverine runoff started increase at ca 18 kyr and ca 136-137 kyr at termination II. For details see text and section 5.

It is now pertinent to discuss the sampling rate of the uranium signal and the averaging degree of flow we estimated with the approximate equation of (26). The required sampling rate and averaging depend on how the time of uranium residence in the lake is related to the characteristic time of flow fluctuations (Q). With ~ τ_Q being the characteristic time of centennial flow change relative to the millennial-scale mean and $\tau_0 \sim V/Q$ being the water renewal time, the uranium residence time τ_U is

$$\tau_U \sim \tau_0 / (1 + S \cdot K_D \cdot \varepsilon \cdot j_{ter} / Q) \qquad (29)$$

If flow does not change much for the time span of uranium residence, i.e., $\tau_Q >> \tau_U$, uranium swiftly tracks all flow variations and the flow magnitude is documented faithfully according to (26). Otherwise, at $\tau_Q << \tau_U$, the approximate reconstruction of (26) records the average flow over the time ~ τ_U rather than its momentary value. Since paleoflow can always change even at a seasonal scale, approximate equation (26) gives its mean over the time τ_U at any high sampling rate. Thus, the present flow record is a mean over ~100 years. As paleoflow decreases, both the numerator and the denominator in (29) increase (to any degree) but the uranium residence time τ_U is quite finite approaching

$$\tau_U \sim V / (S \cdot K_D \cdot \varepsilon \cdot j_{ter}) \qquad (30)$$

For instance, the water residence time during the Karga interglacial was few thousands years and the uranium residence time was as short as τ_U = 60 years. The same τ_U held in glacial periods when flow was vanishing. Therefore, the almost zero flow in those periods is rather a zero mean over 50-100 years than absence of flow at any moment of time. This is in principle compatible with brief high-flow pulses within 50-100-yr periods, and by no means it rules out peaking summer discharge of melt water into the lake.

The discussed conceptual model is obviously just a crude approximation but it appears quite reasonable and stimulating, and can be updated by giving up some initial assumptions.

5.3. Mechanism of Diatom Response in Sediments of Lake Baikal

Our reconstruction of flow history gives a straightforward qualitative explanation for the mystery why diatoms disappeared from Baikal during glacials. The cessation of river flow broke the supply of biogenic elements (and dissolved silica) from the watershed, and Baikal diatoms cannot survive without silica input from an external source because silica production in the lake is low [45]. At present, 76% dissolved silica received by the lake becomes irreversibly buried in diatom frustules [3-7, 93]. Thus the extinction of diatoms in cold periods, which was discovered long ago but remained poorly understood, finally becomes explained. Therefore, the diatom abundance profile in the 200m long BDP-96-1 core (fig. 26) indicates that precipitation in the Baikal watershed in the Pliocene was as a rule sufficient to supply enough dissolved silica to the lake, except for a few spells of so-called Pliocene cooling. It was, especially, at the onset of GNHC (Global Northern-Hemisphere Cooling) 2.5 –2.7 Myr ago [163].

Diatom abundance changed much faster than flow (see diatom patterns for the past 100 kyr in fig. 23). For example, flow during MIS 3 was fife times less than now and the diatom abundance in the sediments of the Akademichesky Ridge was 20-50 times as low as in the Holocene. The diatom response is obviously nonlinear. The key point in understanding the Baikal diatom signal is to explain why diatoms in the pelagic sediments appear and disappear so rapidly. Their sudden inception and extinction, within ~ 400 years, seems really puzzling. Sediments of the respective ages contain nonzero concentrations of authigenic uranium, i.e., its supply had already began or had not ceased yet (see fig 23, 25). Did the lake receive dissolved silica when the supply of soluble uranium began (or did not stop) but diatoms were absent? Of course, it did! See fig. 27 to compare high-resolution authigenic uranium profiles with patterns of sponge spicules and diatoms abundances for the past 100 kyr.

The log- profile of sponge spicule looks very much like the behavior of $A4_0/A_8$ and these, in turn, are similar to the river flow pattern and to the smoothed temperature profiles in the Greenland ice cores. The behavior of diatoms differs from all these patterns as they remained absent for much longer periods than sponges which are accurate tracers of flow dynamics. Since sponges use dissolved silica to build their skeletons, it is reasonable to infer that its input to Baikal was parallel to river flow. Nevertheless, diatoms appeared shortly after and disappeared shortly before the respective onset and break of the silica flux to the lake.

The scenario we suggest provides a simple explanation why diatoms appeared and disappeared so suddenly. Diatoms are sinking. If diatom cells fail to divide while they stay in the illuminated upper water, the population gradually disappears from the photic layer and becomes extinct; otherwise, the diatom population grows [164,165]. Thus, a population survives at

$$\tau_2 < \tau_1$$

and becomes extinct at

$$\tau_2 > \tau_1, \tag{31}$$

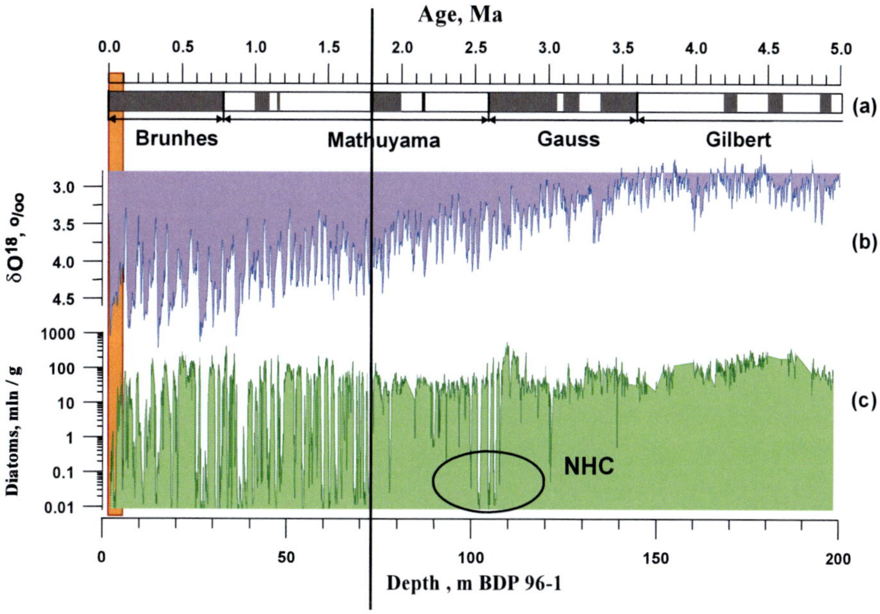

Figure 26. 5 Myr records of diatom history in Lake Baiakl. A – paleomagnetic data from BDP-96 drill core [15,16, 86]; B – marine oxygen isotope record [163] of global ice volume; C – diatoms abundances in BDP 96-1 core [45]. The onset of North Hemisphere Cooling (NHC -2.5-2.7 Ma [163]).

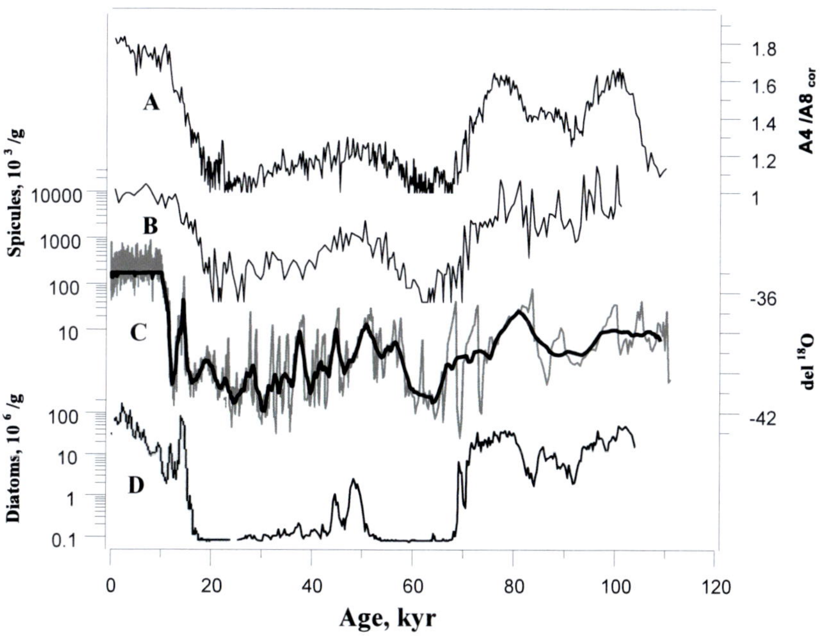

Figure 27. Uranium disequilibrium (A), sponge spicules (B), and diatoms (D) compared to the Greenland temperature record (C). Note that the patterns of uranium disequilibrium and sponge spicules match the Greenland temperature profile. Diatoms appear earlier and disappear later than the respective paleoflow switch on and off which is responsible for the input of dissolved uranium and silica to Lake Baikal.

where τ_1 is the average residence time of a diatom particle in the photic layer and τ_2 is its doubling time. The time τ_1 is controlled *by physical factors only:* the rate of particle settling in water (v) and the turbulence coefficient (D_t) which impedes sedimentation by throwing a part of sinking particles back to the photic layer from deeper water. According to [164,165], τ_1 is found as

$$\tau_1 = 4\ D_t / v^2 \tag{32}$$

The doubling time (τ_2) depends on the kinetics of the uptake of dissolved nutrients, nutrient concentration in the photic layer and on the size of diatom particles. The faster a diatom accumulates the mass of silica necessary for division the shorter the doubling time. A diatom particle of the size - d staying still relative to the water would quickly assimilate all soluble silica within an area of the size ~ 2d and its further growth would be limited by silica diffusion from far away areas. The vicinity of the nearby area becomes exhausted for the time at least ~ d^2/D_m ~ 10 s at the characteristic size of the Baikal diatoms of d ~50 mkm and D_m ~ $3 \cdot 10^{-6}$ cm^2/s, where D_m is the molecular diffusion coefficient of $Si[OH]_4$. The characteristic doubling time of diatoms measured in oceans is about one to seven days [166-168] and is approximately the same in Baikal [4, 93]. Thus, to be able to grow, a diatom particle has to repeatedly change its microenvironment (~ 15,000 times) and, hence, has to move relative to the lake water, as only this motion keeps it in a nutrients-rich environment. This prerequisite, which requires turbulence of the photic layer, what is mentioned in [169, 165]. As for sponges, they need only some current to let water flow them around.

Now we estimate the survival conditions for diatoms in Lake Baikal. If the kinetics of nutrients uptake is fast enough, the effective uptake rate depends on the rate of nutrients supply to the surface of a diatom particle by slow molecular diffusion rather than on the rate of the chemical process [165, 170, 171]. The uptake is limited by diffusion especially for the Baikal diatoms which are much larger than their marine counterparts. With the gross density of skeleton ρ~1 g/cm^3, the size d~50mkm, and concentration of $Si[OH]_4$ in water C ~1.8 mg/l [92], the diffusion flux to a particle is [165, 170, 171]

$$j \sim \pi \cdot C \cdot d^2 \cdot 2D_m / d \sim 2\pi \cdot C \cdot d \cdot D_m, \tag{33}$$

wherefrom the doubling time, supposedly ranging from 1 to 7 days, is

$$\tau_2 \sim \pi \cdot \rho \cdot d^3 / (6\,j) \sim \rho \cdot d^2 / (12 \cdot C \cdot D_m) \sim 6 \text{ days.} \tag{34}$$

This value is compatible with the experiment in Lake Baikal as reported by [4, 93] who also estimated the residence time of diatoms in the photic layer (τ_1) in winter as 15-20 days, i.e., the survival condition is satisfied in the modern Baikal ($\tau_2 < \tau_1$). It follows from (34) that the doubling time increases as dissolved silica decreases in the lake. Then, the survival condition breaks when dissolved silica is of limited supply from the watershed and drops to a critical level. The condition (for C*) when a population becomes extinct, according to (31) with τ_1 and τ_2 found by (32) and (34), respectively, is

$$4\ D_t / v^2 < \rho d^2 / (12 \cdot C^* \cdot D_m) \tag{35}$$

$$C^* < \rho \cdot v^2 \cdot d^2 / (48 \cdot D_t \cdot D_m) \sim \rho \cdot d^n \cdot \Delta^2 / (D_t \cdot D_m) \quad (36)$$

Eq. (36) was obtained using the dependence of the settling rate of a diatom particle on its size and density excess: $v \sim d^k \Delta$ ($2 > k > 1$), where Δ is the density excess of a diatom particle relative to water ($\Delta \sim 0.02\text{-}0.05$ g/cm^3) [e.g., 165, 172] .

Therefore, diatoms never appear right after a glacial ends but only as soon as riverine dissolved silica in the lake grows gradually with increasing river flow to the critical concentration C*. Climate cooling and drying, on the contrary, causes gradual silica decrease from ~2 mg/l typical of interglacials but diatoms persist until it falls below C* and then disappear. Thus, in cold dry periods at limited river flow and silica input, diatoms can survive only near river outlets where dissolved silica remains above the critical concentration. Diatoms use the input silica right at the outlets where they live through the harsh glacial conditions waiting for better times to return to the open lake. Diatoms in deepwater areas appear and disappear abruptly with wetting or drying. This nonlinear effect explains the sharp boundaries (~300 years) of diatom intervals in sedimentary records. Unlike diatoms, sponges respond immediately to $Si[OH]_4$ input as soon as authigenic uranium indicates increase in river flow and in nutrients it carries because they need only silica and bottom fluxes are always available.

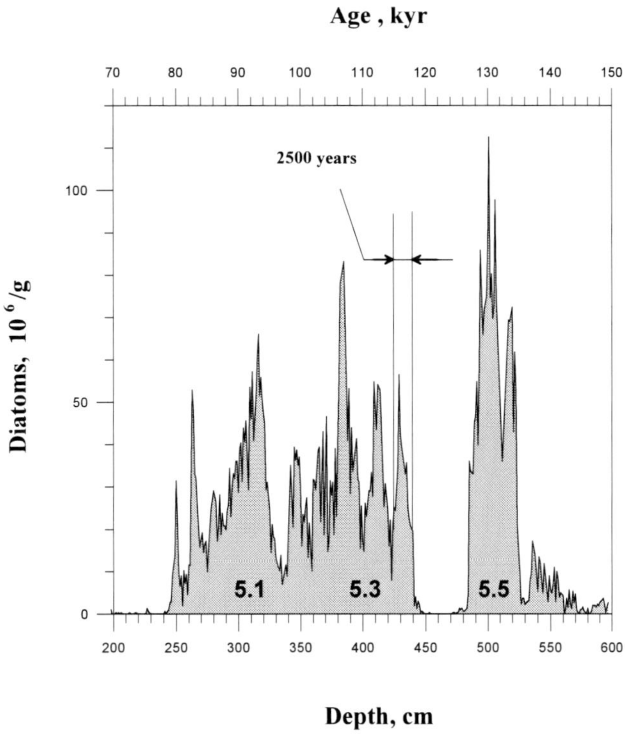

Figure 28 Diatom record at st18 for MIS5, data from Grachev et al [17]. Note climate instability within each warm stage, with an average periodicity of 2-3 kyr. There are several climatic or environmental events over each warm interstitial.

The estimate of C* according to (36), other physical parameters remaining invariable for Lake Baikal environment [3, 4, 93], gives C*~0.7 mg/l (at d = 50 mkm) and is consistent with the empirical evidence that diatoms give way to other algae at dissolved SiO_2 below 0.5 mg/l [172]. The found critical concentration C* also explains why dissolved SiO_2 is never below the minimum of ~ 0.5 mg/l in the photic layer of Baikal even after greatest diatom blooms [3, 4, 93]. In those favorable periods, diatoms are apparently limited by food only and use all available dissolved silica in the photic layer.

Note that C* in (36) decreases together with size as rapidly as d^n (6> n> 4). Therefore, populations of small diatoms have a better chance to survive even at low concentrations of dissolved nutrients. In $Si[OH]_4$-depleted water, other things being equal, small-celled diatoms increase progressively as nutrients decrease. This may be the reason why small diatom species (*C.Minuta*) are dominant in Baikal sediments in the semi-arid Karga interstadial (MIS3), at the very end of warm periods and, occasionally, even during the MIS6 and MIS8 glacials apparently, in brief wetting episodes. This pattern holds throughout the Pleistocene [20, 26].

5.4. Millennial Oscillations of Diatom Abundances in Baikal Sediments

First evidence that Lake Baikal responds to millennial-scale Greenland climate events was reported in [173, 174] where a biological signal was investigated and in [33] where abiotic proxies were studied. Not surprisingly, diatoms, being most sensitive to environmental change, respond to millennial climate instability. See the diatom abundance patterns in sediments from the Akademichesky Ridge (station 18) in fig. 28 [17]. Diatom count, without species division, was carried out using microphotographs rather than by direct microscopy, which provided a good accuracy. Diatom abundances vary in prominent millennial cycles during warm substages of MIS 5 at a mean periodicity of 1-3 kyr: the warm intervals of a ~ 50 kyr total length include at least 15 peaks. If diatoms increased during warm periods, one may expect that the profile in fig.28 documents millennial climate oscillations. Abundances in the patterns of several diatom species from the st2 core (Akademichesky Ridge, fig. 29) likewise oscillate at a millennial scale, both during MIS 5 and in the Holocene. Some diatom species exist throughout the warm intervals (*A. islandica*, *S. grandis*) and others are restricted to short millennial-scale spells (*S.acus*, *S.flabellatus*). The sequence of diatom dominance events shown in fig. 29 is remarkably reproduced in all studied cores from Lake Baikal (see [20, 26, 37]). Note that uranium concentration, an abiotic warm proxy, likewise changes in a millennial rhythm and its peaks are almost in-phase with the diatom peaks (fig. 29c).

Millennial-scale climate events within glacial periods, when diatoms disappear and cannot be used as tracers, are identified from abiotic signals, especially concentrations of some elements and their ratios (see section 2). We expected to track millennial events against geochemical changes by increasing the sampling rate.

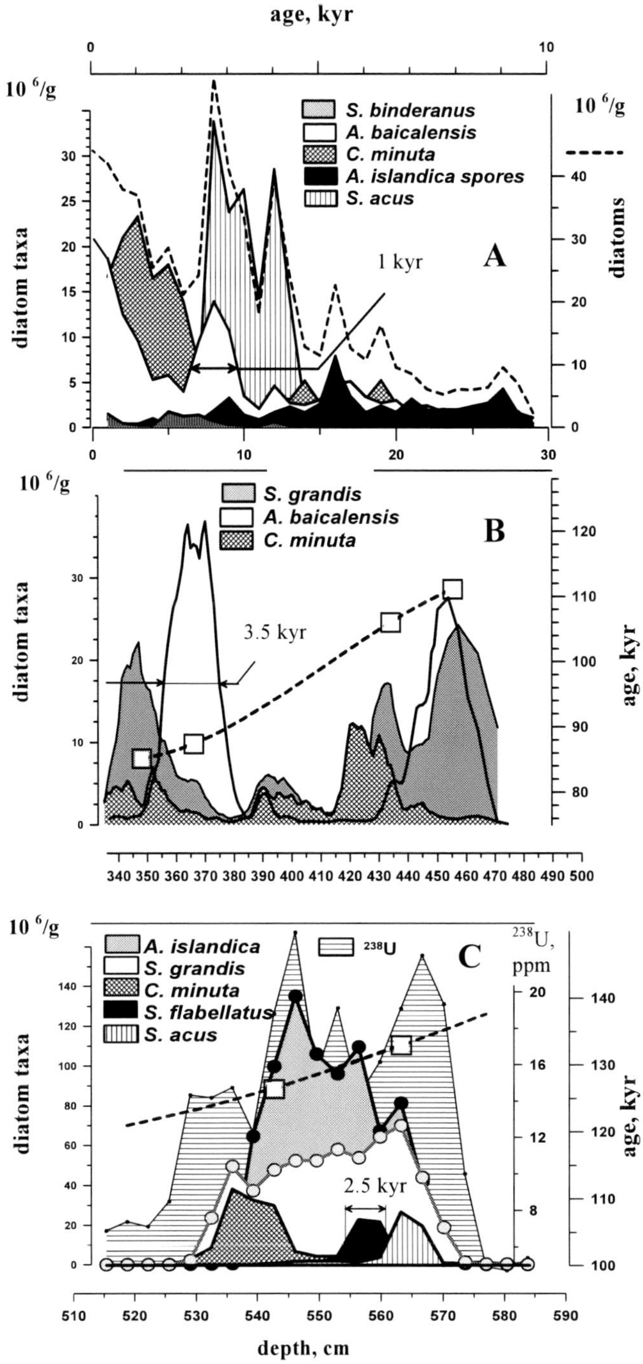

Figure 29. The records of diatom ensembles at st2 (Vorobyeva and Grachev, unpublished data) and uranium profile [44] for comparison. A- for Holocene, B – for MIS 5.1-5.3, C- MIS5.5. See six or seven climate events during Holocene; 6-7 events during MIS 5.1-5.3 and 4-5 events over MIS5.5. Note that the uranium profile is broader than the diatom profile, and all changes in uranium input are accompanied by diatom species changes. Abrupt climate events have an average periodicity of ~ 2-3 kyr. U-Th ages (squares) are after [44].

5.5. SRXRF Scanning the Baikal Records at a Decadal Sampling Interval: Search for Dansgaard-Oeschger Warm Events

To identify ~ 400-yr long events (~ 2 cm at SR ~ 5 cm/kyr), one has to sample the Baikal cores at every 1-2 mm. This sampling rate is hard to achieve with the common wet chemistry as it requires time-consuming preconditioning procedures. High-resolution elemental analysis of the Baikal cores became possible with a scanning device for a SRXRF station designed at the Budker Institute of Nuclear Physics and the appropriate methods of samples treatment and processing of XRF spectroscopy data. See [32, 33, 51, 52] for details and fig. 30 for the layout of the scanning device. The scanned samples were 5–7 mm thick slices of wet sediment scooped from a halved core with 13-30 cm long aluminum trays. The trays were wrapped up in several layers of ~10 micrometer polymer film to avoid dryout of the sediments. The samples were moved vertically, at a controlled scanning interval at least 100 mcm, with reproducibility no worse than 100 micrometers within a 20-cm long sample, and scanned under a sharp monochromatic synchrotron beam collimated to the scanning interval (0.1-1 mm). The measured fluorescence of elements was converted to concentrations by the fundamental parameter method [52]. For details of scanning and data processing see [33,51]. Thus we obtained a continuous record of concentrations for U, K, Ca, Ti, Mn, Fe, Ni, Cu, Zn, As, Br, I, Rb, Sr, Y, Zr, Nb, Mo, Pb, Th, Sn, Sb, Cs, Ba, La, Ce, and some other elements in the st2 core.

Figure 30. SRXRF station with scanning device "Baikal" for elemental and diffraction analyses at Budker Institute of Nuclear Physics (Novosibirsk) [32,33,51]. 1 – scanner with a sediment specimen, 2 – diffraction detector, 3- fluorescence detector channel.

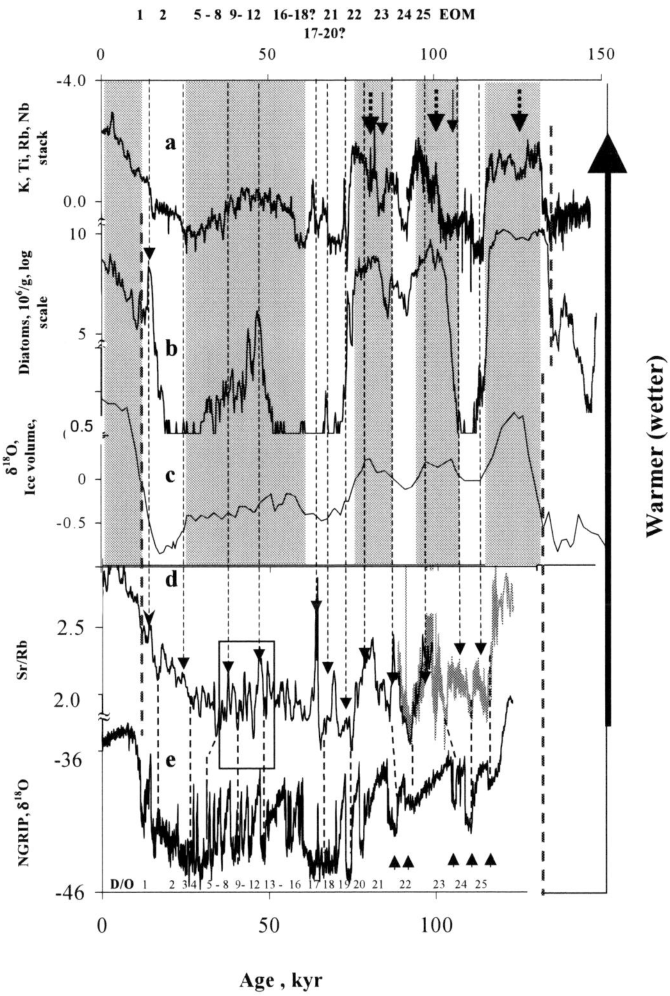

Figure 31. High-resolution records of Lake Baikal responses [46] on abrupt climate shifts in the Greenland and North Atlantic. The resolution of records is ca 20 years. Upper panel shows (A) - the stack of terrigenous elements, B – diatoms (logarithmic scale), compared with volume of global ice (C). Thick dot lines shows Terminations 1 and 2 at 11.5 and 130 ky, respectively. Warm periods (odd MIS) are in grey. Bottom panel shows Sr/Rb ratio – the index of warm (or wet) climates in Lake Baikal -D and indicator of air temperatures under Greenland ($\delta^{18}O$ – ice [182]) – E together with D/O numbers (lower string). Events correlated on Lake Baikal and Greenland are noted by dot lines. The numbers in the upper string shows the correspondence of Baikalian D/O and Greenland's D/O. Short arrows on upper panel show European cooling and arrows on bottom panel show N. Atlantic SST cooling: C21 – C25 events [175, 176]. The rectangular shows interval of rescanning (see fig 32).

In fig. 31 the Baikal profiles of the warm proxies of Sr/Rb ratios and diatom abundances and the clastic stack (K, Ti, Rb, Nb) as tracer of glacial clay (cf. section 1) are compared to the Greenland records of the global ice volume and temperature proxies during the last 140kyr [46]. The plots in fig. 31 are oriented so that the peaks pointed to the direction of warming and/or wetting (right-hand bold arrow). The Baikal watershed obviously responds to all prominent and brief climate events recorded in the North Atlantic cores (D/O events), possibly, due to transport of air and moisture from the Atlantic region to East Siberia. If this is really the cause of the response, it should be immediate and synchronous (inphase). Indeed, most D/O events are synchronous within the accuracy of our ^{230}Th-^{234}U age-depth model. Note that the premise of this synchronicity allows us to update our time scale and use this improved depth-age model for humidity reconstructions in East Siberia (see section 4).

The East Siberian warm substages equated to MIS5 (5.1, 5.3 and 5.5) of the marine oxygen isotope stratigraphy (Kazantsevo interglacial in Siberia) were intervened by notable cooling excursions (fig.31 a, b,d). The three substages of MIS5, that correspond to the precession insolation maximums in the global ice volume curve ($\delta^{18}O$ in marine benthic foraminifers, fig.31c) are represented in our record by series of two or three peaks. Therefore, East Siberian climate was unstable during all warm substages of the last interglacial (cf. fig 28, 29). Note that the abrupt cooling (and/or aridity) events in Siberia are in correspondence with synchronous European cooling episodes of Middle Eemian (at ~ 122-124 kyr BP) and Montaigu (C23 in the North Atlantic region [175, 176], at 104 kyr BP). The warm substage MIS5.1 (73.9-85 kyr BP) likewise was interrupted by cooling in the North Atlantic water [175, 176] during the C21a and C21 events (Melisey-2 in Europe), which are traceable also in the Baikal record (Fig. 31a, b,d).

East Siberia showed a clear response to D/O events 4-17 that fall within the Karga warming of the Siberian stratigraphy equated to MIS3 at 24-59 kyr BP. Two triplet and quartet peaks common to that period (D/O 6-8 and 9-12) are found in the Baikal records. See the results of two scanning experiments in a narrow core interval (fig.32) which show a typical D/O quartet (D/O 9-12 between 34 and 50 kyr BP) containing cosmogenic ^{10}Be anomalies in both Baikal sediments [178] and Greenland ice [177] (the interval of repeated scanning is framed by a box in fig.31d). The two experiments were performed by different methods and in different times. One was applied to sediments in epoxy resin from the stored core (fig. 32,b) [33] and the other to wet sediment from the main (Fig. 32, a) core part. The scanning results show an acceptable reproducibility (fig.31) and the oscillations seen in fig.32 are reproduced to fine details. Therefore, the variability in element compositions is not an instrumental artifact but is realistic and synchronous to the Greenland D/O warm events within the limits of our time scale.

It is worthy of note that diatoms respond in a notable way only to rather long warm episodes (>700 years) within glacials and do not grow much in brief D/O events (fig 31), though temperature in the North Atlantic region during those events rose as much as in long warm spells. The reason is apparently that water renewal in highly inertial Lake Baikal occurred as rarely as every ~ 3000 years on average during MIS 3 (according to the river flow of ~10 km^3/yr, see section 4, fig. 23), or still rarer during the MIS2 and MIS 4 glacial maximums.

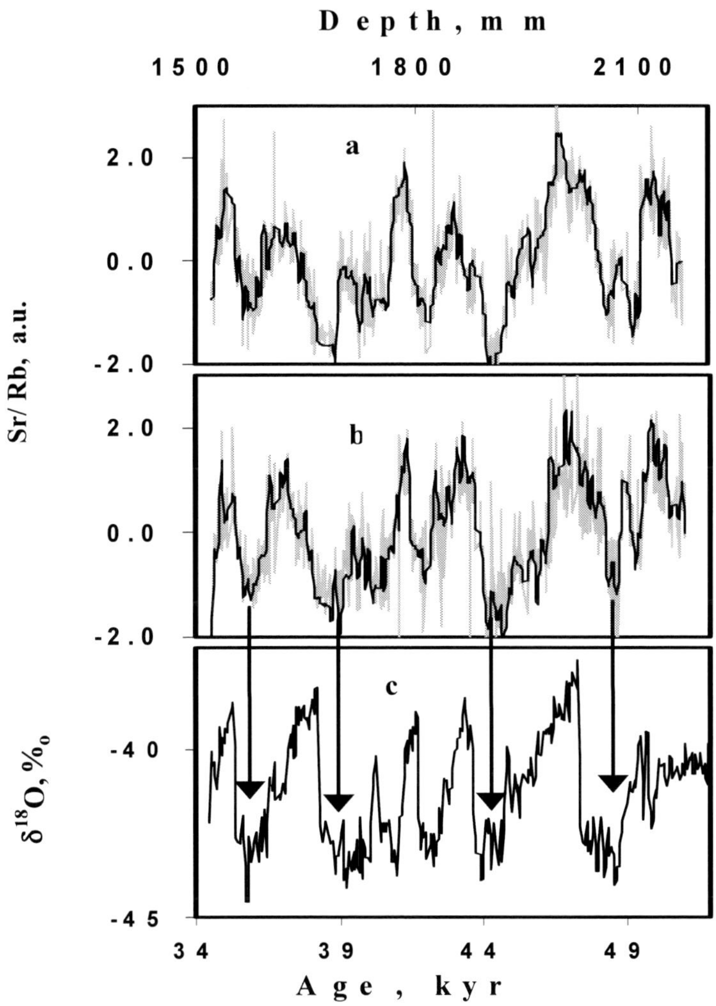

Figure 32. The results of rescanning experiment for the narrow core interval where are typical D/O quartet (D/O 9-12 between 34 and 50 kyr BP; see Figure31). Sr /Rb ratio in the raw sediment (present paper) – A, the sediment of archive core filled in epoxy resin [33] – B, $\delta^{18}O$ ice in NGRIP core [182]- C

5.6. Abiogenic High-Resolution Records on the Last Glacial/Interglacial Transitions in Lake Baikal

We expected to trace the sequence of climate events at the last glacial-Holocene transition in another Baikal core, the one from the Posolsk Bank in the South Baikal basin where sedimentation rate was higher than on the Akademichesky Ridge. The submerged island of Posolsk Bank has its top under 35 m of water; the core was collected at station 6 (52° 05'20"E, 105° 50'26") to a depth of 200 m [31]. Onset of the last glaciation end and transition to interglacial is marked in the core by the peak of *S. flabellatus* at the core horizon corresponding to the B/A (Boelling-Allerod).

The concentrations of 62 elements were measured by ICP-MS in the sediment extracts obtained by dissolution of sediments in concentrated nitric acid for 2 hours at 80°C [31] which fails to open only the most resistant minerals. As in earlier experiments, we revealed a number of typical trends of elements (see the experiment results in figs. 33, 34) identified as a K, U, and Cs-like trends. K, Ca, Mg, and Na fall abruptly at the glacial-Boelling-Allerod boundary (K-like trend), U and Br, on the contrary, rise but more smoothly (U-like trend), while Ti, Nb, Rb, Ba and some other typically terrigenous elements decrease gradually at the Pleistocene-Holocene transition (Cs-like trend). The smooth Rb and Sr decrease is unexpectedly attendant with an abrupt increase in the typically warm signal of Sr/Rb ratio at the glacial-Boelling-Allerod boundary (fig. 33). Note that this Sr/Rb change appears somewhat smoother in the SRXRF pattern of the st2 core (fig.31), possibly, because hot nitric acid selectively extracts Sr and Rb from easily leached feldspar or/and clay minerals, content and composition of which were effected by climate change, too.

Thus, taken together, the available data suggest that the source of terrigenous particles changed at glacial/interglacial transitions for the time of the order of water renewal in modern Baikal (500-1000 years), or even faster. These rapid changes may be triggered either by the onset of river flow (humidity signal) or by glaciers melting (temperature signal). Which of the two worked in the case of Baikal?

Global humidity increased abruptly at the glacial-Boelling-Allerod boundary. See the Greenland temperature, atmospheric CO2 and methane profiles compared with the stack of light elements (Na, Mg, K, Ca) from st6 [31] in fig. 35. The most rapid in these profiles are abrupt humidity changes at onset of B/A, marked by methane forming in peatlands that swiftly appear at climate wetting. The changes in the K- like trends (Sr/Rb ratio and stack of Na, Mg, K, Ca) in the Baikal sediments and the onset of riverine terrigenous influx that replaced glacial input (fig 35) were as abrupt as the increase in atmospheric methane.

One might expect, following Chebykin et al [31], that sudden wetting of global climate caused rapid river flow increase in Siberia which, in turn, changed the composition of terrigenous particles carried to the lake within a short time span of ~ 1000 years. This scenario appears to be consistent with our records (see fig 35). Further brief decrease in atmospheric methane at Y/D (fig. 35), though associated with global climate drying, was not accompanied by formation of large glaciers around Lake Baikal and river flow extinction and thus weakly influenced the chemistry of terrigenous input from the watershed (fig 35). The discussion would end at this point were there no contradiction which emerges on closer examination of data in figures 18, 23, 31, 33, 34 and 35. The abrupt increase in the reconstructed flow occurs at ~ 9-10 kyr and notably lags behind B/A (fig.23). The same follows implicitly from figs. 18 and 33: the U humidity proxy peaks at ~ 9-10 kyr in the st 6 and st 2 cores, whereas element patterns in both cores abruptly changed *earlier,* in onset of global methane increase at B/A ~14.6 kyr BP (see stacks on fig 25 and 35). To cancel the contradiction, one has to take into account two factors. First, methane records humidity in both subtropic and high-latitude regions, i.e., a global average peatland pattern [128-130], while wetting in the Northern Hemisphere may have lagged behind the global wetting event, f.e. [130]. It is reasonable to assume that humidity in Siberia was governed by both global and regional controls. The onset of general climate wetting in Siberia was triggered by global causes, began smoothly at the end of the glacial (fig 23) and peaked at the glacial –B/A transition. Another humidity peak in Siberia was associated directly with later abrupt wetting in the Northern Hemisphere, about 10 kyr BP (see figs. 18, 23, 33). This interpretation agrees fairly well with recent studies of peatlands in Central and Western Siberia [179] and of the methane gradient between the Northern Hemisphere and the tropic zone [130, 180]. From B/A up to ca.10 kyr BP, methane emission by peatlands in the Northern hemisphere was 35-40 Tg per year and grew twice that (60-70 Tg/yr) for ~ 1000 years in the Subboreal period at ~ ca. 10 kyr ago [130, 180]. At that very time, peatlands, which were absent till this moment in Central Siberia, rapidly appeared here and reached the absolute maximum at 9.6 kyr with two clear peaks at 8.6 and 10.2 kyr [179].

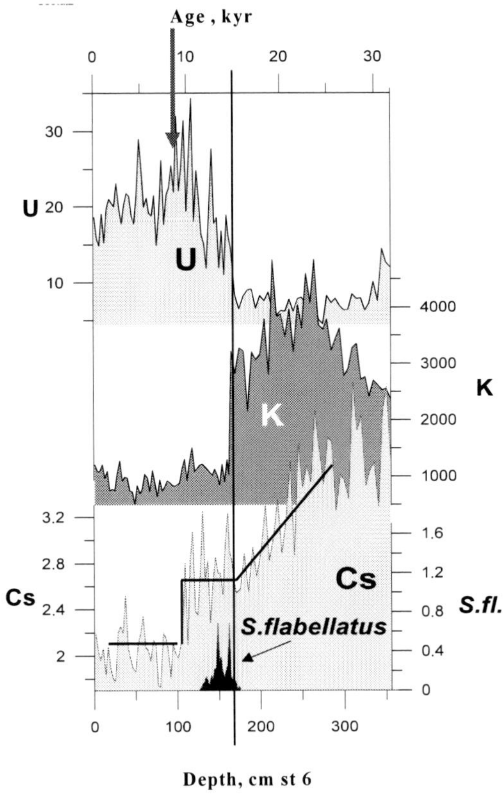

Figure 33. The element records of st 6. *St. Flabelatus* marks the onset of B/A (14.5 kyr).Uranium maximum (red arrow) lags behind the sharp transition during the last deglaciation.

Then, there remains a single question: why the sudden flow increase marked by quite abrupt chemistry changes in the lake suspended load at the B/A transition (15 kyr), both in the st 2 (Akademichesky Ridge, Central Baikal) and st6 (Posolsk Bank, South Baikal) cores (figs. 18, 25, 33, 34 and 35), while the flow change recorded by authigenic uranium occurs about 10 kyr? This controversy can be reconciled by assuming that change in the terrigenous source in onset of B/A was caused by abrupt decrease in glacial input from rapidly degrading glaciers rather than by increase of river runoff. Moreover, runoff apparently increased only in northern tributaries (and, possibly, also in the Barguzin), whereas the arid southern provinces of the Baikal watershed and the Selenga catchment in Northern Mongolia (fig 1) remained almost as dry as in the end of the glacial. The Selenga influx from the southern provinces began later when peatlands grew rapidly in Siberia after the Y/D cooling. Furthermore, moisture transport from the Atlantic region to the southern provinces of the Baikal watershed was facilitated then by thermohaline circulation: TCH resumed ~11 kyr ago after its collapse at Y/D [131-137] and increased the evaporation in the high latitudes and the moisture transport to the continent by the western air flows. The concentration of soluble uranium and the $^{234}U/^{238}U$ ratio in the Selenga being much greater than in other tributary rivers [42], the uranium signal at that transitional (and unstable) time recorded the onset of the Selenga influx since 10 kyr BP. Once the Selenga flow set on, even a minor excess of precipitation over evaporation in its enormous and arid catchment could quickly increase its input to the lake water budget and cause the ensuing abrupt uranium increase in lake sediments.

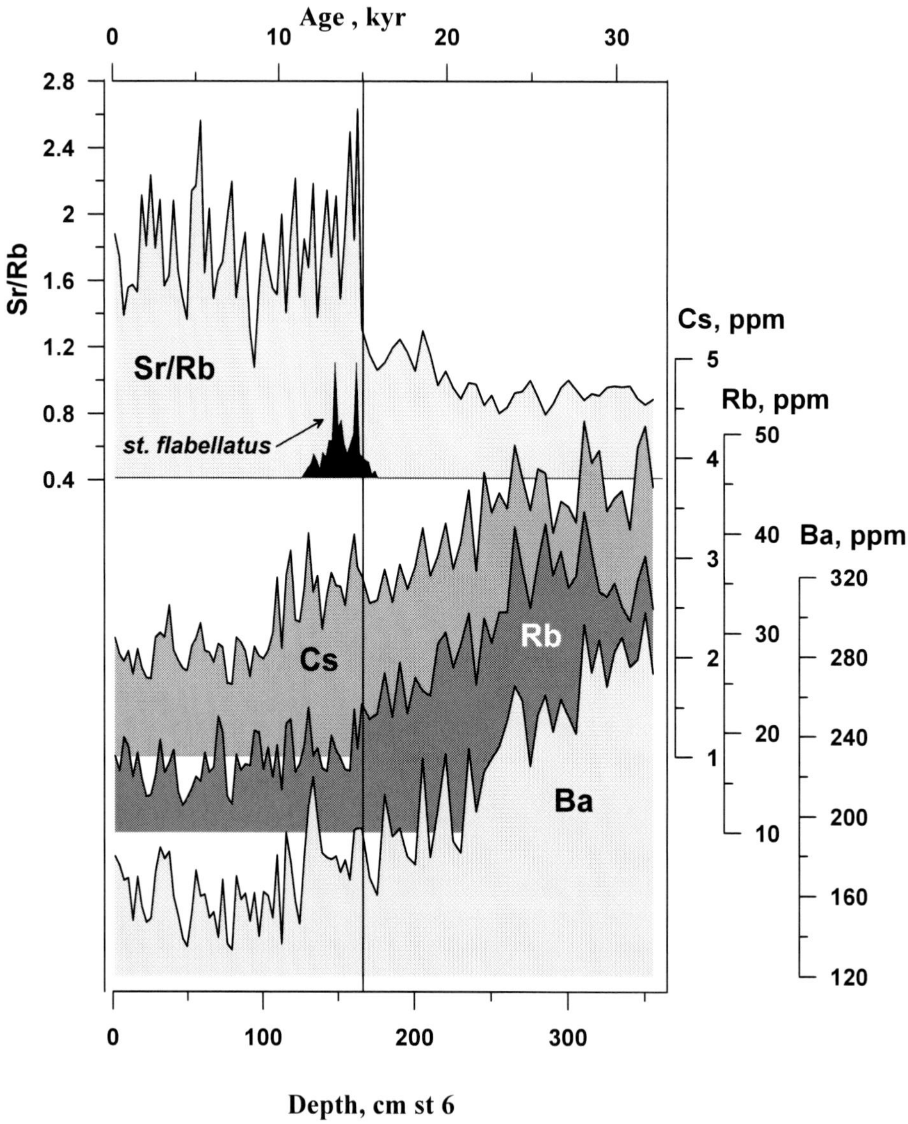

Figure 34. Continuation of fig 33.

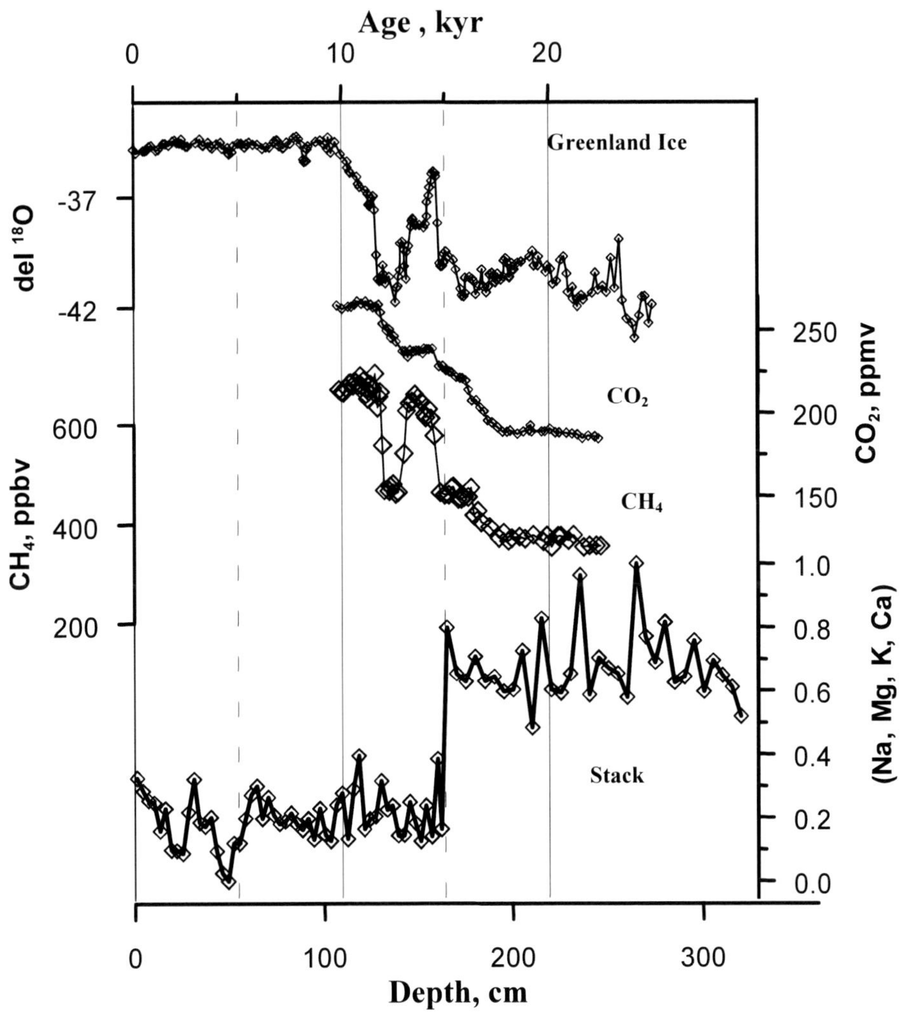

Figure 35. Stack of light elements (Na, Mg, K, Ca) at st6 (data from [31]) compared to global records at last glacial-Holocene transition [181, 182]. The B/A transition is marked by abrupt warming and wetting of global climate (mainly in tropics and SH) and by rapid degradation of glaciers in the mountains around Lake Baikal. Nevertheless, onset of abrupt increasing of climate humidity in Siberia (Lake Baikal U – peak, fig 33) and NH lags behind B/A 5 kyr (see text for details).

Conclusion

In conclusion I summarize the key inferences from the reported results.

We found a number of abiotic proxies that record climate change in the watershed of Lake Baikal and proved that the instability of the East Siberian climate was modulated by orbital cycles. The continental response to orbital forcing differed slightly from that of ice sheets. The nonlinear signals at the eccentricity band (100 kyr) are essentially less prominent in Siberia than the global ice volume variations but the Siberian records bear a strong additional signal at a periodicity of 72 kyr. The Baikal records show a stronger continental response to the precession and obliquity cycles of 23 and 41 kyr which are responsible for most of insolation change. Some geochemical tracers turned out to respond more readily to the 23 and 41-kyr cycles. Therefore, the continental response to insolation is more intense and direct than the global response and bears higher-frequency prominent glacial/interglacial cycles.

There are at least two sources of terrigenous input to Lake Baikal which differ in element compositions and, hence, in mineralogy. One source became active in warm and the other in cold periods. The glacial source consisted almost uniquely of glacial milk as river flow was then vanishing. In interglacial periods, terrigenous particles were carried by rivers and some variability in element abundances was due to redistribution of terrigenous input by the lake tributaries.

We applied direct U-Th dating to BiSi-rich sediments from the Baikal cores and identified the intervals corresponding to all marine isotope stages down to MIS7.1. Quite reliable isotope ages were obtained for the intervals that correspond to MIS5.5 (Kazantsevo interglacial), MIS5.1, MIS 5.3 and MIS7.1. The U-Th dating experience shows that peak-to-peak correlation of biogenic and abiotic records to the SPECMAP $\delta^{18}O$ curve, which is rather an "aesthetic" check, gives generally consistent identification of stages but does not ensure accurate age estimates. Age uncertainty in these correlations can reach 5-10 kyr even for clearly identified horizons.

Both sedimentation rates and terrigenous fluxes to sediments are above the background mean in glacials and below it in interglacials, which is evidence for intense mechanical rock weathering by glaciers in the watershed during glacial periods.

River flow in Lake Baikal was very low or almost zero during glacials and cold periods (MIS2, 4, 6 and MIS5.4). The watershed lost through evaporation all moisture it received. Quantitative reconstruction of paleoflow, precipitation, and evaporation in the Baikal watershed using a conceptual phenomenological model of uranium deposition in sediments indicates that the climate of the Karga interstadial was drier than in the Holocene (300 against 400 mm of mean annual precipitation). The Holocene humidity has been generally similar to that during MIS5.1 and MIS5.3. The low-diatom intervals of the BDP-96-2 core from the Akademichesky Ridge deposited in a relatively dry climates.

The Kazantsevo interglacial in Siberia (MIS5.5) lasted about 20 kyr between 133 and 114 kyr. Warming began immediately with insolation increase unlike the waning of global ice sheets which respond to insolation at a lag of ~ 5-6 kyr.

Humidity at the last glacial-Holocene transition began to grow slowly at the end of the glacial before 18 kyr BP, and abruptly increased about 9.5 kyr, i.e., ~ 2000 years after the onset of the Holocene. The Holocene humidity optimum occurred at ~5 (4-6) kyr, and the climate has been drying for the past 2 kyr. The Holocene wet climate with humidity peaks at ~ 5, ~7 and ~9.5 kyr was intervened by several short dry spells when precipitation fell to 330-350 mm/yr against the maximum of 470-500 mm/yr. One dry episode coincides with a prominent global cooling event at 8.2 kyr BP and the other occurred at the latest Subboreal ~ 2500 -3000 years ago.

The diatom signal in the lake sediments tracks wetting/drying cycles being associated with riverine supply of nutrients (which was limited at reduced flow). Diatoms appear abruptly, later than the flow resumes, and disappear as rapidly before the flow stops. The semi-quantitative model of the diatom response explains why small diatom species are more successful than large (heavy) diatoms. The latter survive only in wettest periods (*S. Grandis*) whereas the former (*C.Minuta*) can survive even during short interstadials within long glacial cycles (MIS3, MIS6, MIS8).

The East Siberian climate experienced centennial-millennial oscillations both in interglacial and glacial periods. Abrupt climate changes intervened each precession cycle of MIS5.1, MIS5.3 and MIS5.5, as well as the Holocene, at every 1.5-3 kyr on average.

The Baikal watershed responded to rapid warming during the last glaciation (Greenland Dansgaard-Oeschger events). Almost each D/O and H-event is marked by the corresponding change in the Baikal sediment chemistry.

The main source of terrigenous input changed swiftly during terminations I and II for the time comparable to the water residence time in Lake Baikal, simultaneously with increase in atmospheric methane caused by abrupt wetting of global climate. The humidity peak in Siberia lagged behind the global peak and occurred at ~ 10 kyr during the last deglaciation, exactly when the Selenga flow began growing rapidly after a decrease to a very low level at the end of the glacial.

Acknowledgements

This study is a synthesis of results I obtained jointly with my colleagues and co-authors of previous publications, and is thus a cooperative contribution. First of all I wish to thank academician M.A. Grachev who turned a physicist engaged in physical chemistry, which I used to be, into a paleoclimate man and involved me into this fascinating science. I greatly appreciate all his help and patience. O.M. Khlystov showed special skill and care in handling the cores and provided valuable geological support. E.P. Chebykin charged himself with much of ICP-MS work and made a part of isotope determinations. M.A. Phedorin contributed to SRXRF analyses and developed the method of processing of fluorescent spectra. S.S. Vorobyeva performed diatom analysis and G.Ziborova measured BioSi and water contents in the sediments. The work would be impossible without V.M. Tsukanov, E.G. Miginskaya, V.I. Kondratiev and, especially, K.V. Zolotarev from the Budker Institute of Nuclear Physics who designed the SRXRF scanner with special purpose for our investigations. The paper benefited from discussions with V.S. Zykin, V.S. Zykina, and L.Z. Granina. At last, I wish to thank T.I. Perepelova who translated my MS from Russian in English. The study was carried out as part of Program 16.8 of the Russian Academy of Sciences and was supported by grant 06-05-91576-JP from the Russian Foundation for Basic Research.

References

[1] Intergovernmental Panel on Climate Change (IPCC). Climate Change 2001. The Scientific Basis, Contribution of Working Group I to the Third Scientific Assessment Report of the Intergovernmental Panel on Climate Change. Houghton J.T., Ding Y., Griggs D.J., Noguer M., van der Linden. and Xiaosu D. (eds.). New York: Cambridge University Press; 2001.

[2] Intergovernmental Panel on Climate Change (IPCC). Climate Change. Summary for Policymakers, Third Assessment Report. Houghton J.T., Ding Y., Griggs D.J.,

Noguer.M, van der Linden P.J. and Xiaosu D (eds). New York: Cambridge University Press; 2001.

[3] Galaziy, G.I. Baikal Atlas. Moscow: Federal Agency for Geodesy and Cartography of Russia; 1993 (in Russian).

[4] Shimaraev, M.N., Verbolov, V.I., Granin, N., Sherstayankin, P.P. *Physical limnology of Lake Baikal: a review*. Irkutsk-Okayma; 1994.

[5] Afanasev, A.N. (1960) Water budget of lake Baikal (Proc. of Baikal Limnolgical Inst), 18, 155-241. Database on website at www.nti.lin.irk.ru/bankzn/catalog.aspv (in Russian).

[6] Afanasev, A.N. Water storage and Water budget of lake Baikal. Novosibirsk: *Nauka*; 1976 Database on website at www.nti.lin.irk.ru/bankzn/catalog.asp (in Russian).

[7] Votintsev, K. K. Hydrochemistry of Lake Baikal. Moscow: Izd. Akad. Nauk. SSSR; 1961. Database on website at www.nti.lin.irk.ru/bankzn/catalog.asp (in Russian).

[8] Martin, P., Granina, L., Martens, K., and Goddeeris, B. (1998). Oxygen concentration profiles in sediment of two ancient lakes: Lake Baikal (Siberia, Russia) and Lake Malawi (East Africa). *Hydrobiologia*, **367**, 163–174.

[9] Granina, L., Muller B., Wehrli, B., and Martin, P. (2000). Oxygen, iron, and manganese at the sediment–water interface in Lake Baikal. *Terra Nostra* , **9**, 87–94.

[10] Lut, B.F. Pribaikal geomorphology of the Lake Baikal depression (Trudy Limnologicheskogo Instituta, v 26(46)). Novosibirsk: *Nauka*; 1978. (in Russian).

[11] Mats, V.D. (1993). The structure and development of the Baikal rift depression. *Earth-Sci. Rev*, **34**, 81–118.

[12] Colman, S.M., Karabanov, E.B., Nelson, C.H. (2003). Quaternary sedimentation and subsidence history of Lake Baikal, Siberia, based on seismic stratigraphy and coring. *J. Sediment. Res*, **73**, 941– 956.

[13] Antipin, V., Afonina, T., Badalov, O., Bezrukova, E., Bukharov, A., Bychinsky, V., Dmitriev, A.A., Dorofeeva, R., Duchkov, A., Esipko,O., Fileva, T., Gelety, V., Golubev, V., Goreglyad, A., Gorokhov, I., Gvozdkov, A., Hase, Y., Ioshida, N., Ivanov, E., Kalashnikova, I., Kalmychkov, G., Karabanov, E., Kashik, S., Kawai, T., Kerber, E., Khakhaev, B., Khlystov, O., Khursevich, G., Khuzin, M., King, J., Konstantinov, K., Kochukov, V., Krainov, M., Kravchinsky, V., Kudryashov, N., Kukhar, L., Kuzmin, M., Nakamura, K., Nomura, S., Oksenoid, E., Peck, J., Pevzner, L., Prokopenko, A., Romashov, V., Sakai, H., Sandimirov, I., Sapozhnikov, A., Seminsky, K., Soshina, N., Tanakah, A., Tkachenko, L., Ushakovskaya, M., Williams, D. (2001) The new BDP-98 600-m drill core from Lake Baikal: a key late Cenozoic sedimentary section in continental Asia. *Quaternary International*, **80**–81,19–36

[14] Colman, S.M., Peck, J.A., Karabanov, E.B., Carter, S.J., Bradbury, J.P., King, J.W., Williams, D.F. (1995). Continental climate response to orbital forcing from biogenic silica records in Lake Baikal Siberia. *Nature*, **378**, 769–771

[15] Kuzmin, M.I., Karabanov, E.B., Kawai, T., Williams, D., Bychinskii, V.A., Kerber, E.V., Kravchinslii, V.A., Bezrukova, E.V., Prokopienko, A.A., Geletii, V.F., Kalmychko, G.V., Goreglyad, A.V., Antipin, V.S., Khomutova, M.Yu., Soshina, N.M., Ivanov, E.V., Khursevich, G.K., Tkachenko, L.L., Solotchina, E.P., Ioshida, N., Gvozdkov, A.N. (2001). Deep drilling on Lake Baikal: main results. *Geol. Geophys*, **42** (1–2), 8 –34

[16] BDP-Members. (1997). Continuous paleoclimate record of last 5 Ma from Lake Baikal, Siberia. *EOS American Geophysical Union Transactions,* **78**, 597–604.

[17] Grachev, M.A., Likhoshwai, E.V., Vorobyova, S.S., Khlystov, O.M., Bezrukova, E.V., Veinberg, E.V., Goldberg, E.L., Granina, L.Z., Kornakova, E.G., Lazo, F.I., Levina, O.V., Letunova, P.P., Otinov, P.V., Pirog, V.V., Fedotov, A.P., Iaskevich, S.A., Bobrov, V.A., Sukhorukov, F.V., Rezchikov, V.I., Fedorin, M.A., Zolotarev, K.V., Kravchinsky, V.A. (1997). Signals of the paleoclimates of upper Pleistocene in the sediments of Lake Baikal. *Russian Geology and Geophysics*, **38**, 957–980.

[18] Peck, J.A., King, J.W., Colman, S.M., Kravchinsky, V.A. (1994). A rock-magnetic record from Lake Baikal, Siberia: Evidence for Late Quaternary climate change. *EPSL*, **122**, 221-238

[19] Williams, D.F., Peck, J.A., Karabanov, E.B., Prokopenko, A.A., Kravchinsky, V., King, J., Kuzmin, M.I. (1997). Lake Baikal record of continental climate response to orbital insolation during the past 5 million years. *Science*, **278**, 1114– 1117.

[20] Grachev, M.A., Vorobyova, S.S., Likhoshway, Y.V., Goldberg, E.L., Ziborova, G.A., Levina, O.V. & Khlystov, O.M.. (1998). A high-resolution diatom record of the palaeoclimates of East Siberia for the last 2.5 My from Lake Baikal. *Quaternary Science Reviews*, **17**, 1101-1106.

[21] Demory, F., Nowaczyk, N.R., Witt, A., Oberhansli, H. (2005). High resolution magnetostratigraphy of late quaternary sediments from Lake Baikal, Siberia: timing of intracontinental paleoclimatic responses. *Global and Planetary Change*, **46**, 167–186.

[22] Carter, S.J., Colman, S.M.. (1994). Biogenic silica in Lake Baikal sediments: results from 1990-1992 American cores. *J.Great Lakes Res* ., **20**, 751-760

[23] Colman, S.M., Peck, J.A., Hatton, J., Karabanov, E.B., King, J.W. (1999). Biogenic silica records from the BDP93 drill site and adjacent areas of the Selenga Delta, Lake Baikal, Siberia. *Journal of Paleolimnology*, **21**, 9 –17.

[24] Colman, S.M., Jones, G.A., Rubin, M., King, J.W., Peck, J.A., Orem, W.H. (1996). AMS radiocarbon analyses from Lake Baikal, Siberia: challenges of dating sediments from a large, oligotrophic lake. *Quaternary Science Reviews*, **15**, 669–684.

[25] Bezrukova, E.V. Palaeogeography of Pribaikalia During Late Glacial and Holocene. Novosibirsk: *Nauka*; 1999 (in Russian).

[26] Khursevich, G.K., Karabanov, E.B., Prokopenko, A.A., Williams, D.F., Kuzmin, M.I., Fedenya, S.A., Gvozdkov, A.A. (2001). Insolation regime in Siberia as a major factor controlling diatom production in Lake Baikal during the past 800,000 years. *Quaternary International*, **80**–81, 47– 58.

[27] Prokopenko, A.A., Karabanov, E.B., Williams, D.F., Kuzmin, M.I., Shackleton, N.J., Crowhurst, S.J., Peck, J.A., Gvozdkov, A.N., King, J.W. (2001). Biogenic silica record of the Lake Baikal response to climate forcing during the Brunhes chron. *Quaternary Research*, **55**, 123–132.

[28] Prokopenko, A.A., Karabanov, E.B., Williams, D.F., Khursevich, G.K. (2002). The stability and the abrupt ending of the last interglaciation in southeastern Siberia. *Quaternary Research*, **58**, 56–59.

[29] Prokopenko, A.A. (2002). On the nature of the link between the Lake Baikal biogenic silica record and regional climate change in Siberia: a reply to M.A. Grachev. *Quaternary Research* **58**, 56–59.

[30] Prokopenko, A.A., Williams, D.F., Kuzmin, M.I., Karabanov, E.B., Khursevich, G.K., Peck, J. 2002. Muted climatic oscillations in continental Siberia during the mid-Pleistocene. *Nature*, **418**, 65–68.

[31] Chebykin, E. P., Edgington, D. N., Grachev, M.A., Zheleznyakova, T.O.,Vorobyova, S.S., Kulikova, N.S., Azarova, I .N., Khlystov, O.M., Goldberg, E.L. (2002). Abrupt increase in precipitation and weathering of soils in East Siberia coincident with the end of the last glaciation (15 cal kyr BP). *EPSL*, **200**, 167-175.

[32] Goldberg, E.L., Grachev, M.A., Chebykin, E.P., Phedorin, M.A., Kalugin, I.A., Khlystov, O.M., Zolotarev, K.V. (2005). Scanning SRXF analysis and isotopes of uranium series from bottom sediments of Siberian lakes for high-resolution climate reconstructions. *Nuclear Instruments and Methods in Physics Res earch.A.*, **543**, 250–254

[33] Goldberg, E.L., Grachev, M.A., Phedorin, M.A., Kalugin, L.A. , Khlystov, O.M., Mezentsev, S.N., Azarova, I.N. , Vorobyeva, S.S. , Zheleznyakova, T.O., Kulipanov, G.N., Kondratyev, V.I. , Miginsky, E.G. , Tsukanov, V.M., Zolotarev, K.V., Trunova, V.A., Kolmogorov, Yu.P., Bobrov, V.A. (2001). Application of synchrotron X-ray fluorescent analysis to studies of the records of paleoclimates of Eurasia stored in the sediments of Lake Baikal and Lake Teletskoye. *Nucl. Instr. and Meth. in Phys. Res. A,* **470** (1-2), 388-395.

[34] Demske, D., Heumann, G., Granoszewski, W., Nita, M., Mamakowa, K., Tarasov, P.E., Oberhansli, H. (2005). Late Glacial and Holocene vegetation and regional climate variability evidenced in high-resolution pollen records from Lake Baikal. *Global and Planetary Change*, **46**, 255–279.

[35] Granoszewski, W., Demske, D., Nita, M., Heumann, G., Andreev, A.A. (2005). Vegetation and climate variability during the last interglacial evidenced in the pollen record from Lake Baikal. *Global and Planetary Change*, **46**, 187–198.

[36] Bradbury, J.P., Bezrukova, Ye.V., Chernyaeva, G.P., Colman, S.M., Khursevich, G.K., King, J.W., Likhoshway, Ye.V. (1994). A synthesis of post-glacial diatom records from Lake Baikal. *Journal of Paleolimnology*, **10**, 213-252

[37] Grachev, M.A., Gorshkov, A.G., Azarova, I.N., Goldberg, E.L., Vorobyva, S.S., Zheleznyakova, T.O., Bezrukova, E.V., Krapivina, S.M., Letunova, P.P., Khlystov, O.M., Levina, O.V., Chebykin, E.P. Pervasive millennial climate oscillations and bio speciation in Lake Baikal, in "The main trends in global and regional climate changes during the late Pleistocene in Siberia", **1,** 107-121. Novosibirsk, IAET, (2002) (in Russian).

[38] Kashiwaya, K., Ochiai, S., Sakai, H. & Kawai, T. (2001). Orbit-related long-term climate cycles revealed in a 12-Myr continental record from Lake Baikal. *Nature*, **410**, 71–74

[39] Kashiwaya, K., Sakai, H., Ryugo, M., Horii, M., Kawai, T., (2001). Long-term climato-limnological cycles found in a 3.5-million-year continental record. *Journal of Paleolimnology*, **25**, 271–278.

[40] Goldberg, E.L., Phedorin, M.A., Grachev, M.A., Bobrov, V.A., Dolbnya, I.P., Khlystov, O.M., Levina, O.V., Ziborova, G.A. (2000). Geochemical signals of orbital forcing in the records of paleoclimates found in the sediments of Lake Baikal. *Nucl. Instr. and Meth. in Phys. Res. A*, **448** (1-2), 384-393.

[41] Fagel, N., Boski, T., Likhoshway, L., Oberhaensli, H. (2003). Late Quaternary clay mineral record in Central Lake Baikal (Academician Ridge, Siberia). Palaeogeography, Palaeoclimatology, *Palaeoecology*, **193**, 159-179

[42] Edgington, D.N., Robbins, J.A., Colman, S.M., Orlandini, K.A. & Gustin, M.-P, (1996). Uranium-series disequilibrium, sedimentation, diatom frustules and paleoclimate change in Lake Baikal. *Earth & Planet. Sci. Lett.*, **142**, 29-42.

[43] Edgington, D. N., Robbins, J. A., Colman, S. M., Orlandidni, K.A., Gustin, M.-P., Klump, J. V. & Granina, L. Z. (1997). Reply to Comment by R. Anderson on "Uranium-series disequlibrium, sedimentation, diatom frustules, and paleoclimate change in Lake Baikal". *Earth Planet. Sci. Lett.*, **148**, 399 – 404.

[44] Chebykin, E.P., Edgington, D.N., Goldberg, E.L., Phedorin, M.A., Kulikova, N.S., Zheleznyakova, T.O., Vorobyova, S.S., Khlystov, O.M., Levina, O.V., Ziborova, G.A. & Grachev, M.A. (2004) Uranium-Series Isotopes as Proxies of Late Pleistocene Climate and Geochronometers in Bottom Sediments of Lake Baikal. *Russ. Geol. Geophys*, **45** (5), 539-556

[45] Goldberg, E.L., Chebykin, E.P., Vorobyova, S.S., Grachev, M.A. (2005). Uranium Signals of Paleoclimate Humidity Recorded in Sediments of Lake Baikal. *Doklady Earth Sciences*, **400** (1), 52–56.

[46] Goldberg, E.L., Phedorin, M.A., Chebykin, E.P., Zolotarev, K.B, Zhuchenko N.A. (2007). Decade–centenary resolution records of climate changes in East Siberia from elements in the bottom sediments of lake Baikal for the last 150 ky. *Nuclear Instruments and Methods in Physics Research A,* **575,** 193–195

[47] Goldberg, E.L., Grachev, M.A., Bobrov, V.A., Bessergenev, A.V., Zolotaryov, K.V., Likhoshway, Ye.V. (1998). Do diatom algae frustules accumulate uranium?. *Nuclear Instruments and Methods in Physics Researsch A.*, **405**, 584-589

[48] Phedorin, M.A., Bobrov, V.A., Goldberg, E.L., Navez, J., Zolotarev, K.V., Grachev, M.A. (2000). SR-XFA as a method of choice in the search of signals of changing palaeoclimates in the sediments of Lake Baikal, compared to INAA and ICP-MS. *Nucl. Instr. and Meth. in Phys. Res.A*, **448** (1-2), 394-399.

[49] Phedorin, M.A., Goldberg, E.L., Grachev, M.A., Levina, O.L., Khlystov, O.M., Dolbnya, I.P. (2000). The comparison of biogenic silica, Br and Nd distributions in the sediments of Lake Baikal as proxies of changing paleoclimates of the last 480 ky. *Nucl. Instr. and Meth. in Phys. Res. A*, **448** (1-2), 400-406.

[50] Phedorin, M.A., Goldberg, E.L., Bobrov, V.A., Khlystov, O.M., Grachev, M.A. (2000). Multi-wavelength synchrotron radiation XRF determination of U and Th in sedimentary cores from Lake Baikal. *J. of Geostandarts and Geoanalysis*, **24**(2), 217-226.

[51] Zolotarev, K.V., Goldberg, E.L., Kondratyev, V.I. , Kulipanov, G.N., Miginsky,E.G., Tsukanov, V.M., Phedorin, M.A., Kolmogorov, Yu.P. (2001). Scanning SRXRF beamline for analysis of bottom sediments. *Nucl. Instr. and Meth. in Phys. Res. A*, **470** (1-2), 376-379

[52] Phedorin, M.A., Goldberg, E.L. (2005). Prediction of absolute concentrations of elements from SRXRF scan measurements of natural wet sediments. *Nuclear Instruments and Methods in Physics Research, A* **543**, 274–279

[53] Goldberg, EL, Grachev, MA, Edgington, DN, Navez ,J, Anre, L, Chebykin, A.P., Shulpyakov, O.I. (2001). Direct U-Th dating of the last two interglacials in sediments of Lake Baikal. *Dolkldy Earth Sci,* **381** (8), 952-955,

[54] Imbrie, J., Imbrie, J.Z. (1980). Modeling the climatic response to orbital variations. *Science*, **207**, 943-953
[55] Imbrie, J., Hays, J.D., Martinson, D.G., McIntyr,e A., Mix, A.C., Morley, J.J., Pisias, N.G., Prell, W.L. and Shackleton, N.J. *The orbital theory of Pleistosene climate*: Support from a revised chronology of marine δ18O record.In Milankovitch and Climate, part 1, A. Berger, J.Imbrie, J.D. Hays, G. Kukla & B.Saltzman eds., pp. 269-305. Norwell Mass: D. Reidel; 1984.
[56] Imbrie, J., Boyle, E.A., Clements, S.C. et al. (1992). On the structure and origin of major glaciation cycles 1. Linear responses to Milankovitch forcing. *Paleoceanography*, **7**(6), 701-738
[57] Imbrie, J., Berger, A., Boyle, E.A. et al. (1993). On the structure and origin of major glaciation cycles. 2. The 100000-year cycle. *Paleoceanography* , **8** (6), 699-735
[58] Hays, J. D., Imbrie, J., and Shackleton, N. J. (1976). Variations in the Earth's orbit: Pacemakers of the ice ages. *Science*, **194**, 1121–1132
[59] Berger, A. and Loutre, M.F. (1991) Insolation values for the climate of the last 10 million years. *Quaternary Science Re*views, **10**, pp 297-317
[60] Berger, A, Loutre, MF., Tricot, C. (1993). Insolation and Earth's orbital periods. *J. Geophys. Res*., **98**, 10341-10362
[61] Paillard, D. (2001). Glacial cycles: toward a new paradigm. *Reviews of Geophysics,* **39**, 325–346
[62] Bassinot, FC, Labeyrie, LD, Vincent, E, Quidelleur, X, Shackleton, NJ and Lancelot, Y. (1994).The astronomical theory of climate and the age of the Brunhes-Matuyama magnetic reversal. *Earth Planet. Sci. Lett*., **126**, 91–108
[63] Martinson, DG., Pisias, NG, Hays, JD, Imbrie, J, Moore, TC, and Shackleton, NJ. (1987).Age dating and the orbital theory of the ice ages - development of a high-resolution 0 to 300000-year chronostratigraphy. *Quat. Res*., **27**, 1–29
[64] EPICA community members. (2004). Eight glacial cycles from an Antarctic ice core. *Nature*, **429**, 623-628
[65] Petit, JR, Jouzel, J, Raynaud, D, Barkov, NI, Barnola, J-M, Basile, I , Bender, M , Chappellaz, J, Davisk, M, Delaygue, G, Delmotte, M, Kotlyakov, VM, Legrand, M, Lipenkov VY, Lorius, C , Pe pin, L, Ritz, C, Saltzman E, Stievenard, M. (1999). Climate and atmospheric history of the past 420,000 years from the Vostok ice core, Antarctica. *Nature*, **399**, 429–436
[66] Bender, ML. (2002) Orbital tuning chronology for the Vostok climate record supported by trapped gas composition. *Earth and Planetary Science Letters*, **204** 275 – 289
[67] Kukla, G, An ZS, Melice, J-L, Gavin, J, Xiao, JL. (1990). Magnetic susceptibility record of Chinese Loess. *Trans. R. Soc. Edinb. Earth Sci*, **81**, 263–88
[68] Kukla, G, Cilek, V. (1996). Plio-Pleistocene megacycles; record of climate and tectonics. *Palaeogeogr. Palaeoclimatol. Palaeoecol*., **120**(1–2), 171–94
[69] Wunsch, C. (2004). Quantitative estimate of the Milankovitch-forced contribution to observed Quaternary climate change. *Quaternary Science Reviews*, **23**, 1001–1012
[70] Hinnov, LA. (2000). New persorctives on Orbitally forced stratigrraphy. *Annu. Rev.Earth& Planet. Sci*., **28**, 419–75
[71] Paillard, D. (1998). The timing of Pleistocene glaciations from a simple multiple-state climate model. *Nature*, **391**, 378-381

[72] Huybers, P, Wunsch, C. (2005). Obliquity pacing of the late Pleistocene glacial terminations. *Nature*, **434**, 491-494.
[73] Saltzman, B. Dynamical Paleoclimatology: Generalised Theory of Global Climate Change. San Diego: Academic; 2002.
[74] Pelletier, J. (2003). Coherence resonance and ice ages. *J.Geophys. Res.*, **108**, doi:10.1029/2002JD003120
[75] Saltzman, B. (1982). Stochastically-driven climatic fluctuations in the sea-ice, ocean temperature, CO2, feedback system. *Tellus*, **34**, 97–112
[76] Wentz, FJ, Ricciardulli, L, Hilburn, K, Mears, C. (2007). How Much More Rain Will Global Warming Bring? Science, DOI:10.1126/science.1140746
[77] Pierrehumbert, RT. (2002). The hydrologic cycle in deep-time climate problems. *Nature*, **419**, 191-198
[78] Khotinsky, NA. Holocene vegetation history. In: Velichko, A.A., Wright, H.E., Barnosky, C.W. (Eds.). *Late Quaternary Environments of the Soviet Union.* Minneapolis: University of Minnesota Press; 1984
[79] Velichko, AA., Andreev, AA & Klimanov, VA. (1997). Climate and vegetation dynamics in the tundra and forest zone during the Late Glacial and Holocene. *Quatern. International*, **41/42**, 71-96
[80] Velichko, AA, Catto, N, Drenova, AN, Klimanov, VA, Kremenetski, KV, Nechaev, VP. (2002). Climate changes in East Europe and Siberia at the Late glacial–holocene transition. *Quaternary International*, **91,** 75–99
[81] Osipov, EYu. (2004). Equilibrium-line altitudes on reconstructed LGM glaciers of the northwest Barguzinsky Ridge, Northern Baikal, Russia. *Palaeogeography, Palaeoclimatology, Palaeoecology*, **209**, 219– 226
[82] Horiuchi, K, Matsuzaki, H, Osipov, E, Khlystov, O, Fujii, S. (2004). Cosmogenic 10Be and 26Al dating of erratic boulders in the southern coastal area of Lake Baikal, Siberia. *Nucl.Instr.&Meth. in Phys. Res. B (NIMB),* 223–224, 633–638
[83] Isaeva, LL. Late Pleistocene Glaciation of North-Central Siberia, Late Quaternary Environments of Soviet Union. Minneapolis: Univ. of Minesota Press; 1984.
[84] Fagel, N, Boski, T, Likhoshway,. L and Oberhaensli, H. (2003). Late quaternary clay mineral record in Central Lake Baikal (Academician ridge, Siberia). *Paleogeogr. Paleoclimatol. Paleoecol*, **193**, 159-179.
[85] Berger, WH, Bickert, T, Yasuda, M.K., Wefer,G. (1996). Reconstruction of atmospheric CO2 from ice core data and the deep-sea record of Ontong Java plateau: the Milankovitch chron. *Geol. Rundschau*, **85**, 466-495
[86] Kravchinsky, VA, Krainov, MA, Evans, ME, Peck, JA, King, JW, Kuzmin, MI, Sakai, H, Kawai, T, Williams, DF. (2003). Magnetic record of Lake Baikal sediments: chronological and paleoclimatic implication for the last 6.7 Myr. *Palaeogeography, Palaeoclimatology, Palaeoecology* **195**, 281-298
[87] Prokopenko, AA, Hinnov, LA, Williams, DF, Kuzmin, MI. (2006). Orbital forcing of continental climate during the Pleistocene: a complete astronomically tuned climatic record from Lake Baikal, SE Siberia. *Quaternary Science Reviews*, **25**, 3431–3457
[88] Dawson, R, Wei, R, Tao, S, Ito, Y, Yamanaka, K, Haraguchi, H. (2004). Analysis of silicon concentration periodicity for the past 2.4Ma in sediments from Lake Baikal site BDP 96-2. *Climate Research*, **26**, 193–197

[89] Granina, L, Muller, B, and Wehrli, B. (2004). Origin and dynamics of Fe- and Mn-sedimentary layers in Lake Baikal. *Chem. Geol.*, **205**, 55–72.
[90] Trunova, VA, Bobrov, VA, Pampura, VD. (1996). Synchrotron Radiation XRFA study of Compositional Homogenity of Sediment Core from Lake Baikal. *X-Ray Spectr.*, **25**, 55-59
[91] Makrygina, VA, Petrova, ZI, Koneva, AA. (1994). Geochemistry of carbonate ores of Olkhon and nearest cites (west baikalian region). *Geochemistry*, **10**, 437-1450 (in Russian)
[92] Falkner, KK, Church, M, Measures, CI, LeBaron, G, Thouron, D, Jeandel, C, Stordal, MC, Gill, GA, Mortlock, R, Froelich, P & Chan, L-H. (1997). Minor and trace element chemistry of Lake Baikal, its trubutaries, and surrounding hot springs. *Limnol. Oceanogr.*, **42**, 329-345.
[93] Grachev, MA. On Modern State of the Ecological System of Lake Baikal. Irkutsk: *Nauka*; 2002 (in Russian)
[94] Mizandronzev, IB. Chemical processes in lake bottom sediments. Novosibirsk: *Nauka*;1990 (in Russian)
[95] Zhmodik, SM, Mironov, AG, Grachev, MA, Bobrov, VA, Nemirovskaya, NA, Khlystov, OM, Zheleznyakova, TO, Titov, AT. (2001). Uraniferous phosphorites in bottom sediments of Lake Baikal. *Doklady Earth Sci.*, **379A** (6), 682-687.
[96] Zhmodik, SM, Verkhovtseva, NV, Nemirovsraya, NA, Titov, AT, Goldberg, EL, Phedorin, MA, Khlystov, OM, Mironov, AG, Karmanov, NS, Zolotarev, KV. *Study of phosphate layers in Lake Baikal sediments by complex of local methods*. v1, 653- 657. In: Water-Rock Interaction (Proceedings 11th International Symposium on Water-Rock Interaction WRI-11, 27 June-2July). Eds R.B.Wanty & R.R.Seal A.A. Balkema Publisher: New York, USA; 2004
[97] Deike, RG, Granina, L, Callender, E, and McGee, JJ. (1997) Formation of ferric iron crusts in quaternary sediments of Lake Baikal, Russia, and implications for paleoclimate. *Mar. Geol.* **139**, 21–46.
[98] Muller, B, Granina, L, Schaller, T, Ulrich, A., and Wehrli, B. (2002). P, As, Sb, Mo, and other elements in sedimentary Fe/Mn layers of Lake Baikal. *Environ. Sci. Technol*, **36**, 411–420.
[99] Chalov, P I, Merkulova, K I. (1966). Comparative rate of oxidation of 234U and 238U atoms in certain minerals. *DAN*, **167**, 146–148 (in Russian).
[100] Chalov, PI. (1969). Mechanism for formation of nonequilibrium ratios between natural compounds containing uranium and thorium. *Soviet Atomic Energy,* **27**(1), 708-716 (in Russian)
[101] Chalov, PI. Isotope fractionation of natural uranium. Frunze: Ilim Publisher; 1975 (in Russian)
[102] Cherdyntsev, VV. Uranium-234. Jerusalem: Israel Program of Sci. Transl; 1971 (Translation from Russian in English of Cherdyntsev, VV, Uran-234. Moscow: Atomizdat; 1969).
[103] Fleischner, RL. (1988). Alpha-recoil damage: Relation to isotopic disequilibrium and leaching of radionuclides. *Geochem&Cosmohim.Acta*, **52**, 1459-1466
[104] Gavshin, VM, Bobrov, VA, Bogdanov, YuA. (1994). Uranium anomalies in bottom sediments of Lake Baikal. *DAN*, **34** (3), 356 – 359 (in Russian).

[105] Winograd, IJ, Coplen, TB, Landwehr, JM, Riggs, AC, Ludwig, KR, Szabo, BJ, Kolesar, PT, and Revesz, KM. (1992). Continuous 500,000-year climate record from vein calcite in Devils Hole, Nevada. *Science* , **258**, 255–260.

[106] Winograd, IJ, Landwehr, JM, Ludwig, KR, Coplen, TB and Riggs, AC. (1997). Duration and Structure of the Past Four Interglaciations. *Quat. Res*., **48**, 141–154

[107] Cutler, KB, Edwards, RL, Taylor, FW, Cheng, H, Adkins, J, Gallup, CD, Cutler, PM, Burr, GS, Bloom, AL. (2003). Rapid sea-level fall and deep-ocean temperature change since the last interglacial period. *EPSL*, **206**, 253-271

[108] Gallup, CD, Cheng, H, Taylor, FW, Edwards, RL. (2002). Direct Determination of the Timing of Sea Level Change During Termination II. *Science*, **295**, 310-313

[109] Eisenhauer, A, Zhu, ZR, Collins, LB, Wyrwoll, KH, Eichstätter, R. (1996). The Last Interglacial sea level change: new evidence from the Abrolhos islands, West Australia. *Geology Rundschau*, **85**, 606–614

[110] Henderson, GM, Slowey, NC. (2000). Evidence from U-Th dating against Northern Hemisphere forcing of the penultimate deglaciation. *Nature*, **404**, 61-66

[111] Clark, PU, Webb, RS, Keigwin, LD. Mechanisms of Global Climate Change at Millennial Time Scales, *Geophysical Monograph Series*. Washington: American Geophysical Union; 1999

[112] Alley, RB, Clark, PU. (1999). The deglaciation of the northern hemisphere: A Global perspective. *Annual Review Earth Planet Science*, **27**, 149-182.

[113] Broecker, WS. (2000). Abrupt climate change: causal constraints provided by the paleoclimate record. *Earth-Science Reviews*, **51** 137–154

[114] Dansgaard, W, Johnsen, SJ, Clausen, HB, Dahl-Jensen, D, Gundestrup, NS, Hammer, CU, Hvidberg, CS, Steffensen, JP, Sveinbjornsdottir, AE, Jouzel, J, Bond, G. (1993). Evidence for general instability of past climate from 250 kyr ice core record. *Nature*, **364**, 218-220.

[115] Heinrich, H. (1988). Origin and consequences of cyclic ice rafting in the northeast Atlantic Ocean during the past 130 000 years. *Quaternary Research*, **29** (2), 142-152

[116] Grootes, PM, Stuiver, M, White, JWC, Johnsen, SJ, and Jouzel, J. (1993). Comparison of oxygen isotope records from the GISP2 and GRIP Greenland ice cores. *Nature*, **366**, 552-554.

[117] Grootes, PM and Stuiver, M (1997). Oxygen 18/16 variability in Greenland snow and ice with 10-3- to 10-5-year time resolution. *Journal of Geophysical Research*, **102**, 26455-26470.

[118] Dahl-Jensen, D, Mosegaard, K, Gundestrup, N, Clow, GD, Johnsen, SJ, Hansen, AW, Balling, N. (1998). Past Temperatures Directly from the Greenland Ice Sheet. *Science*, **82**, 268-271

[119] Bond, G, Heinrich, H, Broecker, W, Laberyie, L, McManus, J, Andrews, J, Huon, S, Jantschik, R, Clasen, S, Simet, C, Tedesco, K, Klas, M, Bonani, G, Ivy, S. (1992). Evidence for massive discharges of icebergs into the North Atlantic ocean during the last glacial period. *Nature*, **360**, 245-249

[120] Bond, GC, Broecker, WS, Johnsen, SJ, McManus, JF, Labeyrie, L, Jouzel, J, and Bonani, G. (1993). Correlations between climate records from North Atlantic sediments and Greenland ice. *Nature*, **365**, 143-147.

[121] Bond, G, Showers, W, Cheseby, M, Lotti, R, Almasi, P, deMenocal, P, Priore, P, Cullen, H, Hajdas, I, Bonani, G. (1997). A Pervasive Millennial-Scale Cycle in North Atlantic Holocene and Glacial Climates. *Science*, **278**, 1257-1266

[122] Bond, G, Kromer, B, Beer, J, Muscheler, R, Evans, MN, Showers, W, Hoffmann, S, Lotti-Bond, R, Hajdas, I, Bonani, G. (2001). Persistent solar influence on North Atlantic climate during the Holocene. *Science*, **294**, 2130 – 2136

[123] Alley, R.B., Meese, D.A., Shuman, C.A., Gow, A.J., Taylor, K.C., Grootes, P.M., White, J.W.C., Ram, M., Waddington, E.D., Mayewski, P.A., and Zielinski, G.A. (1993). Abrupt increase in Greenland snow accumulation at the end of the Younger Dryas event. *Nature*, **362**, 527-529.

[124] McManus, J.F., Oppo, D.W., Cullen, J.L.. (1999). 0.5-Million-Year Record of Millennial-Scale Climate Variability in the North Atlantic. *Science*, **283**, 971-975

[125] McManus, J.F, Bond, G.C., Broecker, W.S, Johnsen, S., Laberyie, L., Higgins, S. (1994). High-resolution climate records from the North Atlantic during the last interglacial. *Nature*, **271**, 326-329

[126] Oppo, D. W., McManus, J. F., Cullen, J. L. (1998). Abrupt Climate Events 500,000 to 340,000 Years Ago: Evidence from Subpolar North Atlantic Sediments. *Science*, **279**, 1335-1338

[127] Severinghaus, J.P., Brook, E. J. (1999). Abrupt climate change at the end of the Last Glacial Period inferred from trapped air in polar ice. *Science*, **286**, 930 – 934.

[128] Brook, E.J., Sowers, T., Orchardo, J. (1996). Rapid Variations in Atmosphere Methane Concentration During the last 110000 years. *Science*, **273**, 1087-1091

[129] Brook, E.J., Harder, S., Severinghaus, J. P., Steig, E.J., Sucher, C.M. (2000). On the origin and timing of rapid changes in atmospheric methane during the last glacial period. *Global Biogeochemical cycles*, **14** (2), 559-572

[130] Chappellaz, J.A., Brook, E.J., Blunier, T., Malaiz, L.B. (1997). CH4 and d18O of O2 records from Antarctic and Greenland ice: A clue for stratigraphic disturbance in the bottom part of the Greenland Ice Core Project and the Greenland Ice Sheet Project 2 ice cores. *Journal of Geophysical Research*, **102**, 26547-26557.

[131] Clark, PU., Pisias, NG, Stocker, TF, Weaver , A J. (2002). The role of the thermohaline circulation in abrupt climate change. *Nature*, **415**, 863–869.

[132] Ganopolski, A., Rahmstorf, S. (2001). Rapid changes of glacial climate simulated in a coupled climate model. *Nature*. **409**, 153-158.

[133] Rahmstorf, S. (2002). Ocean circulation and climate during the past 120,000 years. *Nature*, **419**, 207-214

[134] Vellinga, M., Wood, R. A. (2002). Global climatic impacts of a collapse of the Atlantic thermohaline Circulation. *Clim. Change*, **54**, 251–267.

[135] Vidal, L., Schneider, R. R., Marchal, O., Bickert, T., Stocker, T. F., Wefer, G. (1999). Link between the North and South Atlantic during the Heinrich events of the last glacial period. *Climate Dynamics*, **15**, 909-919

[136] Boyle, E.A. (2000). Is ocean thermohaline circulation linked to abrupt stadial/interstadials transition? *Quat. Sci.Rev.*, **19**, 255-272

[137] Elliot, M., Labeyrie, L., Duplessy, J.C. (2002). Changes in North Atlantic deep-water formation associated with the Dansgaard-Oeschger temperature oscillations (60-10 ka). *Quat. Sci. Rev*, **21**, 1153–1165

[138] Broecker, W.S. (1998). Paleocean circulation during the last deglaciation: A bibpolar seesaw? *Paleoceanography*, **13**(2), 119-121

[139] Bianchi, G.G., McCave, I. N. (1999). Holocene periodicity in North Atlantic climate and deep-ocean flow south of Iceland. *Nature*, **397**, 515–517.

[140] Alley, R. B., Mayewski, P. A., Sowers, T., Stuiver, M., Taylor, K. C., Clark, P. U. (1997), Holocene climatic instability: A prominent, widespread event 8200 yr ago. *Geology*, **25**(6), 483–486

[141] Broecker, W.S. (2000) . Was a change in thermohaline circulation responsible for the Little Ice Age? *PNAS*, **97**(4), 1339–1342

[142] Broecker, W.S. (2001). Was the Medieval Warm Period Global? *Science*, **291**, 1497-1499

[143] Gasse F. (2000). Hydrological changes in the African tropics since the Last Glacial Maximum. *Quaternary Science Reviews*, **19**, 189-211.

[144] Allen, J.R.M., Brandt, U., Brauer, A., Hubberten, H.-W., Huntley, B., Keller, J., Kraml, M., Mackensen, A., Mingram, J., Negendank, J.F.W., Nowaczyk, N.R., Oberhansli, H., Watts, W.A., Wulf, S., Zolitschka, B. (1999). Rapid environmental changes in southern Europe during the last glacial period. *Nature*, **400**, 740–742.

[145] DeMenocal, P., Ortiz, J., Guilderson, T., Adkins, J., Sarnthein, M., Baker, L., Yarusinsky, M. (2000). Abrupt onset and termination of the African humid period: rapid climate responses to the global insolation forcing. *Quaternary Science Reviews*, **19**, 347-361.

[146] DeMenocal, P.B. (2001). Cultural Responses to Climate Change During the Late Holocene. *Science*, **292**, 667-673

[147] Haug, G.H., Guënther, D., Peterson, L.C., Sigman, D.M., Hughen, K.A., Aeschlimann, B. (2003). Climate and the Collapse of Maya Civilization. *Science*, **299**, 1731-1735

[148] Roucoux, K.H., Shackleton, N. J., Abreu, L., Schonfeld, J., Tzedakis, P.C. (2001). Combined Marine Proxy and Pollen Analyses Reveal Rapid Iberian Vegetation Response to North Atlantic Millennial-Scale Climate Oscillations. *Quaternary Research*, **56**, 128–132

[149] Defreitas, C.R. (2002). Are observed changes in the concentration of carbon dioxide in the atmosphere really dangerous? *Bulletin of Canadian Petroleum Geology*, **50**,297-327

[150] Baker, P.A., Rigsby, C.A., Saltze, G.O., Fritz, S.C., Lowenstein, T.K., Bacher, N.P., Veltz, C., (2001). Tropical climate changes at millennial and orbital timescales on Bolivian Altiplano. *Nature*, **409**, 698-700.

[151] Ames, L.L., McGarraf, J.E., Walker, B.A., Salter, P.F. (1983). Uranium and radium sorption on amorphous ferric oxyhydroxide. *Chemical Geology*, **40**, 135-148

[152] Fuller, C.C., Bargar, J.R., Davis, J.A., Piana, M.J. (2002). Mechanisms of Uranium Interactions with Hydroxyapatite: Implications for Groundwater Remediation. *Environ. Sci. Technol.,* **36**,158-165

[153] Froelich, P.N., Bender, M.L., Heath, G.R. (1977). Phosphorous accumulation rates in metalliferrous sediments on the East Pacific Rise. *Earth and Planetary Science Letters*, **34**, 351-359

[154] Lucotte, M., Angelan, B. (1988). Processes controlling phosphate adsorption by iron hydroxides in estuaries. *Chemical Geology*, **67**, 75-83

[155] Porcelli, D., Andersson, S., Wassenburg, G.J., Ingri, I., Baskaran M. (1997). The importance of colloids and mires for the transport of uranium isotopes through the

Kalix River watershed and Baltic Sea. *Geochimica et Cosmochimica Acta*, **61** (19), 4095-4113

[156] Samuelarey, J., Seaman, J., Bertsch, P. (1999). Immobilization of Uranium in Contaminated Sediments by Hydroxyapatite Addition. *Environ. Sci. Technol.*, **33**, 337-342

[157] Segebade, C., Hedrich, M. (2003). Behaviour of uranium in iron-bearing permeable reactive barriers: investigation with U as a radioindicator 237. *Science of the Total Environment*, **307**, 231–238

[158] Simon, F.G. , Biermann, V., Segebade, C., Hedrich, M. (2004). Behaviour of uranium in hydroxyapatite-bearing permeable reactive barriers: investigation using U as a radioindicator . *Science of the Total Environment*, **326**, 249–256

[159] Wite, T., Davis, J.A., Payne, T.E., Waychunas, G.A., Xu, N. (1994). Uranium(VI) adsorption to ferrihydrite: Application of a surface complexation model. *Geochim. Cosmoshim. Acta*, **58**, 5465-5478.

[160] Qin, B. & Yu, G. (1998). Implication of lake level variations at 6 ka and 18 ka in mainland Asia. *Global Planet. Change*, **18**, 59-72.

[161] Urabe, A., Tateishi, M., Inouchi, Y., Matsuoka, H., Inoue, T., Dmytriev, A. and Khlystov, O.M. (2004). Lake-level changes during the past 100,000 years at Lake Baikal, southern Siberia. *Quat. Res.,* **62** (2), 214-222

[162] BDP-99 member. (2004). High-resolution records along the core BDP-99. *Geologia and Geofizika,* **45** (2), 163-193 (in Russian)

[163] Shackleton, N.J. Composite isotope sequence for Cores V19-30, ODP 677 and ODP 846. http://131.111.44.196/coredata/v677846.html

[164] Riley, G.A., Stommel, H., Bumpus, D.F. (1949). Quantitative ecology of the plankton of the western North Atlantic. *Bull. Bingham Oceanogr. Coll.*, **12** (3), 169

[165] Munk, W.H., Riley, G.A. (1952). Absorption of nutrients by aquatic plants. *Journal of Marine Res*., **11** (2), 215-240

[166] Ragueneau, O., Treguer, P., Leynaert, A., Anderson, R.F., Brzezinski, M.A., DeMaster, D.J., Dugdale, R.C., Dymond, J., Fischer, G., Francois, R., Heinze, C., Maier-Reimer, E., Martin-Jezequel, V., Nelson, D.M., Queguiner, B. (2000). A review of the Si cycle in the modern ocean: recent progress and missing gaps in the application of biogenic opal as a paleoproductivity proxy. *Global and Planetary Change*, **26**, 317–365

[167] Nelson, D.M., Treguer, P. (1992). Role of silicon as a limiting nutrient to Antarctic diatoms: evidence from kinetic studies in the Ross Sea ice-edge zone. *Mar.Ecol.Prog. Ser*., **80**, 255–264.

[168] Nelsonm D.M., Dortchm Q. (1996). Silicic acid depletion and silicon limitation in the plume of the Mississippi River: evidence from kinetic studies in spring and summer. *Mar. Ecol. Prog. Ser*, **136**, 63–178.

[169] Kelley, D.E. (1997). Convection in ice-covered lakes: effects on algal suspension. *Journal of Plankton Research*, **19** (12), 1859-1880

[170] Crank, J. *The mathematics of diffusion*. Oxford: Clarendon Press; 1956

[171] Danckwerts, P.V. (1950). Unsteady-state diffusion or heat-conduction with moving boundary. *Trans. Faraday Soc*., **46**, 701-712.

[172] Wetzel, R.G. and Likens, G.E. *Limnological Analyses* (2nd. edition). Springer-Verlag;1991

[173] Prokopenko, A.A., Karabanov, E.B., Williams, D.F., Kuzmin, M.I., Khursevich, G.K., Gvozdkov, A.A. (2001). The detailed record of climatic events during the past 75,000 yrs BP from the Lake Baikal drill core BDP-93-2. *Quaternary International*, **80–81**, 59–68.

[174] Prokopenko, A.A., Williams, D.F., Karabanov, E.B., Khursevich, G.K. (2001). Continental response to Heinrich events and Bond cycles in sedimentary record of Lake Baikal, Siberia. *Global and Planetary Change*, **28**, 217–226

[175] Oppo, D.W., McManus, J.F., Cullen, J.L. (2006). Evolution and demise of the Last Interglacial warmth in the subpolar North Atlantic. *Quaternary Science Reviews*, **25**, 3268–3277

[176] Lehman, S., Sachs, J., Crotwell, A., Keigwin, L., Boyle, E. (2002). Relation of subtropical Atlantic temperature, high-latitude ice rafting, deep water formation, and European climate 130,000–60,000 years ago. *Quaternary Science Reviews*, **21**, 1917–1924.

[177] Wagner, G., Masarik, J., Beer, J., Baumgartner, S., Imboden, D, Kubik, P.W., Synal, H.-A., Suter, M. (2000). Reconstruction of the geomagnetic field between 20 and 60 kyr BP from cosmogenic radionuclides in the GRIP ice core. *Nuclear Instruments and Methods in Physics Research B*, **172**, 597-604

[178] Horiuchi, K., Kobayashi, K., Oda, T., Nakamura, T., Fujimura, C., Matsuzaki, H., Shibata, Y. (2000) Climate-induced fluctuations of 10Be concentration in Lake Baikal sediments, *Nuclear Instruments and Methods in Physics Research B*, **172**, 562-567

[179] Smith, L.C., MacDonald, G.M., Velichko, A.A., Beilman, D.W., Borisova, O, Frey, K.E., Kremenetski, K.V., Sheng, Y. (2004). Siberian Peatlands a Net Carbon Sink and Global Methane Source since the Early Holocene. *Science*, **303**, 353-356

[180] Diillenbach, A., Blunier, T., Fliickige,r J., Stauffe,r B. (2000). Changes in the atmospheric CH4 gradient between Greenland and Antarctica during the Last Glacial and the transition to the Holocene. *Geophys. Res. Let.*, **27** (7), 1005- 1008

[181] Monnin, E., Indermuhle, A., Dallenbach, A, Fluckiger, J., Stauffer, B., Stoecker, T.F., Raynaud, D., Barnola, J-M. (2001). Atmospheric CO2 concentrations over the Last Glacial Termination. *Science*, **291**, 112-114.

[182] North Greenland Ice Core Project members. (2004). High-resolution record of Northern Hemisphere climate extending into the last interglacial period. *Nature*, **431**, 147-151

[183] Edgington, D.N., Klump, J.V, Robbins, J.A., Kusner, Yu.S., Pampura, V.D.& Sandimirov, I.V. (1991). Sedimentation rates, residence times and radionuclide inventories in Lake Baikal from 137Cs and 210Pb in sediment cores. *Nature*, **350**, 601-604.

In: Climate Change Research Progress
Editor: Lawrence N. Peretz, pp. 87-125

ISBN: 978-1-60021-998-6

Chapter 2

THE ECONOMIC LITERATURE ON CLIMATE POLICY RESEARCH: THE STATE OF THE ART AND SOME DIRECTIONS FOR FURTHER RESEARCH

Pablo del Río[1a], Felix Hernández[2b] and Alejandro Caparrós[2c]

[1]Faculty of Social Sciences and Law, Department of Spanish and International Economics, Econometrics and History and Economic Institutions, University of Castilla-La Mancha, C/Cobertizo de S. Pedro Mártir s/n, Toledo-45071, Spain

[2]Instituto de Economía y Geografía, Consejo Superior de Investigaciones Científicas, C/Pinar 25, Madrid 28006, Spain

Abstract

Awareness of the problem of climate change has boosted the implementation of climate policy instruments at the international (i.e., Kyoto Protocol) and national levels in order to tackle greenhouse gas (GHG) emissions mitigation. In turn, this has stimulated research by various social science disciplines, including economics, on the analysis of policies and measures aimed at GHG mitigation. This literature has focused on the functioning and design of the instruments, on the assessment of different instruments according to several criteria (environmental effectiveness and economic efficiency in emissions reductions) and on a comparison of their respective merits and drawbacks. This paper provides a critical overview of the economic literature on climate policy research, indicating the main lines of research and key results from this fruitful literature. It also proposes several avenues for further research, particularly concerning issues to be tackled and methods to be applied. In this regard, the analysis of the interactions and complementarities between different climate and energy policy instruments, the inclusion of technological change in climate change mitigation and the ancillary benefits are particularly worth undertaking.

[a] E-mail address: pablo.rio@uclm.es.(Corresponding author)
[b] E-mail address: felix1@ieg.csic.es
[c] E-mail address: acaparros@ieg.csic.es

1. Introduction. Aim, Scope and Methodology

Climate policy is currently an important topic in international policy debates. Its relevance lies in the evidence that a climate change problem exists, which requires the adoption of measures to tackle it. Awareness of the problem of climate change has boosted the implementation of climate policy instruments at the international (i.e., Kyoto Protocol) and national levels in order to tackle greenhouse gas (GHG) emissions mitigation. In turn, this has stimulated research by various social science disciplines, including economics, on the analysis of policies and measures aimed at GHG mitigation. This literature has focused on the functioning and design of the instruments, on the assessment of different instruments according to several criteria (environmental effectiveness and economic efficiency in emissions reductions) and on a comparison of their respective merits and drawbacks.

This paper provides a critical overview of the economic literature on climate policy research, indicating key results from this fruitful literature and proposing several avenues for further research, particularly concerning issues to be tackled and methods to be applied.

The focus of this review is specifically on the economic literature on climate policy research. This means that other aspects have not been considered. For example, we have not considered the existing economic literature on climate change research more generally. Although it is obvious that good climate policy should be based on good research on the economics of climate change, there are some aspects of climate change economics that are more closely related to what can be done to mitigate emissions and, thus, concentration levels and do so at the minimum possible social cost (cost-effectiveness)[1]. On the other hand, although relevant for policy-making, non-economic perspectives on climate policy have been omitted (including those pertaining to policy-sciences and sociology). However, we do not claim it is a complete and comprehensive review of climate policy. We have focused on those issues which are deemed most relevant for climate policy, but other aspects which might be worth analysing have not been included in the analysis.

In order to identify the current state of the issue, and in addition to the own experience of the authors with this topic, several sources of information have been consulted, including books, journals and internet searches. The most recent issues (last six years) of the most relevant and cited international journals dealing with climate policy issues (including Climate Policy, Energy Economics, Energy Policy, Energy Journal, Energy and Environment and Mitigation and Adaptation Strategies for Global Change and Climatic Change) as well as other more general journals on environmental and ecological economics have been consulted (including Ecological Economics, Environmental and Resource Economics, Journal of Environmental Economics and Management and Resource and Energy Economics). Finally, an overview of the papers included in the last two congresses of the International and the European Society for Ecological Economics (ISEE, ESEE) and the International and European Association for Environmental and Resource Economics (EAERE) has been undertaken in order to identify those related to environmental technological change. Not all papers relevant for a topic are mentioned, but only those we consider to be the most relevant ones although, of course, this is a subjective judgement.

[1] For instance, a major topic in the literature on climate change economics is the issue of economic valuation of climate change impacts. Since an analysis of the costs and benefits of climate change mitigation is certainly very important for the timing of implementation of policy measures, it would be very difficult to argue that this research is not policy relevant.

Accordingly, this paper is structured as follows. The next section is dedicated to the overview of the state of the art of climate policy research. Section 3 considers relevant lines of future research. The paper closes with some concluding remarks and policy recommendations.

2. The State of the Art of Climate Policy Research

i. Economic impact of Kyoto and the EU Directive.

One of the tasks economists have been engaged in has been the analysis of the potential costs of emissions mitigation, for example, through the Kyoto Protocol and the EU Directive. This has been done with the help of econometric models.

i.1. Concerning the impact of the Kyoto Protocol.

Many economists have defended the cost-effectiveness of the Kyoto Flexible Mechanisms (International Emissions Trading, Clean Development Mechanism and Joint Implementation) to reach the Kyoto Protocol targets. These mechanisms reduce overall compliance costs because their flexibility allows countries with high marginal abatement costs to buy Kyoto units from cheap abatement countries, reducing their compliance costs, whereas the sellers of those units obtain a stream of revenue. Econometric models have been built to simulate the impact of those mechanisms on the cost of reaching the Kyoto Protocol targets and the savings that these instruments allow with respect to alternative situations in which countries reach their targets with direct regulation measures or with domestic (but not international) emissions trading schemes. This model can be classified according to the following table, although the basic difference is between top-down and bottom-up models (Hernández 1999).

Table 1. Types of models.

Type of models	General description	Advantages (ADV)/Drawbacks (DRA)	Example
Integrated assessment models	They represent physical and social processes.	ADV: They allow the joint analysis of physical and economic phenomena. DRA: Very data-intensive. (the economic component of these models belongs to the following categories)	AIM, GRAPE, RICE, IGSM, MERGE
Computable General Equilibrium models (CGE)(*top-down*)	They determine the new equilibrium in the economic system after an exogenous shock (for instance, the implementation of a carbon tax).	ADV: They pick up the influence of energy policy on other industrial sectors, a well as sectoral and feedback effects. DRA: They assume perfect markets, they do not represent the adjustment process after an exogenous shock, they undervalue transaction costs and the imperfections of labour and monetary markets are ignored. The energy sector is better represented in bottom-up models (i.e., CGE models are unable to consider all technological options).	EPPA, GEM-E3, GREEN, WORDSCAN

Table 1. Continued

Type of models	General description	Advantages (ADV)/Drawbacks (DRA)	Example
Emissions trading models	They use marginal abatement costs (MAC) to analyse emissions trading. MACs are generated through CGE models with different emission targets.	ADV: Robust and appropriate method to analyse emissions trading. DRA: They can not capture effects such as carbon leakage.	MACGEM, PET, ZHANG, CICERO, ENEA, R&S
Neo-Keynesian macroeconomic models (*top-down*)	Aggregate production is modelled as a function of capital and labour inputs. Input-output tables are used to represent transaction between sectors.	ADV: Same as CGE models. They take into account monetary policy, imperfect competition and unemployment. DRA: Limited use for long-term analysis. Very complex.	G-CUBED*, OXFORD
Energy system models (*bottom-up*)	They use data of specific Technologies and the optimal technological mix is determined through lineal programming.	ADV: Great technological detail. DRA: -Energy demand is exogenous and independent of prices. Human behaviour is not endogeneized. -Only the energy sector is represented, without any possibility to consider linkages of this sector with the rest of the economy.	POLES MARKAL

Source: Own elaboration from Springer (2003). *Note: some of these models belong to two categories. This is the case, for example, of G-CUBED, which is a neo-Keynesian macroeconomic model and a CGE model. Many models are really hybrids belonging to two categories.

These models lead to different results. In general, the results of top-down models show greater compliance costs than bottom-up models, especially because the later include technologies with negative costs (i.e., no-regret measures). However, there are important differences even between models within the same categories, due to (Banuri et al 2001, Springer 2003, Viguier et al 2003, Hernández 1999):

-Differences between their business as usual scenarios. The business as usual (BAU) scenarios of each model contain different assumption on the future economic policies at the sector and macroeconomic levels as well as other features, including the sectoral structure, the resource intensity, prices and, thus, technological choices. These assumptions cause different emissions growth projections across models in the BAU scenarios.

-Different assumptions about costs and availability of low-cost alternative technologies in the energy sector. This leads to different slopes in the marginal abatement cost curves of the different models.

-Different treatment of short-term macroeconomic shocks.

-Projections of the rate of technological progress, GDP and demographic growth are also very different across models.
-Finally, some models only take CO2 into account (for example, POLES), whereas others include other GHG (GTEM or EPPA).

On the other hand, these models assume an ideal functioning of the emissions trading market. The cost-effectiveness of emissions trading takes place as long as several conditions (numerous participants, low transaction costs etc...). Furthermore, models can not deal with all the aspects in detail.

Notwithstanding, a common result of all these models is that the emissions trading scheme reduces compliance costs for countries. The following table shows the marginal abatement costs in three different scenarios for 17 different models: no emissions trading scenario, emissions trading scenario between Annex I countries and global emission trading (all countries).

Table 2. Main results from the Energy Modelling Forum. Marginal abatement costs (in $1990/tCarbon; Kyoto target in 2010)*.

Model	No emissions trading				Emissions trading between Annex I	Global emissions trading
	USA	OCDE-E	Japan	CANZ		
ABARE-GTEM	322	665	645	425	106	23
AIM	153	198	234	147	65	38
CETA	168				46	26
Fund					14	10
G-Cubed	76	227	97	157	53	20
GRAPE		204	304		70	44
MERGE3	264	218	500	250	135	86
MIT-EPPA	193	276	501	247	76	
MS-MRT	236	179	402	213	77	27
Oxford	410	966	1074		224	123
RICE	132	159	251	145	62	18
SGM	188	407	357	201	84	22
WorldScan	85	20	122	46	20	5
Administration	154				43	18
EIA	251				110	57
POLES	135.8	135.3	194.6	131.4	52.9	184

Source: IPCC (2001). OCDE-E = OCDE Europe. CANZ = Canada, New Zealand and Australia.
* 100$/tCarbon = 27.3 $/tCO2

The results show that marginal abatement costs significantly differ per region and, most importantly, that emissions trading significantly reduces the costs of compliance with the Kyoto Protocol and that costs savings are greater the higher the number of countries participating in emissions trading.

The next table provides estimates on the loss of GDP in 2010 in the different regions if they comply with their Kyoto Protocol targets. USA, Canada and New Zealand are the most negatively affected countries in all scenarios. Again, the shift from a no-trading scenario to the trading scenarios significantly reduces abatement costs.

Table 3. Main results from the Energy Modelling Forum. Loss of GDP (% over GDP levels; Kyoto target in 2010)*.

Model	No trading				Emissions trading between Annex I				Global emissions trading			
	USA	OCDE	Japan	CANZ	USA	OCDE	Japan	CANZ	USA	OCDE	Japan	CANZ
ABARE-GTEM	1.96	0.94	0.72	1.96	0.47	0.13	0.05	0.23	0.09	0.03	0.01	0.04
AIM	0.45	0.31	0.25	0.59	0.31	0.17	0.13	0.36	0.20	0.08	0.01	0.35
CETA	1.93				0.67				0.43			
G-CUBED	0.42	1.50	0.57	1.83	0.24	0.61	0.45	0.72	0.06	0.26	0.14	0.32
GRAPE		0.81	0.19			0.81	0.10			0.54	0.05	
MERGE3	1.06	0.99	0.80	2.02	0.51	0.47	0.19	1.14	0.20	0.20	0.01	0.67
MS-MRT	1.88	0.63	1.20	1.83	0.91	0.13	0.22	0.88	0.29	0.03	0.02	0.32
Oxford	1.78	2.08	1.88		1.03	0.73	0.52		0.66	0.47	0.33	
RICE	0.94	0.55	0.78	0.96	0.56	0.28	0.30	0.54	0.19	0.09	0.09	0.19

Source: IPCC (2001). OCDE-E = OCDE Europe. CANZ = Canada, New Zealand and Australia.

Apart from more general models, there are "sectoral" models, i.e., models which analyse the impact of the Kyoto Protocol on specific sectors using other instruments apart from emissions trading (i.e., taxes). In this context, the work of Mathiesen and Maestad (2004) is worth mentioning. They analyse the impact of emissions reductions in the steel sector and find out that a carbon tax on emissions in industrialized countries may cause relocation of steel production to non-industrialized countries. Using a partial equilibrium model of the steel industry, this paper finds that global emissions from this industry are likely to decline substantially due to factor substitution within the integrated steel mills in the industrialized countries, and to a shift between steel making technologies. The advantage of this type of sectoral models is that the aforementioned effects are not well accounted for in economy wide models, which only provide aggregate numbers.

i.2 Concerning the economic impact of the EU emissions trading scheme (EU ETS).

The implementation of the EU ETS has encouraged the realisation of studies which analyse the impact on specific countries and sectors, using econometric models for this task.

For example, Klepper and Peterson (2006) assess the likely allocation effects of the current climate protection strategy as it is laid out in the National Allocation Plans (NAPs) for the European Emissions Trading Scheme (ETS) with the help of the multi-regional, multi-sectoral CGE-model DART, which is used to simulate the effects of the current policies in the year 2012 when the Kyoto targets need to be met. Different scenarios are simulated in order to highlight the effects of the grandfathering of permits to energy-intensive installations, the use of the project-based mechanisms (CDM and JI), and the restriction imposed by the supplementarity criterion.

Böhringer et al (2005) adopt a wider perspective in which the impact of a hybrid system is analysed: Member States allocate their national emission budgets under the EU Burden Sharing Agreement (BSA) between energy-intensive sectors that are eligible for European emissions trading and the remaining segments of their economies that will be subject to complementary domestic emission regulation. They show that such hybrid emission regulation may lead to substantial excess costs compared to a comprehensive emissions trading system covering all segments of the economy. Therefore, a key issue is how to treat the sectors not covered by the EU ETS. Böhringer et al (2006) note that, as the EU ETS covers only part of domestic carbon emissions, it implies a segmented environmental regulation scheme: Each Member State must specify additional domestic abatement policies for the sectors outside the EU ETS in order to meet its emissions budget under the BSA. Based on numerical simulations for Germany, they quantify the excess costs of segmented carbon regulation and conclude that inefficiencies can be much better explained by lobbying of influential EU ETS sectors than by information problems, i.e., the authors go beyond modelling issues to provide a political economy perspective. This issue is related to, both, the issue of allocation of allowances in the EU ETS and to political economy considerations (see below)[1].

In line with Böhringer et al (2005), several authors have shown that Member States have tended to allocate too many allowances to the covered sectors compared to the non-covered sectors. This increases the effort that the non-covered sectors have to make in order to put the European economies in a Kyoto-path so that they can comply with their Kyoto Protocol

[1] The terms "permits" and "allowances" will be used interchangeably throughout this chapter.

targets (see Buchner and Ellerman 2006; Betz et al 2006). Despite the efforts made by the European Commission, this also seems to be a problem in the second period of the EU ETS, although to a lesser extent (see Neuhoff et al 2006).

Based on the NAPs, Kemfert et al (2006) analyse the effects of the EU ETS compared to fixed quotas with the GTAP-E model. It is found that the welfare gains from regional emissions trading are largest for some countries (Belgium, Denmark and Great Britain) smaller for others (Finland and Sweden) and smallest for Germany and other regions. When the economy-wide and terms-of-trade effects of emissions trading are taken into account, however, (negative) terms-of-trade effects can offset the (positive) allocative efficiency gains for the cases of the Netherlands and Italy, while all other regions end up with positive net welfare gains. All regions, however, experienced increases in real GDP as a result of regional emissions trading.

Demailly and Quirion (2007) quantify the competitiveness impact of the EU ETS on the iron and steel industry, which is arguably the most sensitive among those covered by the EU ETS, since it is both highly CO_2-intensive and relatively open to international trade. The analysis is carried out with a partial equilibrium model of the steel market and the results show that competitiveness impacts are minor.

With respect to the indirect impacts of the EU ETS in the different scenarios, recent studies have shown that these will be limited. For example, Grubb and Neuhoff (2006) suggests that this issue could be a source of concern only after 2012 for those sectors which are electricity-intensive, including aluminium, as long as allocation of allowances is very restrictive, is based on auction and cost-pass through to electricity prices exists[2]. Cement and steel could also be affected in this context. They show that, considering the allocation for 2008-2012 and the likely prices in such period, net costs will not be high with respect to the sectors' value added.

On the other hand, the impact of the EU ETS on the electricity sector and on electricity consumers has received much attention by the literature. For example, Kara et al (2006) use an electricity market and an energy system model (VTT and TIMES, respectively) to analyze the likely impact of the EU ETS on electricity generators, energy intensive industries, other industries and consumers in the Nordic countries. The annual average electricity price is found to rise by 0.74 €/MWh for every 1 €/tCO_2 in the Nordic area. Large windfall profits are estimated to incur to electricity producers in the Nordic electricity market.

In this context, Linares et al (2006) analyse the impact of the EU ETS on the Spanish electricity sector, using a highly detailed model of this sector. Results show that the implementation of the Directive will result in a significant increase of electricity prices, and also, due to the Spanish pricing system, in a large increase in the revenues of generating firms, unless the regulator intervenes. Results also show the different implications of different allocation methods.

---

[2] These authors analyse the potential competitiveness effects of the EU ETS, considering also the impact of a higher electricity price on sectors not covered by the EU ETS. It is assumed that such effects depend on two major factors: the potential impact of the EU ETS on input costs relative to sector value-added (before any mitigation or pass-through of costs into product prices) and the trade exposure of these sectors. In general, no sector scores badly in both factors (except non-ferrous metals including aluminium), i.e., those sectors potentially suffering a potentially high competitiveness impact according to the first criteria do not have a great trade exposure index (electricity generation and cement), whereas those with large trade exposure, are unlikely to suffer from a large impact on input costs (metal manufacturers and textiles).

Sijm et al (2006) analyse the impact of free allocation on the price of electricity and the profitability of the electricity generation business. They use an energy system model (COMPETES)[3], which simulates the impact of the behaviour of the largest electricity generators on the wholesale electricity market, considering different scenarios (which relate to different structures of the electricity market) in order to identify how firms in Germany and the Netherlands pass through their greater costs as a result of the EU ETS into electricity prices. The results show that such pass-through could be as large as 60% to 100% of CO2 costs, depending on the carbon intensity of the marginal electricity generation plant, market factors and technology-specific features. Therefore, according to this study, electric utilities would obtain significant *windfall profits* due to the EU ETS, a result in line with Kara et al (2006) and Linares et al (2006).

Again, a highly relevant issue in the context of the impacts of the EU ETS has been the allocation of emissions allowances. The relevance of this issue deserves that a specific section is devoted to it (see below).

ii. Emissions trading.

Emissions trading schemes (ETS) have been analysed according to a triple perspective:

1) On the one hand, the analysis of the ETS has focused on the criteria of environmental effectiveness and cost-effectiveness. Other criteria (such as dynamic efficiency, wider macroeconomic impacts, distributive impacts, political feasibility and transaction costs) are increasingly being considered.
2) On the other hand, the focus of the analysis of the schemes has shifted from the national to the international level (EU ETS and Kyoto Protocol). The pioneering empirical analyses of emissions trading schemes have been carried out in the U.S. (see Ellerman et al 2003, Schmalensee et al 1998) and they show that significant cost savings in complying with targets were attained, compared to a traditional regulation instrument in the form of emissions standards[4]. Initial analysis of the functioning of the EU ETS have recently mushroomed (see several issues of the Journals Carbon Market Europe and Climate Policy). A substantial volume of papers have been dedicated to what design elements should be chosen for an appropriate functioning of the schemes. In this case, the works of Tietenberg (2005, 2006) and Grubb et al (1998) clearly stand out.
3) Finally, from the initial theoretical/qualitative analyses, which put the emphasis on the demonstration of the cost-efficiency of the scheme and the appropriate design elements, the focus has shifted to quantitative analysis. In this context, macroeconomic simulation models play a very relevant role. Econometric models have played a key role concerning different aspects of the analysis of the functioning of the systems and, particularly, regarding the cost-effectiveness of an international

[3] See http://www.electricitymarkets.info

[4] For an overview of the functioning of the emissions trading schemes in the U.K., Denmark, Norway and New South Wales Greenhouse Gas Abatement Scheme in Australia, see Hernández and del Río (2007).

ETS. We will focus on the major implications of these analyses for international climate policy, particularly in the context of the Kyoto Protocol.

Simulation models have allowed the identification of the impacts of different alternatives and, particularly, the compliance costs with the Kyoto Protocol in several situations:

-Hot air. If Business-as-Usual emissions in a country are below their emissions target, then emissions reductions can not be considered to be the result of mitigation efforts, i.e., they are not "intentional". This may reduce the environmental effectiveness of the Kyoto Protocol if Annex I in this situation are allowed to buy and sell this "hot air", and this may happen through the international emissions trading system. The main argument against allowing hot air is that it would lead to greater emissions than would otherwise occur. Hot air projections for 2010 (middle of the period 2008-2012), cover a wide range (between 257 and 1371 million tons of CO2), depending on emissions projections in the business-as-usual scenario (Zhang 2000). The amount of hot-air available will depend on the economic growth rates of Eastern European countries, on the evolution of energy efficiency, on the fuel mix and on how changes in the use of land are accounted for (Brander 2003). Simulations show the impact of this decision on the carbon market: it increases the supply of permits in the international market and leads to a reduction of the market price. For instance, Haites (2004) shows that Russia and Ukraine would maximize their revenues if they behaved strategically and to sold only 40% of their surplus permits, banking the rest for the next period.

-Impact of non-ratification by the U.S. The impact on the international emissions trading market of the non-ratification of the Kyoto Protocol by the U.S. has been analysed with the help of econometric models. The main conclusion is that, since the U.S. would be a net buyer of Kyoto units, its non-ratification would lead to a reduction in the demand for permits and, thus, its price, indirectly benefiting other buyer countries.

-Supplementarity. The Protocol states that the use of the Kyoto Mechanisms will be supplementary to domestic emissions reductions by Annex I countries, i.e., that these countries may comply with their emissions reduction targets by buying permits in the international market and carrying out emissions reductions projects abroad (CDM and JI projects), but it is also required that a share of the emissions be reduced at home. The impact of this supplementarity rule is a greater price for Kyoto units than if such rule was non-existent. In turn, this greater price would encourage the realisation of domestic emissions reductions in Annex I countries, which would reduce the demand for permits and its price (see Matschoss and Welsch 2006). The net final impact can be expected to be a greater price for Kyoto units than in the absence of supplementarity. With respect to cost-effectiveness in complying with emissions reductions targets, the supplementary provision reduces the potential cost-effectiveness of emissions trading. Matschoss and Welsch (2006) show that restrictions to trading raise the investments in carbon productivity in permit demanding regions while reducing them in permit supplying regions. In terms of per capita consumption, permit demanding regions lose and permit supplying regions gain from restrictions. In scenarios that involve "hot air," restrictions to trade lower overall emissions, which results in reduced climate damage for most regions. Reduced damage, in turn, reduces the incentive to invest in carbon productivity.

-Impact of banking and borrowing. According to the Kyoto Protocol, Kyoto units issued in a given year of the first compliance period (2008-2012) can be used in a later year within such period (intraperiod banking) and even in a later period (interperiod banking). This increases the intertemporal efficiency in emissions reduction, as well as the liquidity of the permit market, possibly leading to a less volatility of the permit price. However, in the Protocol borrowing is not allowed. In other words, permits to be issued in one later period can not be used to comply with emissions targets in a previous period. Stephan and Müller (2004) analyse banking and borrowing of carbon emission rights within the framework of a simple, integrated assessment model. Breaking the world economy in just two regions it is shown that: (1) Increasing intertemporal flexibility in greenhouse gas abatement through banking and borrowing of carbon emission permits has a positive effect on welfare for regions with a poor endowment in emission permits, but negatively affects rich-endowed regions. (2) Intergenerational fairness advocates intertemporal flexibility in greenhouse gas abatement, irrespectively of the initial allocation of permits. (3) Optimal carbon accumulation is not independent of the initial permit allocation. In contrast, Stevens and Rose (2002) argue that there no large benefits from banking and that the great cost-savings and cost-efficiency in the emissions trading scheme result from the spatial difference of marginal abatement costs. This paper presents a generalized dynamic model of greenhouse gas emissions trading under constraints on the volume of transactions. The results indicate the greatest gains would stem from extending permit trading spatially among industrialized nations, although sizable gains would also emanate from the inclusion of developing countries. Gains from intertemporal trading (banking and borrowing) are meagre. Restrictions on the volume of permit purchases have by far the most costly effect when developing countries are incorporated into trading[5].

-Other design elements in the Protocol. Based in the simulations carried out with the integrated model SAP12 (Stochastic Assessment of Climate Change Policy), Hourcade and Ghersi (2002) consider several options to improve the Kyoto Protocol. More specifically, the study shows that a "safety valve" is a good tool to avoid uncertainty on costs[6]. The study concludes that had this approach been adopted in the negotiations in the Conference of the Parties in 2001 (when the US decided to reject the Protocol), an agreement on payments between 35 and 100 $/tC could have been reached.

-The issue of market power. The literature has paid attention to the functioning of the emissions trading instrument in specific situations, including the case of imperfect competition in the product market or the issue of market power in the emissions trading market.

"Market power" has normally been taken to mean a company's use of its position in a market to raise prices above competitive levels. Theoretical models of emissions trading often assume perfect competition such that no firm can exercise power on the selling or buying side of the market. Although this may be appropriate for large markets (like the US SO2), permit

---

[5] Godal and Klaasen (2006) reach a similar conclusion.

[6] The "safety valve" refers to the setting of a maximum price for emission permits. When the price of permits exceeds the safety valve, the government agrees to sell permits at the safety valve price.

markets are sometimes imperfectly competitive and may be thin, concentrated and dominated by large firms (Cason et al 2003)[7].

An essential component of market power in the emissions trading market is the quantity of allowances that any participant may acquire or transfer. If allowances are grandfathered to the existing producers (particularly in the electricity sector), the concentration of allowances on a few large producers can create a significant risk of market power. Theoretical and empirical studies of emissions trading markets show that a dominating source (or a coalition of sources) may attempt to manipulate the price as a monopolist (seller) or monopsonist (buyer). This market power depends on the firm's size relative to the market in which it is operating. If a dominating source offers too few allowances, this increases the prices artificially and, therefore, the company can benefit from monopoly power by withholding allowances. If it buys too few allowances, this can cause the prices to fall to a lower than normal level. The company thus exerts monopsony power. In both cases, the dominant source may achieve a net gain at the expense of a cost-effective outcome (Burniaux 1999).

Empirical studies carried out on this issue can be classified in four different categories. The first three would correspond to the analysis of the conditions under which a market power situation may occur. The fourth category refers to the environmental and macroeconomic impacts of such a situation.

1) Ex-ante simulations. The simulations show that market power should be above 90% of the potential trade volume in order for a firm or group of firms to be able to manipulate the permit market (Cason et al 2003). However, this might be the case at an international level with "hot air" in the Kyoto Protocol. Russia and Ukraine may exercise market power on the permit supply side (see the result of Haites (2004) above), but other factors may work against this exercise. For example, Hagem and Maestad (2006) explore the interactions between the revenues for Russia from permit sales and fossil fuel exports. Their CGE model suggests that, since Russia is a large fossil fuel exporter, it has a small incentive to exercise market power in the emissions permit market, because this would increase the price of fossil fuels and, thus, reduce their exports[8]. The numerical simulation of Hagem and Maestad (2005) suggests that Russia would benefit from coordinating its permit exports with its oil and gas exports during the Kyoto commitment period.

2) Experimental studies. Laboratory tests provide an additional source of information on market power[9]. Most studies show that market power would not be a great problem and only one study on market power (Godby 2000) shows that this could be a problem that could offset the cost savings of emissions permits markets A relatively recent study is that of Carlén (2003), whose experiment tries to mimic the conditions of an international permit market. It includes one large buyer of permits (USA) and ten other countries. The paper shows that trade volumes are efficient and prices most of the time competitive, i.e., there are no negative effects on costs.

[7] According to this author, however, empirical studies find that the extent of market power would need to be great (up to 90% of the potential trading volume) for a firm or group of firms to manipulate the market.

[8] Notwithstanding, Russia is a large gas exporter. If a greater permit price favours the substitution of gas for coal, then Russia would benefit twice if it exercised market power in the permit market.

[9] See Cason et al (2003), Carlén (2003) and Burniaux (1999), among others.

3) Analysis of real-world cases. Case studies have been undertaken to identify the existence of market power in permit markets. The emphasis has been put on the US Acid Rain programme and, recently, on the EU ETS. No studies have been able to identify a market power situation.

4) Impact of market power. The impact of market power may be considered at least at two levels: compliance with emissions reduction targets (and compliance costs) and the macroeconomic effects of market power. Concentration of permits in a few hands may reduce the efficiency of the permit market. Particularly negative would be the case in which market power in the permit market is used to exercise power in the product market.

Tietenberg (2005) observes that market power is a problem not because it affects the environmental effectiveness of the ETS, but because it increases the costs of the scheme. Furthermore, the results from the literature show that, despite market power, ETS are clearly superior to command-and-control (CAC) regulation regarding the cost-effectiveness criteria (Tietenberg 2006). For example, Hahn and Noll (1982) reported very small increases in compliance costs in the case of market power (between 0% and 10% of the total control costs, depending on the assumptions made). Maloney and Yandle (1984) show that, even in the case of a cartel of firms with a monopoly over 90% of permits, the costs of control would increase by 41%. However, and even in this case, an ETS would show cost savings compared to direct regulation.

ii.1 The allocation discussion in emissions trading schemes.

The issue of allocation of emissions permits (or allowances) in the emissions trading schemes has been a major focus of the literature in recent years. This has been spurred by the implementation of the EU ETS and other schemes elsewhere and by the entry into force of the Kyoto Protocol. These analyses were initially mostly theoretical, comparing the advantages and drawbacks of different allocation methods. Several methods are compared between each other (auction, grandfathering and benchmarking), mostly leading to the conclusion that the best method in terms of economic efficiency is making polluting sources pay for those allowances (i.e., auction) and recycle the revenues to reduce pre-existing distortionay taxes (particularly, labour taxes). This would reduce pollution and, at the same time, increase employment levels, a double benefit called "the double dividend" (see Harrison and Radov 2002 for an overview).

A recent set of empirical studies have concentrated on allocation in the EU ETS: 1) whether allocation has been generous 2) possible negative effects of such generosity. Several papers stand out in this context, including Bruchner and Ellerman (2006) and Betz et al (2006). The conclusion is that allocation has been too generous to the covered sectors, with potential negative effects on the overall efficiency of the scheme[10]. The reason is that a greater effort than it is optimal has been put on non-covered sectors. Betz and Sato (2006) provide a summary of the impact of the national allocation plans in 2005-2007 according to

[10] Buchner and Ellerman (2006) calculate that over-allocation in 2005 was as much as 100 million EUAs. The cap is not expected to be very demanding either for the 2008-2012 period. Betz et al (2006) show that, on average, the intended allocation for the sectors covered by the EU ETS in 2008-2012 are only 3% lower than the budget in 2005-2007, historical emissions in 2005 and projected emissions in 2010. In other words, no significant reductions will be required.

three key criteria (effectiveness, efficiency and distributional consequences) as well as the prospects for 2008-2012.

iii. Project-based Flexible Mechanisms.

The so-called Flexible Mechanisms are instruments envisaged in articles 17, 12 and 6 of the Kyoto Protocol to help Annex I countries comply cost-effectively with targets. These are international emissions trading (IET), clean development mechanism (CDM) and Joint Implementation (JI). The latter two entail the realisation of emissions reduction projects in either an Annex I (JI) or a non-Annex I country (CDM). Here we concentrate on the CDM, which is currently the most developed project-based Kyoto Mechanism[11]. The realisation of CDM projects has recently experienced an impressive dynamism (see Fenhann 2007). However, this should not lead us to forget that there have been (and there are) problems related to carrying out this type of projects. The economic literature has concentrated on several of these barriers (del Río 2007a):

(A). Transaction costs.

These transaction costs include search, negotiation, baseline determination, approval, validation, registration, monitoring verification, certification, enforcement, transfer and registry costs. Some of them are fixed and, thus, the larger the project, the lower the impact of fixed transaction costs. The existence of an accurate institutional capacity to host CDM projects in the host countries is crucial to keep some transaction costs at reasonable levels.

(B). Additional risks for investors.

The realisation of a CDM project entails several risks, additional to those linked to an investment project in a developing country:

(i) Political risks. If either the investor or the host country does not comply with the eligibility criteria, then the CERs generated can not be used to comply with the mitigation targets. Particularly relevant is the lack of post-Kyoto targets which, combined with the long lead times for most CDM projects, may restrict the implementation of CDM projects.

(ii) Performance/technical risks. Less polluting technologies may lead to lower reductions of emissions than previously expected, leading to less CERs earned.

(iii) Market risks. The expected market price of the CERs generated may vary considerably, affecting the revenue of the project. This price depends on many variables, some of which are difficult to predict (see below).

[11] All these mechanisms lead to the issuing of so-called Kyoto units, which are transferable: AAUs (Assigned Amount Units to Annex I countries), ERUs (Emissions Reductions Units from JI projects in Annex I countries), RMUs (removal units from afforestation and reforestation projects in non-Annex I countries) and CERs (Certified Emissions Reductions from CDM projects). They all represent one metric ton of CO2 equivalent.

(iv) Risk of duration of the crediting period. If it is decided to renew the crediting period, then the total duration of the crediting period is not fixed with total certainty[12].

(v) Risks on costs. Abatement and transaction costs may be higher than initially considered.

(vi) Contract risks. The seller of CERs may not hand the CERs to the buyer at the date previously agreed.

(vii) Other risks. For example, there might be an unwillingness of the financial community to recognise CERs to manage financial risk (Salter 2004). The Executive Board may not approve the methodology proposed by the project developers or the validation and verification report. The Designated National Authority (DNA) of the host country may consider that the project does not contribute to its sustainable development.

(C). Other barriers.

-Barriers related to the financing of projects. Bhardwaj et al (2004) observe that, in general, financial institutions are not eager to finance projects (especially when these are small).

-Institutional/organisational barriers, i.e., lack of institutional capacity by non-Annex I countries to identify, attract and develop CDM projects.

-Low CER prices. A major factor affecting CER prices is the "hot air" issue. As shown by econometric models (see above), if Russia and Ukraine decide to sell all their surplus AAUs in the international market in the first commitment period, the price of Kyoto units and the demand for CERs will be low. Somehow paradoxically, strategic behaviour by Russia and Ukraine would help the CDM and, particularly, the most expensive CDM projects.

-Size matters. Given the fixed transaction cost component, small projects have relatively more transaction costs in relation to their revenues and might be less attractive for the investor.

A particular difficult issue has been proving the additionality of projects and calculating baselines (i.e., emissions in the absence of the project)[13]. This problem is being mitigated through the elaboration of standardised methodologies, although the problem remains in some sectors (particularly, in transport).

Finally, a stream of the literature has concentrated on the discussion of the type of projects that the CDM is encouraging (see, among others, Ellis et al 2004 and Cosbey et al 2005). It is a source of concern that the share of those projects which are expected to contribute more to the sustainability of host countries (like renewable energy projects) is going down[14] and that CDM projects show a geographical concentration in four countries,

[12] CDM project developers may choose between a crediting period of 10 years or a crediting period of 7 years with two possibilities for renewal (i.e., a maximum of 21 years).

[13] The emissions reductions from a CDM project are measured as the difference between the emissions in the absence of the project (the so-called "baseline") and the emissions when the project is implemented. The former have to be above the later.

[14] Fenhann (2007) shows that, as of May 2007, the share of renewable energy projects in total projects was 59%, but it represented only 25% of the CERs created. In contrast, the figures is, respectively, 3% and 38% for low-cost but low-sustainability projects (i.e., reduction of HFCs, PFCs and N_2O).

whereas some areas are underrepresented[15]. Several papers have tried to assess why this geographical concentration is taking place, showing that it is related to emissions reduction potentials, but also to the institutional capacity and the general investment climate in host countries (see Jung 2006, del Río and Hernández 2007a).

iv. Transaction costs of different mitigation instruments.

Transaction costs in emission permit markets[16] lead to a functioning of those markets which differs from the dynamism that is usually assumed in theoretical models and empirical simulations. As previously mentioned, this is true not only with respect to ETS, but also with respect to the two Kyoto-project mechanisms (CDM and JI). If this is so, then the cost-efficiency advantages attributed to emissions trading would be overvalued. If the transaction costs of implementing an ETS are high, then the instrument would not be as cost-effective as envisaged, if cost-effectiveness is defined as the costs of reaching a certain emissions target, i.e., abatement costs and transaction costs.

Kerr and Maré (1999) can be considered as the pioneering study on transaction costs in permit markets (although unrelated to climate policy). These authors econometrically estimated the impact of transaction costs on the cost-effectiveness of the US lead phase-down programme. They showed that those transaction costs were a significant barrier for firms to trade permits. Transaction costs reduce the potential gains from trade by 10–20%.

However, most analysis have focused on the transaction costs of the project-based Kyoto mechanisms (CDM and JI)(Michaelowa and Jotzo 2005, Liski and Virrankoski 2004, Woerdmann 2004). As a result of these costs, the deployment of CDM projects will actually be lower than that foreseen according to the emissions reduction potentials and marginal abatement costs and, thus, emissions reduced with this mechanism will be lower than previously envisaged in simulation models.

The most relevant study in this context is probably that of Michaelowa and Jotzo (2005), which tries to integrate the transaction costs in the simulations. These authors use a model of the global carbon market based on marginal abatement cost curves per country and sector in order to analyse the impact of transaction costs on the realisation of CDM projects. Such impact is modelled as an upward shift of the marginal abatement cost curve. This reduces the number of CERs exchanged in the market and increases their price.

In turn, Liski and Virrankoski (2004) also show the negative effect of transaction costs of CDM and JI projects on the cost-effectiveness of the Kyoto architecture. Their model combines the search market for projects with a frictionless permit market to quantify the supply-side frictions in the CO_2 market. They also decompose the effects of frictions into the effects of search friction, bargaining, and bilateralism. A calibration using previous cost estimates of CO_2 reductions illustrate changes in cost savings and allocative implications.

Some authors observe that, if the costs of implementation, information and monitoring are taken into account, this could significantly change the cost-effectiveness of different

[15] According to Fenhann (2007), 4 out of 5 CDM projects are being implemented in four countries: Mexico, Brazil, China and India. In contrast, out of 696 CDM projects registered in early June 2007, only 20 were implemented in Africa (see web of the UNFCCC: http://.www.unfccc.de).

[16] These include search costs, negotiation, control, verification and insurance costs, among others. See Woerdman (2001) for further details.

mitigation instruments. This is the main conclusion from Rousseau and Proost (2005), who aim to include rule making, implementation, monitoring and enforcement costs into the cost comparison of policy instruments. They use a simple partial equilibrium model and apply it to the textile industry. The model includes discrete abatement functions and costly monitoring and enforcement. The case study uses individual firm data to simulate the differences in abatement costs and compliance decisions between firms. Combinations of regulatory instruments (emission taxes, emission standards and technology standards) and enforcement instruments (criminal fines, civil fines and transaction offers) are compared. It is shown that the inclusion of the aforementioned costs indeed alters the relative cost efficiency of the different instruments. This reinforces the conclusion that the assessment of the mitigation instruments should be carried out considering several criteria.

v. Interactions.

The literature on environmental policy instruments has traditionally focused on the analysis of the implementation and use of an instrument to tackle one environmental problem. However, it could be appropriate to use a combination of instruments to tackle one environmental problem. For example, Bennear and Stavins (2006) defend such combination in specific situations, in order to increase the environmental effectiveness or reduce the costs of complying with an environmental target.

Furthermore, the analysis of single instruments is at odds with the real-world practice, which is characterised by the coexistence of different environmental and energy policies. This coexistence generates synergies and conflicts between instruments that should be analysed.

The interaction between emissions trading and other environmental and energy policies may be either positive, negative or non-existent. In the first case, there would be synergies between the considered instruments, whereas conflicts would result in the second case. Conflicts may lead to a less-than-optimal functioning of emissions trading schemes, in contrast to the ideal functioning of those schemes normally assumed in the simulations carried out with econometric models.

Notwithstanding, an interesting conclusion of the literature is that such combination may be justified if the instruments which coexist with the emissions trading system aim to tackle problems which can not be effectively dealt by the emissions trading scheme. This is the case with local environmental problems and other local issues and the fact that complementary instruments may be needed to eliminate the different barriers which prevent new mitigation technologies from being developed, adopted and diffused and for which the emissions trading system provides a limited incentive.

The combination of instruments may lead to the use of several policies to achieve one single objective (i.e., "to kill several birds with one stone") or several policies to achieve several objectives (i.e., "to kill several birds with several stones")(Sorrel and Sijm 2005).

Summing up, the coexistence of instruments leading to CO2 emissions reductions with an emissions trading scheme may make sense in several situations and, particularly:

1) When both types of instruments are applied to different sectors and/or firms.

2) When, even if applied to the same sectors and firms, the non-emissions trading instruments have different objectives or lead to additional benefits which can not be achieved with an emissions trading scheme.

In the rest of situations, the coexistence is, in the best of cases, redundant and, in the worst, inefficient. All in all, the transaction costs of such coexistence have to be considered.

A significant degree of attention has recently been paid to the relationship between renewable energy promotion schemes and emissions trading schemes, given the relevance of both instruments in real policy practice[17].

There are two major streams in the literature on the interaction between emissions trading and renewable electricity (RES-E) support schemes[18]. One focuses on the theoretical interactions resulting from the simultaneous application of both instruments. Papers following this tradition include Jensen and Skytte (2003), Skytte (2006), Boots (2003), Morthorst (2001, 2003), Del Río et al (2005) and Del Río and Hernández (2007b) as well as publications from the EU-funded InTraCert project (Boots et al 2001), among others.

The other stream of the literature analyses the possible interactions in several countries, using case studies (see Sorrell 2003 and Boemare and Quirion 2003, among others). These studies do not consider the simultaneous interaction between instruments in order to reach a target but, instead, the analysis is made separately for each target (Skytte 2006).

At a national level, the interactions between RES-E deployment, emissions trading and the electricity market can be summarised as follows (del Río 2007b; Skytte 2006):

[1] *Electricity market and ETS.* An ETS increases final consumer electricity prices, since conventional electricity generators tend to pass the extra costs of GHG control to electricity prices.

[2] *Electricity market and RES-E deployment.* The final effects on the consumer of the increase in RES-E deployment are ambiguous. On the one hand, the introduction of RES-E promotion schemes tend to reduce the electricity price, since they displace and reduce the demand for conventional (fossil-fuel based) electricity (assuming that marginal generation is based on fossil fuels). On the other hand, the costs of the RES-E policy are passed to the consumer.

[3] *ETS and RES-E deployment.* The ETS makes RES-E deployment more competitive, since it increases the costs of conventional electricity. In turn, RES-E deployment reduces the demand for emission permits in an ETS, and, thus, their price.

In the existing literature, three possible situations in the interaction between ETS and RES-E support schemes are discussed: i) reaching the RES-E quota with two instruments; ii) reaching the carbon emissions target with two instruments and; c) Simultaneous reaching of RES-E and carbon emissions targets with two instruments.

The most relevant (and realistic) case is the later one, whereby it might be difficult to reach both goals while at the same time minimising consumer costs. Jensen and Skytte (2003) show that attaining the two goals with a minimum consumer price depends on the correlation of the RES-E quota and the consumer price. With negative correlation, it is better to choose a

[17] This has been particularly so in the European context, where compliance with the Kyoto targets and renewable energy promotion are key objectives of European environmental/energy policy.

[18] The term "RES-E" stands for "electricity from renewable energy sources".

RES-E quota and no emissions quota. In the case of positive correlation, it is optimal to choose a combination of the RES-E and the emission quota. It will never be optimal to regulate only through the emission quota no matter the correlation. The major policy implication is that, in the case of positive correlation and attainment of two targets, there should be coordination between the ministerial departments responsible for the setting of the respective quotas, while in the case of negative correlation there are no gains from coordination. Therefore, an important finding is that the two instruments are to some extent substitutes for each other, i.e. there is a positive synergy in their use[19].

vi. Post-Kyoto (design elements).

The literature on post-Kyoto has discussed several issues. One has been the extent to which a global agreement is possible or, rather, whether regional agreements are more likely, i.e., what type of regime is more likely to be implemented. This has been mostly a theoretical issue, which has been mainly analysed with the help of cooperative versus non-cooperative games[20]. This literature is certainly relevant for the discussion of which post-Kyoto regime is possible/likely, although it is still framed in quite theoretical terms.

Another issue that has received significant attention in the literature has been the different possible proposals for a post-Kyoto regime. Phillibert (2005) and Bodansky et al (2004) provide a complete overview of these proposals[21]. The main conclusion is that several institutional arrangements are possible but that the alternatives fall broadly within two groups: quantitative and qualitative, i.e., those in line with the Kyoto-type architecture (with emissions targets and emissions trading) and those that involve a radical departure from Kyoto, including the "policies and measures" options (such as an international carbon tax, technological cooperation etc…).

Quantitative and non-quantitative approaches are really the two extremes of a continuum with many possibilities in between and even combination of approaches[22]. This distinction illustrates the degree of potential change of a future regime with respect to the current Kyoto Protocol architecture based on targets and emissions trading. Quantitative approaches involve no change or only incremental institutional change with respect to the current regime. In contrast, non-quantitative approaches involve a radical departure from the current Kyoto regime. The current international negotiations show that some countries prefer a quantitative approach, while others prefer a non-quantitative one (for example, in the form of technological cooperation). The post-Kyoto regime might be a fragmented one, at least in the short/medium term, with a coexistence of different approaches instead of a single, comprehensive regime.

---

[19] See del Río (2007b) for further details on these interactions and for an overview of the literature.

[20] For example, Kemfert et al (2004) analyse games of GHG emission reduction in which the emissions and the emission reduction costs of one country depend on other countries' emission abatement. Baker (2004) develops a model of the impacts of uncertainty and learning in a non-cooperative game, showing that the level of correlation of damages across countries is crucial for determining optimal policy.

[21] See also Torvanger (2004) and Storey (2002), among others.

[22] The literature on quantified objectives distinguishes options relative to the nature of targets and options with respect to timing, differentiation and possible allocation schemes (including Multi-stage approaches, Contraction and Convergence, Global Triptych approach, Immediate Equal Per Capita, Brazilian proposal,

An interesting focus of this literature has been on the (theoretical) analysis of different proposals according to different criteria and, particularly, with respect to effectiveness in emissions reductions and cost-effectiveness in those reductions, i.e., the literature has tried to answer the question of what are the consequences for reduction of concentration levels of different schemes and what are their potential mitigation costs. The conclusion is that a Kyoto-type architecture (i.e., with emission targets and permit trading) is probably the best alternative in terms of effectiveness and cost-effectiveness, although it is proving unfeasible in the short term, given the rejection of some countries and academics on the basis of their "high cost" for some countries (since the level of the permit price and, thus, such cost, is unknown beforehand).

vii. Ancillary benefits of climate change mitigation.

In addition to mitigating GHG emissions, climate policy instruments may lead to other social (environmental and socioeconomic) benefits, called "ancillary benefits", which include reduction of local pollution, increasing security and diversification of energy supply, increasing employment levels and increasing access to clean and reliable energy sources. In other words, climate change mitigation and local sustainability may have synergistic effects. However, climate policies are generally evaluated with respect to their costs and benefits and ancillary benefits are usually omitted.

The analysis of these synergies has been carried out at the theoretical level and at the empirical level with the help of case studies (see Markandya and Halnaes 2002). An example of these studies is Rübbelke (2003), who analyse countries' decisions on greenhouse gas abatement in an impure public good model which considers ancillary benefits. If ancillary benefits are integrated into the analysis, environmental regulations independent of the climate problem may influence countries' GHG control. Consequently, different environmental programs should be coordinated carefully and should not be considered as being independent. In turn, Burtraw et al (2003) empirically analyse the ancillary benefits of GHG emissions reductions in the U.S. electricity sector. A key conclusion is that a high tax can be justified if those ancillary benefits are taken into account: Actions to slow atmospheric accumulation of greenhouse gases also would reduce conventional air pollutants yielding "ancillary" benefits that tend to accrue locally and in the near-term[23].

Cost-based allocations). These alternatives are all theoretically compatible with a world-wide ETS and, thus, would also be compatible with recognition of the value of Kyoto units.

[23] Using a detailed electricity model linked to an integrated assessment framework to value changes in human health, the authors find that a tax of \$25/tCarbon would yield NO_x-related health benefits of about \$8 per ton of carbon reduced in the year 2010 (1997 dollars). Additional savings of \$4–\$7 accrue from reduced investment in NO_x and SO_2 abatement in order to comply with emission caps. Total ancillary benefits of a \$25 carbon tax are \$12–\$14, which appear to justify the costs of a \$25 tax, although marginal benefits are less than marginal costs. At a tax of \$75, greater total benefits are achieved but the value per ton of carbon reductions remains roughly constant at about \$12.

viii. The issue of carbon lock-in. How should it be overcome (what combination of instruments could achieve this).

Carbon lock-in is a major issue concerning climate change mitigation and, thus, climate policy. It refers to the observed fact that key technologies and infrastructures (roads, buildings, power plants...) in several sectors of the economic system suffer from inertia and are embedded for decades in the economic structure. If those have been recently implemented, it is very difficult for alternative, low-carbon technologies to be adopted, since the costs of scrapping non-obsolete technologies and infrastructures could be enormous. In this context, a great policy effort is necessary to bend existing infrastructures and the usual receipt of "economic instruments" may not be suit for the task and other measures (including technology policy instruments) may be needed. Authors discussing this type of problem use an analytical framework combining environmental economics, industrial economics and (evolutionary) economics of technological change (Unruh 2001, 2002, Jaffe et al 2005).

However, although empirical work on the barriers to the penetration of abatement technologies has been recent but fruitful (see Report from the III Working Group of the Intergovernmental Panel on Climate Change), to our knowledge not much has been written about the empirical effectiveness and economic efficiency of different instrument combinations on the breaking of carbon lock-in.

ix. Choice of instruments.

Traditionally, comparisons between the different environmental policy (and, thus, mitigation) instruments have been carried out in different contexts (different structure of product markets, uncertainty on the costs and benefits of emissions reductions), paying increasing attention to some criteria (dynamic efficiency, transaction costs) and considering new instruments (voluntary agreements and project-based mechanisms). Comparisons between instruments in the context of Environmental Economics have taken place between direct regulation instruments (emissions and technology standards, among others) and the so-called economic instruments (i.e., taxes and emissions trading), mostly focusing on the cost-effectiveness criteria. The initial theoretical studies were carried out in a partial equilibrium framework and showed the superiority of economic instruments in this regard, because they allow the reaching of the environmental target at a lower cost. However, the wider impacts of compliance on basic macroeconomic variables (employment and production) did not receive a comparable attention.

Econometric models of the top-down type have recently been used in order to compare GHG emissions mitigation instruments both in a general and a partial equilibrium framework. For instance, Gerlagh and van der Zwaan (2006) carry out a comparative analysis of various mitigation instruments with the help of simulations in a top-down model[24]. One of their main conclusions is that a carbon intensity portfolio standard, involving the recycling of carbon taxes to support renewables deployment, is the most cost-efficient way to address the problem

[24] These instruments are: carbon taxes, fossil fuel taxes, non-carbon (renewable) energy subsidies, a portfolio standard for the carbon intensity of energy production, and a portfolio standard for the use of non-carbon (renewable) energy resources.

of global climate change. A comprehensive introduction of the capture and storage of carbon dioxide would contribute to reducing the costs of climate change control, but would not obviate the large-scale need for renewables.

However, in many instances the issue is not so much one of choosing between different instruments, but that of selecting an appropriate basket of measures. A combination of instruments can be more optimal than the use of a single system. It should be thus explored how this optimal mix can be obtained, an issue that overlaps with the interaction question mentioned in a previous section. Furthermore, the design features of specific instruments are probably as relevant as the choice between different instruments.

x. Harmonisation of national policies.

Another issue on which researchers have concentrated is that of the appropriate level of policies, and whether they should be applied internationally and/or coordinated between countries. For example Amir and Nannerup (2005) show that there might be gains from the non-harmonisation of climate change policies. These authors claim that studies of optimal second-best environmental regulation of identical polluting agents have invariably ignored potentially welfare-improving asymmetric regulation by imposing equal regulatory treatment of identical firms at the outset. Yet, cost asymmetry between oligopoly firms may well give rise to private as well as social gains. A trade-off is demonstrated for the regulator, between private costs savings and additional social costs when asymmetric treatment is allowed. Padilla and Roca (2004) analyse the advantages and implications of the implementation of a European tax on carbon dioxide emissions and it focuses on its effects on intercountry distribution. In contrast to a harmonized tax, which would only have distributive effects within each member state, a tax collected at European scale would also have important distributive effects among different countries. These effects would also depend on the use of tax revenues.

The issue has received recent attention in the context of the EU ETS, where several authors have observed that a greater harmonisation of the allocation decisions at EU level should be achieved, leading to benefits in terms of economic efficiency and equal treatment of similar installations (see Grubb and Neuhoff 2006, del Río 2006).

xi. Political economy considerations.

Given that theoretical analysis and econometric simulations have shown that market-based mitigation instruments have proven to be superior to other measures, several authors have tried to provide an answer to the question of why has incentive-based climate policy, such as emissions trading and taxes, which are the most efficient, not been adopted in national climate policies and why has there been a preference to implement other instruments (specially subsidies and, to a much lesser extent, command-and-control regulation).

This political economy discussion has been analysed mostly with the help of a public choice approach (see Svendsen 2000, Schneider and Volker (1999) or with institutional path-dependency theory (see Woerdmann 2004).

Scheneider and Volker (1999) take into account the behaviour and interests of voters, politicians, interest groups and bureaucracies to show that there are conflicts with other policies and that individual rationality is a powerful obstacle in implementing taxes and emissions trading[25]. They show that, when choosing a climate policy instrument, policy makers try to minimise conflicts with firms, while passing the costs to the great majority of ignorant and uninformed voters (i.e., consumers and/or taxpayers). There is thus an incentive by policy makers to implement "soft" climate policy measures (such as subsidies, voluntary agreements, tax exemptions or rebates) or less visible measures (i.e., CAC instruments) which minimise the short-term conflicts with affected firms and are not unpopular concerning voters.

Svendsen (2000) bases the design of taxes and tradable permit systems on a Public Choice analysis of the main stakeholders' interests in the US and Europe, suggesting that a grandfathered CO_2 permit market is a more effective policy than a CO_2 tax in relation to organised interests such as industry, electric utilities and environmental organisations.

Dijkstra (2004) and Damania (1999) analyse the impact of pressures from companies and NGOs on the elections of candidates from political parties which defend the use of a specific instrument (regulatory standards or taxes) and the degree of ambition of such instrument.

Another example is the study of the U.K. Climate Change Levy carried out by David Pearce (see Pearce 2006), who notes that energy taxes designed to control energy consumption, and as a climate change mitigation instrument are common in EU countries, although they bear little resemblance to the design guidance that is given in economics textbooks. Political economy analysis, in which the interaction of economics and political reality is emphasised, explains the gap between theoretical ideals and practical reality. Pearce argues that "a closer look at the factors that influence real-world policy design should help policy-oriented economists in designing measures that have a greater chance of adoption".

In contrast, Woerdmann (2004) uses a neo-institutional economics theoretical framework (combining elements of Oliver North's institutional theory and Brian Arthur's evolutionary economics of technological change[26]) to show that political transaction costs, depending and building incrementally on the path of earlier choices and events and informal constraints in the form of cultural barriers to the implementation of market-based climate policy (for instance, regarding equity) explain the rejection of emissions trading.

xii. Dynamic efficiency of mitigation instruments.

An additional relevant criteria to judge and compare instruments is their capacity to generate a continuous incentive for technological change (dynamic efficiency). Theoretical studies argue that market-based instruments (including emissions trading and taxes) have a greater dynamic efficiency than regulatory instruments in the form of emissions or technology standards (see Field and Field 2003 and Galeotti and Carraro 2003)[27].

However, there is a surprisingly small amount of empirical research on the dynamic efficiency (incentive for technological change) of mitigation instruments, which points out to

[25] This discussion is on environmental policy in general and not specifically focused on mitigation instruments.

[26] This follows North's suggestion to apply Arthur's self-reinforcing mechanisms to institutions.

[27] However, other authors doubt this conclusion, and argue that the alleged superiority of market-based instruments is still an open issue even in theoretical terms (see Requate 2005).

a fertile ground for further research (see next section). It is particularly surprising that there has not been any research we are aware of on the impact of the EU ETS on environmental technological change, more than two years after the scheme started to function. Although putting a price on carbon emissions is likely to create a greater incentive for technological change than when such price is implicitly cero, we do not know what has been the degree of incentive created (supposedly very small, given the low EU ETS allowance prices and their excessive volatility), the impact of the scheme on the different stages of technological change, the types of technologies being adopted and how the incentive for technological change affects different sectors.

xiii. Sinks.

Large scale land use changes have a significant global impact. Land use change (especially deforestation) has been historically responsible for a large part of the cumulative human induced CO_2 emissions (Watson et al 2000). Hougton (2003) has estimated that about one-third of total CO_2 emissions since 1850 come from changes in land use, and two thirds from fossil fuels. In addition, between 1970 and 2004 carbon dioxide (CO_2), methane (CH_4) and nitrous oxide (N_2O) emission from deforestation and other forest related activities have grown by 40%, whilst direct emission from agriculture grew by 27% (IPCC 2007, p.4-5).

Taking into account these figures, it is not surprising that almost two decades ago Sedgo and Solomon (1989) suggested that it would be possible to substantially offset the world's emissions of carbon dioxide by expanding world's forest areas and that a long list of paper have analyzed additional ways in which land use, land use change and forestry (LULUCF) alternatives can contribute to mitigate climate change.

Early studies reported a large potential to sequester carbon by increasing the forest area, at a relatively low cost per ton of carbon (Sedjo and Solomon 1989). Later, more detailed analyses revealed costs to be higher than anticipated. Surveys of the studies that have analyzed the costs of different LULUCF alternatives can be found in Richards and Stokkes (2004) or in Bosello et al (2007), and Van Kooten et al (2004) present a meta-analysis of these studies. The bottom-line is that, although different studies are difficult to compare due to different carbon accounting methods, different geographical areas and different estimation techniques, these alternatives remain competitive and have to be taken into account in the mix of alternatives necessary to mitigate climate change.

The cost-effectiveness shown in these studies were, both, the reason why LULUCF activities were include in the Kyoto Protocol and the reason why so many limitations to the use of LULUCF were introduced in the Protocol and the Marrakech Accords, since it was feared that they were too cheap and would preclude any other abatement measure. The result was that "sinks" were one of the crunch issues that resulted in the failure to reach an agreement at COP 6 in The Hague in November 2000 (still with the US on board). The Parties were not able to decide if sinks activities should be limited in quantity, and if so, what this limit should be. At COP 6bis an agreement was reached (without the US), which was similar in terms of sinks to the one rejected at COP 6. This agreement was then included in the Marrakech Accords in November 2001 at COP7 (UNFCCC 2001). Rules regarding sinks under the Clean Development Mechanism were later adopted at COP9 in November 2003

(UNFCCC 2003). A summary of the complicated system resulting can be found in Caparrós and Jacquemont (2003 and 2005).

There are at least four mitigation strategies in which forest and agricultural lands are involved: conversion of non-forestland to forest, preserving and/or increasing carbon in existing forest and agricultural soils, altering agricultural and forestry fossil fuel usage patterns and growing biomass to substitute for fossil-fuel based products. Although we will make some occasional references to the two latter alternatives, we will focus in this section on the two first alternatives since they constitute what is usually known as "sinks".

Since the literature about "sinks" is too vast to cover it in this heading, we will focus only on the various approaches used to estimate the cost of these alternatives, since they have been of significant political relevance over the entire international negotiations on climate change, and on two potential problems related to LULUCF that are also relevant from a policy point of view: ancillary benefits and leakage.

Cost estimates

There are three general approaches used to estimate the land cost associated to carbon sequestration practices: bottom-up technical and engineering oriented costs studies, partial or general computable equilibrium models, and the econometric analysis of revealed preferences of landowners (Richards and Stokes 2004).

Most studies fall in the first category, the bottom-up studies. Those determine exogenously the value of inputs to production and derive an estimate of costs, normally based on the assumption that land use change decisions are driven by profit maximisation. This approach can rely on highly detailed databases at the farm level, but this technical orientation often forces to restrict the estimates to the sector and/or plot level (Moulton and Richards 1990). Mathematical programming has also been used to examine alternative agricultural management at the farm, regional, or sectoral level (e.g. De Cara et al 2005; Alig et al 1997).

Partial equilibrium studies concentrate on the agricultural and/or forestry sector; while the rest of the economic system remains outside of the picture, assuming that the market of interest is insignificant for the rest of the economy. Agriculture models are more abundant (e.g. Ignaciuk et al 2004) than forestry ones (e.g. Sohngen et al 1999; Solberg et al 2003). Unfortunately, only a handful of models account for both sectors (e.g. Alig et al 1997).

Computable general equilibrium models (e.g. Burniaux and Lee 2003, Pohjola et al 2003) analyse the global economy and are built top-down from the world level to regional or sub-regional levels, guaranteeing the market closure. However, most current models lack necessary data for fine-tuned disaggregating and hence do not provide much region, crop, or management-specific detail (e.g. De Cara and Javet 2000, Antle et al 2003). A global model with a more developed intertemporal component can be found in Sonhgen and Mendelson (2003) or Tavoni et al (2007).

Agriculture and forestry sectors outcomes are strongly dependant on site specific environmental conditions. Thus, some recent models have tried to integrate the link of both sectors to the overall economy with a detailed description of the different options available to reduce emissions. Some examples of those studies are Antle et al (2002). However, "uncertainties surrounding the various integrated components and difficulties of managing those models still limit the potentials of integrated models to determine reliably relative effects" (Bosello et al 2007).

Finally, other studies have used a revealed preference approach in which past land use changes are analysed for estimating the econometric relationship between land use choice and relative returns in forest and agriculture (Stavins 1999, Plantinga et al 1999, Ludowski et al 2006). These studies examine the relationship between observable historic events (changes in timber and agricultural product prices) and landowner responses (land use changes), and statistically estimate a response function. The models then assume a carbon subsidy and estimate how landowners would respond (Ludowski et al 2006). The econometric approach has many advantages since it can account for decision-making inertia or other non-profit maximization reasons that explain land use (Stavins 1999). However, it has also some drawbacks since it is not well suited to analyze long term changes, especially because of technological change. Recent developments in this approach (e.g. Ludowski et al 2006) have used fine-tuned micro-data, extending both, the potential land uses considered, and the geographic scope (the US). Unfortunately, no model is available outside of the US.

As said above, the general conclusion to be drawn from this literature is that more complex and realistic models tend to increase the cost of the different "sink" alternatives, but that they still remain cost-effective compared to other emission reduction strategies.

Non-market ancillary costs and benefits of carbon sequestration programs

In addition to the direct costs, carbon sequestration programmes have a number of indirect market and non-market costs and benefits, which are generally ignored by modelling studies. For instance, reduced tillage decreases soil erosion and increased fertilization to promote biomass growth may have adverse environmental impacts. Other potential ancillary effects can be derived from the forest species (fast vs. slow growing; alien vs. native) and stand type (mono-specific plantation vs. mixed plantation) selection. Incentives to increase carbon sequestration may favour the use of fast-growing alien species, which can potentially be negative for biodiversity, as indicated by Caparrós and Jacquemont (2003). In this line, Caparros et al (2007) show that if biodiversity-scenic values of reforestations with fast-growing alien species (eucalyptus) and with native slow-growing species (cork oak) were internalized, this would favour reforestations with slow-growing native species (with lower carbon sequestration potential, but higher biodiversity-scenic values).

Few carbon cost studies have included ancillary costs and benefits. However, Plantinga and Wu (2003) show that secondary benefits could be as great as land conversion costs; Matthews et al 2002 suggests that reforestations might negatively affect certain types of birds' habitat; and Feng and Kling (2005) provide evidence of positive ancillary benefits.

Leakage

In the context of carbon sequestration, leakage may take place by shifting activities from the project site to areas outside the boundary of the project or by changing market prices. Leakage effects have been addressed by Alig et al. (1997) using sectoral models and by Stavins (1999) and Ludwoski et al. (2006) using econometric models. Sectoral models analysed the countervailing pressure from rising agriculture land prices and its influence over the accomplishment of specific carbon abatement targets. Alig et al (1997) does not specifically estimate the size of leakage, but suggest that the pressure from rising agriculture land prices might be strong enough to dissipate the potential accomplishments of US carbon sequestration programs over several decades. This study, however, does not provide for incentives to keep existing land in forests. More recent studies for the US have analysed specifically the leakage effect of forest carbon sequestration programs (Murray et al 2002), showing that empirical estimates for the US suggest leakage ranges from minimal (<10%) to very large (>90%), depending on the different programs (afforestation, avoided deforestation and old forest set aside program). At the international level, Sonhgen et al (2001) suggest that forest conservation strategies in North America and Europe might increase harvesting pressure in subtropical areas.

3. Main Lines of Future Research

In this final section we aim to provide some directions for further research on the topic of climate policy. Of course, this should not be regarded as a "closed list" of proposals, but rather some which we believe have a great potential for relevant analysis concerning global warming mitigation.

i. Economic impact of Kyoto and the EU Directive.

As explained in section 2, the economic analysis of the likely impacts of the Kyoto Protocol and the EU ETS in the different countries has been carried out mostly with two types of simulation models (top-down versus bottom-up). However, both types of models have their weaknesses and advantages and it could be interesting to build new models which integrate to the greater extent possible both perspectives. Some attempts in this direction have been made (see Wing 2006, Böhringer and Rutherford 2007), but much remains to be done in this context.

ii. The allocation discussion in emissions trading schemes.

Although the allocation of allowances in the EU ETS has received much attention in the literature (see Buchner and Ellerman 2006; Betz et al 2006[28]), some issues are still worth exploring. These are related to ex-post empirical analysis on the impact of allocation on sectoral competitiveness, including the effects on sales and market quotas with respect to countries which do not have carbon restrictions. Furthermore, the impact of allocation decisions taken in non-EU ETS should also be explored (like the Australian North South Wales ETS). On the theoretical front, the implications of a top-down allocation in the EU ETS should be analysed further, as well as a greater harmonisation of other elements of the EU ETS (including treatment of new entrants, closures, banking...)(see del Río 2006 in this regard). These implications should be considered according to several criteria (effectiveness in reducing emissions, cost-effectiveness, dynamic efficiency, equity and political feasibility).

[28] See also the rest of papers included in the special issues of the journal Climate Policy devoted to allocation and national allocations plans in the EU ETS (volume 6, issue 4 and volume 6, issue 1).

iii. Transaction costs.

The empirical analysis of the transaction costs of mitigation instruments and, particularly, of emissions trading schemes has not received much attention in the literature. This discussion (and calculations) has taken place only with respect to the Kyoto Flexibility Mechanism (see Michaelowa and Jotzo 2005, for example), but the impact of transaction costs on the wider international carbon market has not been analysed. Neither has there been any analysis we know of in the context of the EU ETS, although at least the administrative costs of implementing and running the system could be significant. Firm-level transaction costs should also be identified. Policy measures in the future should aim to reduce these costs and more research should be devoted to how this could be done[29].

Therefore, an empirical analysis of the impact of those transaction costs on the cost-effectiveness and emissions reductions of the mitigation instruments is worth undertaking. This would be particularly so in the case of the implementation and functioning of the EU ETS, the International Emissions Trading under the Kyoto Protocol, the Clean Development Mechanism and renewable energy promotion schemes. The analysis of the impact of those transaction costs on the cost-effectiveness and emissions reductions of the scheme could be carried out through the inclusion of those costs in existing econometric models. For instance, in general equilibrium top-down models, this could be done without significantly altering the model, assuming that those costs represent an additional cost over the market price of permits or credits (in the case of project-based Kyoto Mechanisms) or an additional cost over the tax rate in the case of carbon taxes. The impact on key macroeconomic variables resulting from the inclusion of those transaction costs could then be analysed.

The identification of those costs could be done through interviews with experts (regarding transaction costs for firms) or with public officials engaged in the implementation and/or functioning of the scheme (with respect to the administrative costs of the system). In the case of the EU ETS, the administrative costs for the European Commission should also be taken into account. Furthermore, the costs charged by financial intermediaries can be taken as a *proxy* of the transaction costs for companies.

Finally, it could also be interesting to analyse what are the transaction costs of implementing (or changing) RES-E promotion schemes, currently one of the main mitigation instruments. Traditionally (and still today predominantly), countries have supported renewable electricity with feed-in tariff schemes[30]. However, some countries have recently moved to an alternative, more market-based instrument, such as Tradable Green Certificates (TGCs). While this is not the place to discuss the respective merits and demerits of these systems, the transaction costs of the respective systems and the costs of moving from one system to another should be explored.

---

[29] The reference for the calculation of transaction costs should not be the situation without a policy, but rather the transaction costs should be compared across different instruments. This is only possible to a certain extent, however, given that many countries have only recently implemented mitigation instruments.

[30] These are subsidies per kWh generated paid in the form of guaranteed premium prices (either a total payment per kWh of RES-E electricity or a payment per kWh on top of the electricity wholesale-market price) and combined with a purchase obligation by the utilities. The costs are usually borne by consumers.

iv. Interactions[31]

A general challenge of the interaction literature is how to combine the different emissions reduction instruments (Kyoto Mechanisms, taxes, voluntary agreements...) at different territorial levels (international, national, regional and local) in a way that optimises the environmental effectiveness and the cost-effectiveness of the coexistence of such instruments while, at the same time, a substantial number of countries participate in the emissions reduction effort and commit to comply with emissions reduction targets[32]. The difficulty lies in the interactions between those flexible measures, which may lead to synergies but also to conflicts and, thus, to trade-offs between those goals. The result is a wider complex system which should be assessed according to the fulfilment of those goals.

Concerning the specific interaction between RES-E promotion and emissions trading schemes, some issues are worth investigating at, both, the theoretical and empirical levels. Concerning the theoretical literature, the analysis of the interactions according to the dynamic efficiency criteria has been neglected in the literature. Furthermore, trade-offs between different criteria should be analysed, particularly the possible conflict between static and dynamic efficiency in RES-E promotion.

Although the existing literature has been relatively successful in providing a theoretical framework and doing initial empirical research through case studies, we believe that, at an empirical level, it is necessary to carry out a formal analysis of the interactions between emissions trading (whether in the context of the Kyoto Protocol or the Emissions Trading Directive) and other environmental and/or energy instruments (renewable energy promotion schemes and measures to encourage energy efficiency). Further research should focus on the empirical analysis of the interactions in national and international settings with the help of econometric models closely approaching the conditions of real-world markets. Finally, efforts should also be made to extend the analysis of interactions through model simulations and case studies to other non-EU regions (i.e., the US, Japan and Australia).

v. Post-Kyoto (design elements).

Further research should be devoted to the economic analysis of the different alternatives which are possible in a post-Kyoto world and their compatibility with emissions trading, which is currently the main instrument to achieve reductions, given the adoption of the EU emissions trading scheme, the features of the Kyoto Protocol architecture and the implementation of gradual national and subnational trading schemes around the world (including the regional schemes in the US and Australia). This analysis of which post-Kyoto regime is more appropriate should be undertaken according to different criteria (abatement effectiveness, cost-efficiency in reducing emissions, equity etc...). In this regard, the analysis should take into account which is the post-Kyoto option most likely to result in a broader coalition of countries shouldering up the mitigation effort.

---

[31] This discussion heavily draws from del Río (2007b).

[32] Obviously, given the global character of the climate change problem, the broader the coalition of countries shouldering up the mitigation effort, the better.

In this context, but also related to the issue of "dynamic efficiency", the issue of intertemporal efficiency is also worth investigating. This refers to the post-Kyoto policy regime which best achieves long-term reductions at the lowest possible costs. In order to achieve this, a post-Kyoto regime should not only achieve reductions in the short and medium terms at low costs, but also be dynamically efficient: i.e., create sufficient stimulus for technological change so that technologies will be available in the future which will be able to reduce emissions at moderate costs.

vi. Ancillary benefits of climate change mitigation.

More research should be devoted to the ancillary benefits of climate change mitigation. For example, RES-E investments reduce CO2 emissions but also local pollutants while contributing to the diversification of energy supply (and, thus, to security of supply) and to improve regional and rural development opportunities. If those ancillary benefits are high then, from a broader "social efficiency" perspective, those investments could be defended, even if their abatement costs were relatively high. Although some studies have analysed separately the employment effects and also the local externalities avoided by the substitution of RES-E for conventional electricity, an integrated analysis of this issue is lacking.

Although reducing all ancillary benefits to a monetary figure is a difficult task (particularly concerning employment benefits), an effort in this direction should be made. This could not only justify the implementation of mitigation measures, but also help discriminate between different types of measures.

vii. The issue of carbon lock-in. How it should be overcome (what combination of instruments could achieve this).

Some experts have recently put the emphasis on the difficulty to change the economy towards a low-carbon path in the presence of long-term infrastructures and socioeconomic habits and institutions which are "adapted" to the existing high-carbon technological regime (see Unruh 2001 and 2002, del Río 2004). These elements create a self-reinforcing economic structure which is difficult to change. Although those theoretical papers have discussed what should be the combination of instruments aimed at breaking such lock-in, an empirical analysis of how a combination has worked in practice (as might be the case in the Dutch context) is lacking. However, the (un)available evidence points out that this combination is rare. Also in this context, an analysis of the factors leading to technological change and, particularly, the sources of learning effects (learning by doing and learning by using) should be explored.

viii. Political economy considerations.

Further research should be devoted to explain the paradox of why market-based instruments, which are theoretically superior to other instruments (CAC regulation, voluntary agreements, subsidies...) have not been adopted in national policy context. Although Public Choice perspectives have provided some theoretical explanation on this (see Scheider and Volker

1999), empirical research has neglected this issue, with some notable exceptions (see Svendsen 2000, Pearce 2006).

ix. Dynamic efficiency of mitigation instruments.

Mitigation instruments should be assessed not only with respect to their static economic efficiency (i.e., achieving targets in the short-term at low costs) but according to whether they stimulate technological changes which will allow countries to reach future targets at reasonable costs. In this context, two avenues for research stand out. On the one hand, theoretical research is still inconclusive on which instrument is more dynamically efficient (see Requate 2005). It is necessary to carry out a deep analysis on how environmental policy and technological policy instruments should be combined in order to encourage the development, adoption and diffusion of mitigation technologies in each of these stages of technological change. We believe that this issue should not be explored in general, however, but sector by sector. It is highly likely that the features and structure of sectors influence which instrument is better in this regard. It is particularly interesting to explore what are the most appropriate instrument mix to be implemented in the transport sector, given the great increase of the emissions of this sector.

On the other hand, more empirical research should be devoted to the analysis of how much dynamically efficient current systems have been. In this context, it is striking that a deep analysis of how the most relevant ETS currently in place (the EU ETS) has influenced technological change has been absent. This analysis should explore two interrelated issues: First, if the EU ETS has effectively promoted technological change and the type of technological change being promoted. Since allowance prices in the first phase of the EU ETS have been very volatile, with very low figures by the end of the period, this issue could possibly be more interesting in the second EU ETS period (2008-2012), when allocations are expected to be less generous and, thus, a greater allowance price is likely to result, leading to a greater incentive for technological change.

ix.1. Policy measures to improve the functioning of the project-based Kyoto Flexible Mechanisms.

The project-based Flexible Mechanisms envisaged in the Kyoto Protocol (and, particularly, the CDM) suffer a number of barriers to their effective deployment, including delays in their approval (related to the excessive work overload of the CDM Executive Board, to the practical difficulties to assess additionality and to establish baselines and to the lack of institutional capacity of the host countries), transaction costs (see above), risks (related to the absence of a post-Kyoto regime, uncertainty on the future value of CERs and unclear, vague or restrictive definition of sustainability), to the lack of information on the CDM, to market problems (including a low CER price, low share of CDM projects leading to sustainability benefits) and access to financing[33]. Measures to tackle and mitigate these barriers should be identified. It should also be investigated further how different types of projects can be encouraged (specially the ones contributing more to the sustainability of host countries).

[33] For further details on these barriers, see Ellis et al (2004), Cosbey et al (2005), Sugiyama et al (2005), Bineau (2004), Michaelowa (2003) and Haites (2004).

x. Sinks.

Estimates of LULUCF practices or biomass potentials should be made on a geographic explicit bases and taking into account all the alternatives available. Most studies analyse certain GHG land-base mitigation options, but almost no study takes into account all the alternatives available. Thus, it is not clear if the estimates for potential reforestations proposed by one study are not assumed to be done on the same area that another study proposes to use for bioenergy production.

More research is also needed in order to analyse the environmental consequences from adopting some carbon sequestration practices, given that the "positive" or "negative" labelling of externalities depends on specific project design and site conditions. Enhancing carbon sequestration (and growing bioenergy crops for carbon substitution) might be set in conflict to other mayor environmental goals as biodiversity conservation, depending on practices design and scope. There are different ways to sequester carbon, and more cost-effective ones (in financial terms) might be related to fast growing mono-specific forest plantations, due to their higher short-term sequestration potential. Thus, carbon sequestration costs should be extended to include secondary costs or benefits (Caparros et al 2007).

Finally, a very important topic that needs more attention is the most appropriate way to account for carbon to avoid leakage and to make it compatible with the existing or emerging emission trading markets (Richards et al 2006). There are a large number of different accounting methods, which makes it difficult to compare between different studies. Although there are studies comparing these methods (e.g Maréchal and Hecq 2006), there is no systematic assessment of the performance of these methods regarding leakage, temporality, ancillary benefits and compatibility with emission trading systems.

4. Conclusion

The aim of this chapter has been to provide an overview of the main topics in the climate policy literature and to identify issues for further research. The overview has shown that there is a fertile ground for further analysis at least in the following topics: simulations of the economic impact of Kyoto and the EU ETS Directive, the economic impacts resulting from the adoption of different types of allocation methods, the empirical calculation of transaction costs of different mitigation instruments, interactions between different instruments, implications of different Post-Kyoto regimes design elements, empirical analysis of the ancillary benefits of climate change mitigation, how climate policy measures should be combined in order to avoid carbon lock-in, political economy considerations explaining the choice for certain mitigation instruments, empirical study of the dynamic efficiency of mitigation instruments, policy measures to improve the functioning of the project-based Kyoto mechanisms and further research associated to sinks.

References

Alig, R., Adams, D., McCarl, B., Callaway, J.M., Winnett, S. (1997). Assessing effects of mitigation strategies for global climate change with an intertemporal model of the U.S. forest and agriculture sectors. *Environmental and Resources Economics* **9**, 259-274.

Antle, J., Capalbo, S., Mooney, S., Elliott, E., Paustian, K.H. (2002). Sensitivity of carbon sequestration costs to soil carbon rates. *Environmental Pollution* **116**(3), 413-422.

Antle, J., Capalbo, S., Mooney, S., Elliott, E., Paustian, K. (2003). Spatial heterogeneity, contract design, and the efficiency of carbon sequestration policies for agriculture, *Journal of Environmental Economics and Management,* **46**(2), 231–250.

Baker, T. (2004). Uncertainty and learning in a strategic environment: global climate change. *Resource and Energy Economics*, **27**(1), 19-40

Banuri, T. et al (2001). Technical Summary. Climate Change 2001: Mitigation. Report of Working Group III of the Integovernmental Panel on Climate Change. http://www.ipcc.ch

Bennear, L. and Stavins, R. (2006). The use of multiple policy instruments for environmental protection: an economic perspective. *World Congress of EAERE World Congress. July 2006,* Kyoto

Betz, R., Rogge, K. and Schleich, J. (2006). EU emissions trading: an early analysis of national allocation plans for 2008-2012. *Climate Policy,* **6**(4), 361-394.

Betz, R. and Sato, M. (2006). Emissions trading: lessons from the first phase of the EU ETS and prospects or the second phase. *Climate Policy,* **6**(4), 351-360.

Bhardwaj N. et al (2004). *Realising the potential of small-scale CDM projects in India.* ECN-C—04-084. Petten (the Netherlands)

Bineau, V. (2004). *Practical guide to the Clean Development Mechanism and its application to rural electrification using renewable energy sources.* Funded by the European Commission under the Sinergy Programme. Foundation Énergies pour le Monde. Paris.

Bodansky, C. (2004). *International Climate Efforts Beyond 2012: A Survey of Approaches.* Pew Centre on Global Climate Change (USA).

Boemare C, Quirion P (2003). *Interaction in EU Climate Policy. France policy brief.* INTERACT project. June 2003. Paris

Böhringer, C., Hoffmann, T., Lange, A., Löschel, A. and Moslener U. (2005). Assessing Emission Regulation in Europe: An Interactive Simulation Approach. *Energy Journal,* **26**(4), *1-22*

Böhringer, C., Hoffmann' T. and Manrique-de-Lara-Peñate, C. (2006). The efficiency costs of separating carbon markets under the EU emissions trading scheme: A quantitative assessment for Germany. *Energy Economics,* **28**(1), *44-61*

Böhringer, C. and Rutherford, T. (2007). Combining bottom-up and top-down. Energy Economics (forthcoming).

Boots, M. (2003). Green certificates and carbon trading in the Netherlands. *Energy Policy,* **31**(1), 43–50.

Boots, M. et al (2001). *The interaction of tradable instruments in renewable energy and climate change markets.* Final Report Intracert project. Energie n°5. European Commission (Directorate General for Energy and Transport). ECN-C—01-048. 2000, Bruselas.

Bosello, F., Giupponi, C., Povellato, A. (2007). *A Review of Recent Studies on Cost Effectiveness of GHG Mitigation Measures in the European Agro-Forestry Sector. FEEM Working Paper* No. **14**.2007.

Brander, L. (2003). The Kioto mechanisms and the economics of the design. In Faure, M.; Gupta, J. and Nentjes, A. (Eds.). *Climate Change and the Kioto Protocol* (25-44). Cheltenham (U.K.): Edward Elgar.

Burniaux, J.M. (1999). *How important is market power in achieving Kyoto? An assessment based on the GREEN model.* OECD paper. OECD. Paris.

Burtraw, D., Krupnick, A., Palmer, K., Paul, A., Toman, M. and Bloyd, C. (2003). Ancillary benefits of reduced air pollution in the US from moderate greenhouse gas mitigation policies in the electricity sector. *Journal of Environmental Economics and Management*, **45**(3), 650-673

Caparrós, A., Jacquemont, F. (2003). Conflicts between biodiversity and carbon offset programs: economic and legal implications. *Ecological Economics*, **46**, 143-157.

Caparrós, A. and Jacquemont, F. (2005). Biodiversity and carbon sequestration in forests: economic and legal issues. In M. Bothe and E. Rehbinder (Eds.) *Climate Change Policy (149-182).* Utrecht: Eleven International Publishers.

Caparrós, A., Cerdá, E., Ovando. P., Campos, P. (2007). *Carbon Sequestration with Reforestations and Biodiversity - Scenic Values. FEEM Working Paper* No. **28**.2007.

Carlén, B. (2003). Market Power in International Carbon Emissions Trading: A Laboratory Test. *Energy Journal*, **24** (13), 1-26.

Cason, T., Gangadharan, L. and Duke, C. (2003). Market power in tradable emission markets: a laboratory testbed for emission trading in Port Phillip Bay, Victoria. *Ecological Economics*, **46**, 469-491

Cosbey, A et al (2005). *Realizing the Development Dividend: Making the CDM Work for Developing Countries.* International Institute for Sustainable Development. Winnipeg (Canada).

Damania, R. (1999). Political Competition, Rent Seeking and the Choice of Environmental Policy Instruments. *Environmental and Resource Economics,* **13**, 415–433.

De Cara, S., Jayet, P.A. (2000). Emission of greenhouse gases from agriculture: the heterogeneity of abatement cost in France. *European Review of Agricultural Economics,* **27**(3), 281-303.

De Cara, S., Houze, M., Jayet, P.A. (2005).′ Methane and Nitrous Oxide Emissions from Agriculture in the EU: A Spatial Assessment of Sources and Abatement Costs.
Environmental and Resource Economics, **32**, 551–583.

Del Río, P. (2004). Public policy and clean technology promotion. The sinergy between environmental economics and evolutionary economics of technological change. *International Journal of Sustainable Development*, **7**(2), 200-216.

Del Río, P. (2006). Harmonisation versus decentralisation in the EU ETS. An economic analysis. *Climate Policy*, **6**(4), 457-475.

Del Río, P. (2007a). Encouraging the implementation of small renewable electricity CDM projects. An economic analysis of different options". *Renewable and Sustainable Energy Review,* **11**(7), 1361-1387.

Del Río, P. (2007b). The interaction between emissions trading and renewable electricity support schemes. An overview of the literature. *Mitigation and Adaptation Strategies for Climate Change* (forthcoming) DOI: 10.1007/s11027-006-9069-y.

Del Río, P. and Hernández, F. (2007a). Benefits and barriers to the implementation of renewable electricity clean development projects: The case of the South Mediterranean basin. *World Review of Science, Technology and Sustainable Development*, **4**(1), 14-37.

Del Río, P. and Hernández, F. (2007b). How do energy & environmental policy goals and instruments affect electricity demand?. A framework for the analysis. *Renewable and Sustainable Energy Reviews*, **11**(9), 2006-2031.

Demailly, D and Quirion, P. (2007). European Emission Trading Scheme and competitiveness: A case study on the iron and steel industry. *Energy Economics* (forthcoming). Available online at http://www.sciencedirect.com/

Dijkstra, B. (2004). Political Competition, Rent Seeking and the Choice of Environmental Policy Instruments: Comment. *Environmental and Resource Economics*, **29**(1), 39-56.

Ellerman, D., Joskow, P. and Harrison, D. (2003). *Emissions trading in the US. Experience, Lessons, and Considerations for Greenhouse Gases*. Pew Center (USA).

Ellerman, D. and Buchner, B. (2006). *Over-Allocation or Abatement? A Preliminary Analysis of the EU ETS*. FEEM. Nota di lavoro 139.2006, Milano.

Ellis, J, CorfeeMorlot, J, Winkler, H. (2004). *Taking stock of progress under the Clean Development Mechanism.* Paris: COM/ENV/EPOC/IEA/SLT(2004)4/FINAL.

Feng, H. and Kling, C. (2005). The Consequences of Co-benefits for the Efficient Design of Carbon Sequestration Programs. *Canadian Journal of Agricultural Economics,* **53**, 461–476

Fenhann, J. (2007). Overview of the CDM pipeline. Available at: http://www.cd4cdm.org/ Publications/CDMpipeline. Last accessed: 20 March 2007.

Field, B. and Field, M. (2003). *Environmental Economics. An introduction.* New York: McGraw Hill.

Galeotti, M. and Carraro, C. (2003). The future evolution of the Kyoto Protocol: costs, benefits and incentives to ratification and new international regimes. In Carraro, C. and Egenhoffer, C. (Eds). *Firms, Governments and Climate Policy* (267-316). Cheltemham (U.K.): Edward Elgar.

Gerlagh, R. and van der Zwaan, B. (2006). Options and Instruments for a Deep Cut in CO2 Emissions: Carbon Dioxide Capture or Renewables, Taxes or Subsidies?. *Energy Journal,* **27**(3), 25-48

Godal, O. and Klaasen, G. (2006). Carbon trading across sources and periods constrained by the Marrakesh Accords. *Journal of Environmental Economics and Management,* **51**, 308-322.

Godby, R. (2000). Market Power and Emission Trading: Theory and Laboratory Results, *Pacific Economic Review,* **5**(3), 349-363.

Grubb, M., Michaelowa, A., Swift, B., Tietenberg, T., Zhang, X. (1998). *Greenhouse Gas Emissions Trading. Defining the Principles, Modalities, Rules and Guidelines for Verification, Reporting & Accountability*. UNCTAD.

Grubb, M. and Neuhoff, K. (2006). Allocation and competitiveness in the EU Emissions Trading System: Policy Overview. *Climate Policy* **6**(1), 7-30.

Gunter, S. and Müller-Fürstenberger, G. (2006). Does Distribution Matter? Efficiency, Equity and Flexibility in Greenhouse Gas Abatement. Environmental and Resource Economics, **27**(1), 87-107.

Hagem, C. and Maestad, O. (2006). Russian exports of emission permits under the Kyoto Protocol: The interplay with non-competitive fuel markets. *Resource and Energy Economics*, **28**(1), 54-73.

Hahn, R. and Noll, R. (1982). Designing a Market for Tradable Emissions Permits. In Magat, W. (ed.). *Reform of Environmental Regulation*. Cambridge, Massachusetts.

Haites. E. (2004). *Estimating the Market Potential for the Clean Development Mechanism: Review of Models and Lessons Learned*. PCFplus Report 19. Washington D.C.

Harrison, D. and Radov, D. (2002). *Evaluation of Alternative Initial Allocation Mechanisms in a European Union Greenhouse Gas Emissions Allowance Trading Scheme*. London: National Economic Research Associates.

Hernández, F. (Ed.)(1999). *Global warming in Spain. An analysis of its socioecomomic and environmental impacts*. Madrid: Editorial CSIC (in Spanish).

Hernández, F. and del Río, P. (2007). *The Kyoto Protocol and its impact on Spanish companies*. Madrid: Editorial CSIC (forthcoming, in Spanish).

Houghton, R.A. (2003). Revised estimates of the annual net flux of carbon to the atmosphere from changes in land use and management: 1850-2000. *Tellus Series B-Chemical and Physical Metereology*, **55**, 378-390.

Hourcade, C. and Ghersi, F. (2002). The Economics of a Lost Deal: Kyoto - The Hague - Marrakesh. *Energy Journal*, **23**(3), 1-26

Ignaciuk, A., Vöhringer, F., Ruijs, A., van Ierland, E.C. (2006). Competition between biomass and food production in the presence of energy policies: a partial equilibrium analysis. *Energy Policy,* **34**(10), 1127-1138.

Intergovernmental Panel on Climate Change (IPCC)(2001). *Third Assessment Report. Mitigation*. Geneve.

IPCC (2007). Working Group III contribution of the IPCC Fourth Assessment Report. Climate Change 2007: Mitigation of Climate Change. *Summary for Policymakers*. IPCC, Geneve.

Jaffe, A., Newell, R. and Stavins, R. (2005). A tale of two market failures: Technology and environmental policy. *Ecological Economics*, **54**(2-3), 164-174.

Jensen, S.G. and Skytte, K. (2003). Simultaneous attainment of energy goals by means of green certificates and emission permits. *Energy Policy,* **31**, 63–71.

Jung, M (2006). Host country attractiveness for CDM non-sink projects. *Energy Policy*, **34** (15), 2173-2184.

Kara, M., Syri, S., Lehtilä, A., Helynen, S., Kekkonen, V., Ruska, M. and Forsström, J. (2006). The impacts of EU CO_2 emissions trading on electricity markets and electricity consumers in Finland. *Energy Economics* (in press).

Kemfert, C., Kohlhaas, M., Truong, T. and Protsenko, A (2006). The environmental and economic effects of European emissions trading. *Climate Policy,* **6**(4), 441-455.

Kerr, S, y Maré, D. (1997). Transaction costs and tradable permit markets: the United Status Lead Phasedown. http://www.arec.umd.edu/skerr/transact.htm.

Klepper, G. and Peterson, S. (2007). Emissions Trading, CDM, JI, and More: The Climate Strategy of the EU. *Energy Journal* **27**(2), 1-26

Linares, P., Santos, F.J. Ventosa, M., Lapiedra, L. (2006). Impacts of the European Emissions Trading Scheme Directive and Permit Assignment Methods on the Spanish Electricity Sector. *Energy Journal* **27**(1), 79-98

Liski, M. and Virrankoski, J. (2004). Frictions in Project-Based Supply of Permits. *Environmental and Resource Economics,* **28**(3), 347-365.

Lubowski, R.N., Plantinga, A.J., Stavins, R.N. (2006). Land-use change and carbon sinks: Econometric estimation of the carbon sequestration supply function. *Journal of Environmental Economics and Management,* **51**, 135-152.

Maloney, M. and Yandle, B. (1984). Estimation of the Cost of Air Pollution Control Regulation. *Journal of Environmental Economics and Management,* **11**, 244-263.

Markandya, A. and Halnaes, K. (2002). *Climate Change and Sustainable Development. Prospects for Developing Countries.* London: Earthscan.

Matthews, S., O'Connor, R., Plantinga, A. (2002). Quantifying the impacts on biodiversity of policies for carbon sequestration in forests. *Ecological Economics,* **40**(1), 71–87.

Marechal, K. and Hecq, W. (2006). Temporary credits: a solution to the potential non-permanence of carbon sequestration in forests?. *Ecological Economics*, **58**, 699-716.

Mathiesen, L. and Maestad, O. (2004). Climate Policy and the Steel Industry: Achieving Global Emission Reductions by an Incomplete Climate Agreement. *Energy Journal,* **25**(4), 91-114

Matschoss, P. and Welsch, H. (2006). International Emissions Trading and Induced Carbon-Saving Technological Change: Effects of Restricting the Trade in Carbon Rights. Environmental and Resource Economics, **33**(2), 169-198

Michaelowa, A. (2003). CDM Host country institution building. *Mitigation and Adaptation Strategies for Global Change*, **8**, 210-220.

Michaelowa, A. and Jotzo, F. (2005). Transaction costs, institutional rigidities and the size of the clean development. *Energy Policy,* **33**(4), 511-523.

Morthorst, P.E. (2001). Scenarios for the use of GHG-reduction instruments—How can policy-instruments as carbon emission trading and tradable green certificates be used simultaneously to reach a common GHG-reduction target?. *Energy and Environment,* **11** (4), 423–438.

Morthorst, P.E. (2003). National environmental targets and international emission reduction instruments. *Energy Policy,* **31**(1), 73–83.

Moulton, R., Richards, K. (1990). *Costs of Sequestering Carbon through Tree Planting and Forest Management in the United States*, General Technical Report WO-58. Washington, D.C.: U.S. Department of Agriculture

Murray, B.C., McCarl, B.A., Lee, H-C., 2002. *Estimated Leakage from Forest Carbon Sequestration Program.* RTI International. Working paper 02-06.

Padilla, E. and Roca, J. (2004). The Proposals for a European Tax on CO_2 and Their Implications for Intercountry Distribution. *Environmental and Resource Economics,* **27**(3), 273-295

Pearce, D. (2006). The political economy of an energy tax: The United Kingdom's Climate Change Levy. *Energy Economics*, **28**(2), 149-158.

Plantinga, A.J. and Wu, J. (2003). Co-Benefits from Carbon Sequestration in Forests: Evaluating Reductions in Agricultural Externalities from an Afforestation Policy in Wisconsin. *Land Economics,* **79**(1), 74-85.

Plantinga, A., Mauldin, T., Miller, D. (1999). An econometric analysis of the costs of sequestering carbon in forests. *American Journal of Agricultural Economics,* **81**(4), 812-24.

Philibert, C. (2005). *Approaches for Future International Co-Operation.* Paris: IEA.

Pohjola, J., Kerkelä, L., Mäkipää, R. (2003). Credited forest carbon sinks: how the cost reduction is allocated among countries and sectors. *Climate Policy,* **3**, 445–461.

Requate, T. (2005). Dynamic incentives by environmental policy instruments-a survey. *Ecological Economics* **54**, 175-195

Richards, K.R. and Stokes, C. (2004). A Review of Forest Carbon Sequestration Studies: A dozen Years of Research. *Climate Change,* **63**(1-2), 1-48.

Richards, K.R, Sampson, R.N., Brown, S., 2006. *Agricultural and Forestlands: US Carbon Policy Strategies*. Pew Center on Global Climate Change.

Rousseau, S. and Proost, S. (2005). Comparing Environmental Policy Instruments in the Presence of Imperfect Compliance – A Case Study. Environmental and Resource Economics, **32**(3). 337-365

Rübbelke, D. (2003). An analysis of differing abatement incentives. *Resource and Energy Economics,* **25**(3), 269-294.

Salter, L. (2004). A Clean Energy Future? The role of the CDM in promoting renewable energy in developing countries. *19th World Energy Congress. Sidney: September 5-9, 2004.*

Schmalensee, R., Joskow, P.L., Ellerman, D., Montero, J.P. (1998). An interim evaluation of sulphur dioxide emissions trading. *Journal of Economic Perspectives,* **12**(3), 53-68.

Schneider, F. and Volkert, J. (1999). No chance for incentive-oriented environmental policies in representative democracies? A Public Choice analysis. *Ecological Economics,* **31**, 123-138

Sedgo, R. and Salomon, A. (1989). Greenhouse warming: abatement and adaptation. In Crosson P., Darmstadter, J., Easterling, W. and Rosenberg, N. (Eds.). *RFF Proceedings* (110-119)., July 1989.

Sijm, J., Neuhoff, K. and Chen, Y. (2006). CO2 cost pass through and windfall profits in the power sector. *Climate Policy,* **6**, 49–72.

Skytte, K. (2006). *Interplay between Environmental Regulation and Power Markets.* San Domenico di Fiesole (Italy): European University Institute Working Papers nº2006/04.

Solberg, B., Moiseyev, A., Kallio, M.A.I. (2003). Economic impacts of accelerating forest growth in Europe. *Forest Policy and Economics,* **5**, 157–171

Sohngen, B., Mendelsohn R. (2003). An optimal control model of forest carbon sequestration. *American Journal of Agricultural Economics,* **85**(2), 448-457.

Sohngen, B., Mendelsohn R., Sedjo, R. (1999). Forest Management, Conservation, and Global Timber Markets. *American Journal of Agricultural Economics,* **81**(1), 1-13.

Sohngen, B., Mendelsohn R., Sedjo, R. (2001). A Global Model of Climate Change Impacts on Timber Markets. *Journal of Agricultural and Resource Economics*, **26**(2), 326-343.

Sorrell, S. (2003). *Back to the Drawing Board? Implications of the EU Emissions Trading Directive for U.K. Climate Policy.* Executive summary EU-funded project Interaction in EU Climate Policy (INTERACT). University of Sussex at Brighton (U.K.).

Sorrell, S., and Sijm, J. (2005). Carbon Trading in the Policy Mix. In Helm, D. (Ed.). *Climate Policy* (194-217). Oxford (U.K.): Oxford University Press.

Springer, U. (2003). The market for tradable GHG permits under the Kyoto Protocol: a survey of model studies. *Energy Economics,* **25**, 527-551.

Stavins, R.N. (1999). The Costs of Carbon Sequestration: A Revealed-Preference Approach. *The American Economic Review,* **89**(4), 994-1009.

Stevens B and Rose, A. (2002). A Dynamic Analysis of the Marketable Permits Approach to Global Warming Policy: A Comparison of Spatial and Temporal Flexibility. *Journal of Environmental Economics and Management,* **44**(1), 45-69.

Storey, M. (2002). *Kyoto and Beyond, Issues and options in the global response to climate change.* Stockholm (Sweden): Swedish Environmental Protection Agency,. www.internat.environ.se/documents/issues/ climate/report/Kyoto.pdf

Sugiyama, T, Yamaguchi, K, Yamagata, H. (2005). *CDM in the Post-Kyoto Regime: Incentive mechanism for developing countries to promote energy conservation and renewable energies.* Workshop Issue Paper. March 22-23, 2005. Mitsubishi Research Institute (Japan)

Svendsen, G.T. (2000). *Public Choice and Environmental Regulation. Tradable Permit Systems in the United State and* CO_2 *Taxation in Europe.* Cheltenham (U.K.): Edward Elgar.

Tavoni, M., Sohngen, B., Bosetti, V., (2007). Forestry and the carbon market response to stabilize climate *FEEM Working Paper* No. **15**.2007. Milano.

Tietenberg, T. (2005). The Tradable Permits Approach to Protecting the Commons: Lessons for Climate Change. In Helm (Ed.). *Climate Change Policy* (167-193). Oxford: Oxford University Press.

Tietenberg, T. (2006). *Emissions Trading Principles and Practice.* 2nd edition. Washington: Resources for the Future.

Torvanger, A., Twena, M. and Vevatne, J. (2004). *Climate policy beyond 2012. A survey of long-term targets and future Frameworks.* Oslo (Norway): CICERO Report 2004:02.

Unruh, G.C. (2000). Understanding carbon lock-in. *Energy Policy,* **28**(12), 817-830.

Unruh, G.C. (2002). Escaping carbon lock-in. *Energy Policy,* **30**, 317-325.

Van Kooten, G.C.; Eagle, A., Manley, J., Smolak, T. (2004). How costly are carbon offsets? A meta-analysis of carbon forest sinks. *Environmental Science and Policy,* **7**(4), 239-251.

Viguier, L., Babiker, M., Reilly, J. (2003). The costs of the Kyoto Protocol in the European Union. *Energy Policy,* **31**, 459 –481.

Watson, R.T., Noble, I.R., Bolin, B., Ravindranath, N.H., Verardo, D.J., Dokken, D.J., (2000). *Land use, land-use change and forestry. A special report of the IPCC,* Cambridge: Cambridge University Press.

Wing, I.S. (2006). The synthesis of bottom-up and top-down approaches to climate policy modelling: Electric power technology detail in a social accounting framework. *Energy Economics* (forthcoming). Available online at http://www.sciencedirect.com/

Woerdman, E. (2001). Emissions trading and transaction costs: analyzing the flaws in the discussion. *Ecological Economics,* **38**, 293-304.

Woerdman, E. (2004). *The Institutional Economics of Market-Based Climate Policy.* Amsterdam: Elsevier.

Zhang, X.Z. (2000). Estimating the size of the potencial market for the Kyoto mechanisms. http://ideas.uqam.ca/ideas/data/Articles

In: Climate Change Research Progress
Editor: Lawrence N. Peretz, pp. 127-159

ISBN: 978-1-60021-998-6

Chapter 3

MULTIPROXY ANALYSIS OF CLIMATE VARIABILITY FOR THE LAST MILLENNIUM IN THE TROPICAL ANDES

V. Jomelli [1,2], J. Argollo[3], D. Brunstein[1], V. Favier[4], G. Hoffmann[5], M. P. Ledru[2] and J. E. Sicart[2]

[1] CNRS Laboratoire de Géographie Physique de Meudon UMR8591 France.
[2] IRD UR GREAT ICE MSE Montpellier, France
[3] IGEMA, University Mayor San Andrés, La Paz, Bolivia
4 CEAZA, Benavente 980, La Serena, Chile
[5] CEA Laboratoire des Sciences du Climat et de l'Environnement, Gif sur Yvettes, France

Abstract

Reconstructing climate over the last millennium is one major target of climate research. The extent of climate variability over recent centuries sets the scale on which ongoing climate changes can be measured. The quality of reconstructions of millennial temperatures in the northern hemisphere has greatly improved over the last decade (*IPCC*, 2007). Nevertheless there is still a lack of precisely dated, high-quality climate reconstructions in the tropics, a lack which is particularly disturbing since this region is a major climate engine for the global water cycle. In recent years, a significant effort has been made to improve our knowledge on climatic variability in the tropical Andes by developing new proxy records, new transfer functions and better time resolution of available proxy records. Hence, the aim of this paper is to review the different proxies with multi-decadal to annual resolution available in the tropical Andes (Bolivia, Peru, Ecuador) such as glacial moraines, pollens, isotopic variations in ice core records, and tree rings. For each proxy, we focus on recent major methodological progress and on new results revealing climatic shifts over the last millennium.

1. Introduction

The last millennium is obviously the period which most resembles (and actually includes) the pre-industrial period in the early 19th century. The amplitude of ongoing anthropogenic climate change should therefore be compared with this most recent period. Presumably "natural" climate variability, - similar to that which human societies experienced from the medieval period up to the first years of industrialisation also defines the range of climate to which humans have successfully adapted. Early meteorological observations are available since about the late 18th century but mainly in Western Europe and North America (Auer, et al., 2007). South America offers a wealth of historical information and natural climate proxies that have only recently been discovered and analysed.

The last millennium enables us to investigate the influence of different forcing factors. It includes a relatively cold period, called the Little Ice Age (LIA, ~1500-1900 yr AD), when solar activity was low and volcanoes were active, and a relatively warm Medieval Warm Period (MWP: before 1200 yr AD) both of which were characterised by relatively stable greenhouse gas forcing. Hence, the last 1000 years serve as a "laboratory" where the effect of climate mechanisms can be tested, and provide a basis for improving our understanding of climatic change. Indeed, temperature changes that occurred during the LIA are in the same order of magnitude as those estimated by some Global Circulation Models for the middle of the 21st century. On the other hand, despite a small amplitude temperature change during the last millennium (less than 1°C), considerable environmental consequences have been observed. Numerous catastrophic events related to temperature and hydrological cycles (freezes, floods, droughts, heat waves, etc.) as well as dramatic events with serious consequences for human societies (diseases, famines, etc.) have been documented (Ladurie, 1987; Grove, 1988). Consequently, long climatic series are needed to validate ocean-atmosphere models over a long period of time and to study the impact of change on pre-modern societies.

Another important aspect concerns the heterogeneous geographical distribution of available climatic data. Most of these data are based on observations or multi-proxy reconstitutions (dendroclimatology, historical documents, weather data, etc.) at mid and high latitudes. However the inter-tropical band, considered a key area for the energetic exchanges that play a key role in global climate (Mc Gregor and Nieuwolt, 1998), is also the region where data are the most rare (Jones, et al., 2001). The surface of tropical oceans is the main source of heat for atmosphere–ocean circulation (Broecker and Denton, 1989; Wang et al., 2004) and the impacts of tropical climate reach far beyond the limits of the tropics (Mc Gregor and Nieuwolt, 1998). In addition, intra-annual and abrupt changes in tropical climate (like ENSO) have a major influence beyond the actual tropics, including in mid-latitudes.

Why focus on the tropical Andes?

- The first reason is geographical. The Tropical Andes cover about 3800 km from 10°N to 23°S. The shape and setting of the tropical Cordilleras means they form a barrier for the dominant and persistent easterly atmospheric flow separating the wet Amazon side from the dry Pacific side, producing pronounced windward and leeward effects (Kaser and Georges, 1997). Many summits reach elevations of over 6000 m asl. For these reasons, the tropical Andes are characterized by contrasted climates and steep climatic and ecological gradients over short spatial distances that significantly influence biological, geomorphological, and hydrological processes, and form a system that is extremely sensitive to environmental changes.

- Second, climatic changes are frequently inferred to explain either the demise of some historic or older civilisations and their economy, or the development of subsistence strategies to ensure their survival. For example, the decline of Tiwanaku field agriculture from 950 AD (Middle Horizon) in the Titicaca Basin (Ortloff and Kolata, 1993; Binford et al., 1997) has been attributed to a failure of the corresponding societies to respond to short-term climatic deterioration. Hence, the better our knowledge of past climatic changes the better we understand the relationships between old civilisations and climate changes. In addition, climate variability analysed either using observed data or output modelling (Aceituno, 1988; Vuille *et al.*, 1998b; Vuille and Bradley, 2000; Garreaud *et al.*, 2003) has a strong impact on the availability of water resources in most of the Andean capitals such as Lima, Arequipa or La Paz, which are experiencing a high migratory flux from the countryside (Kaser et al., 2003; Mark and Seltzer, 2003; Pouyaud et al. 2005). All these cities are located in dry areas where drinking water and the dam water that supplies electric power is expected to run short in the near future.
- The last reason is the recent improvements in our understanding of the relationships between the climate and the proxies via transfer functions, and to the discovery of new proxies that help to better document past climatic conditions.

The last question is why do we need a multiproxy approach? The answer is that each proxy has a specific relationship with climatic conditions, for example some dendrochronological records are more sensitive to changes in precipitation than others. On the other hand, some proxies do not enable high-frequency climate variability (less than a decade) to be documented either because of their low time resolution or to their specific relationship with climatic shifts. In addition, each of the proxies has its own specific problems that can mostly be overcome by combining approaches.

In this context, the aims of this paper are to 1) highlight recent improvements in the continental proxies in the tropical Andes, 2) document climatic conditions during the last millennium inferred from different proxies in the tropical Andes.

This paper is organized as follows: in the next section, we describe high resolution proxies i.e. ice core and tree ring records; in section three, we describe low resolution proxies i.e. pollen and moraine records; and in section 4, we compare climatic conditions inferred from these different proxies.

2. High Resolution Proxies

2.1. Ice Core Records

Over the last 30 years, numerous high-altitude ice cores have been drilled and successfully analysed. In the Andes, these ice cores were located within a huge latitudinal strip that reaches between tropical Ecuador and the dry central Cordillera Real in Bolivia. Ongoing drilling activities are extending the geographical range even farther towards the south of Chile and Patagonia. Here we give only a short overview of centennial scale variability (basically during the last Millennium) archived in these cores. In addition, we refer to two other recent review papers (Thompson and Davis, 2006; Vimeux, et al., 2007) that raised some of the questions we discuss here.

All together, five main deep ice core sites have been explored in the Andes and many short and deep cores have been drilled and analysed (figure 1) from North to South: in Chimborazo (1.3°S, Ecuador), Huascaran (9°S, Cordillera Blanca, Peru), Quelccaya (13.5°S, Peru), Illimani (16.4°S, Cordillera Real, Bolivia) (figure 2), and Sajama (18°S, Bolivia). At least two additional cores (Corupuna, Peru; and San Valentin, Chile) have been drilled and partly analysed, but results are not yet published. For the time period of interest here, i.e. the last millennium, the Illimani and the Quelccaya core will be analysed in most detail.

When analysing low-latitude ice cores, many practical and theoretical difficulties have to be overcome, most of which are due to their exceptional location and due to the specific impact of tropical climate on the various proxies archived in the ice. In contrast to polar ice cores, the preservation and transport of the ice is a major issue in the tropics. The ice is typically carried down from the glaciers in commercial ice boxes and afterwards put into boxes cooled with dry ice. Due to the rapidly rising temperatures during the descent, the time the ice is preserved without additional cooling must be kept as short as possible. Transport by balloons was envisaged, but was unsuccessful due to the difficult wind conditions at the typical drilling sites located between 5500 and 6400 m asl. (L.Thompson, pers. communication).

A further problem is the dust load and the general contamination of tropical ice with different organic compounds which makes analysis of the gas content very difficult. To measure the $\delta^{18}O_{Air}$ composition of atmospheric oxygen (a proxy for general biological activity and sea level change on glacial/inter-glacial cycles), or the greenhouse gases CH_4 and CO_2, high "quality" ice is required. Only snow that was relatively free of impurities when it was deposited under very cold conditions enables satisfactory reproducibility of the measurements and ensures that post-depositional reactions do not alter the respective gas signals (Tschumi and Stauffer, 2000). To our knowledge, only a few gas measurements have been successful in low latitude ice cores ($\delta^{18}O_{Air}$ in the Sajama and Illimani core (Ramirez, et al., 2003; Thompson, et al., 1998). The dust records from Andean ice cores are commonly used in a multi-proxy dating approach by counting the sequence of dry dusty layers within the firn and ice. Although different sources of dust in the vicinity of the glaciers may cause problems, dust records are nevertheless considered as a valuable qualitative drought proxy of at least regional significance (Thompson, et al., 1988). However, the most widely used climatic proxy in ice cores is the isotopic composition of the ice (expressed as ‰ deviation, δ 18O ice and δ D ice, from a global standard, SMOW= Standard Mean Ocean Water, (Craig, 1961). Below we present a detailed climatic interpretation of the isotope records in the Andes, and in particular, of ice cores located in the dry core of the tropical/sub-tropical subsidence zone which undergo substantial annual precipitation losses by sublimation. Detailed field studies have determined that annual mass losses due to sublimation are in the order of 30%.

Two questions are of importance with regard to the interpretation of isotopic records, 1) is the mean annual precipitation altered by fractionation processes during the transition from snow to atmosphere? 2) Does a combination of seasonally varying sublimation and a strong cycle in the isotope signal lead to a "selection" bias in the ice cores. The first question has been investigated in a detailed study carried through on the Cerro Tapado (Stichler, et al., 2000). In fact each phase change leads to fractionation which necessarily enriches a micro layer at the snow surface. However, diffusive processes combined with a continuous condensation/re-evaporation are too slow to produce mixing of the enriched isotope signal in the fractionated micro-layer down into the firn. Stichler, et al (Stichler, et al., 2000) assume that layers affected by sublimation penetrate the firn to a depth of 5-10 cm at best. Net fractionation therefore is close to zero. The second question is unfortunately much more difficult to answer. Interannual variations in temperature and humidity due to the ENSO phenomenon can produce significant variations in the Evaporation/Precipitation relationship up to complete sublimation of an entire annual snow layer. Obviously there will be a tendency to stronger sublimation in the dry season, typically several tens of percents of yearly precipitation (Wagnon, et al., 2003). This should therefore lead to significant under-representation of the isotope signal of precipitation during or close to the dry period. Potentially a major change in rainfall patterns could lead to the loss of snow layers and their associated water isotopes over even longer periods, decades or more.

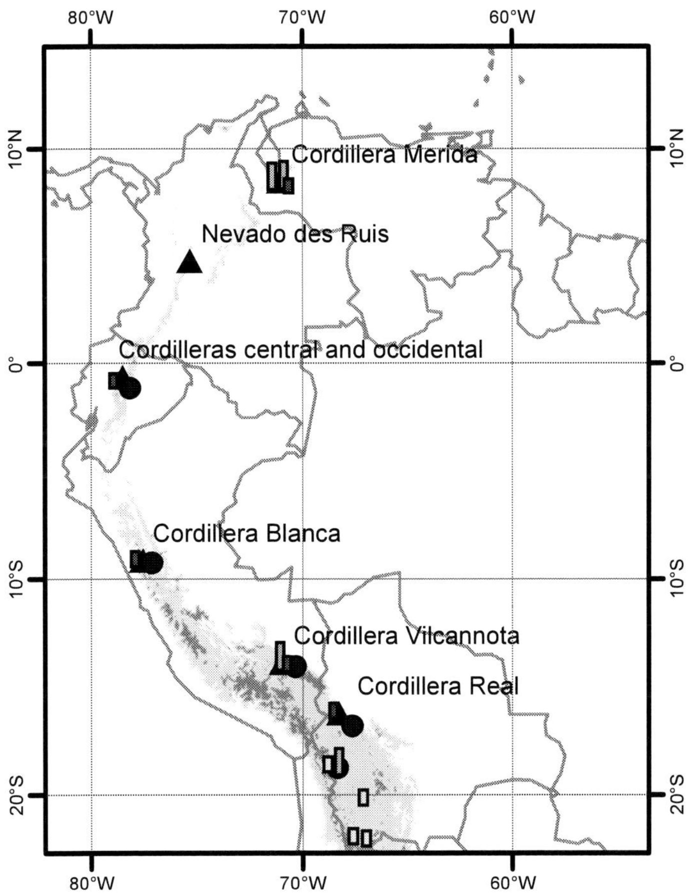

Figure 1. Location of the different proxies available in the tropical Andes.

A further difficulty might arise from the extreme daily insolation amplitude in the tropics and/or from hidden heat fluxes in the bedrock in areas with ongoing volcanic activity. Both processes might lead to water percolating through the firn, even though the mean annual temperatures apparently guarantee cold glacier conditions. Percolating water resulted in the abandonment of ice drilling at the saddle position on the Chimborazo in Ecuador.

Quelccaya was the first high-altitude low-latitude ice core drilled to the bedrock. High accumulation at Quelccaya allows annual dating over the last 1000 years and provides excellent resolution for the entire period from the Little Ice Age to the medieval warm period. Relatively "warm" conditions due to its low altitude (5670 m asl) cause strong basal melting and limit the temporal range covered. Their counting accuracy enabled the authors to publish the first high altitude climate reconstruction of the last millennium in the tropical Andes (Thompson et al., 1986). The Illimani core is the most recent ice core covering, as Huascaran and Sajama, the entire Holocene and the final stages of the last glacial. Due to the marked thinning of ice layers, annual counting of the Illimani was only possible back to 1772 (see figure 13). For the two other cores, we consider only the annually dated part of the record and refer to a recent analysis of the smoothed record (Thompson et al., 2006).

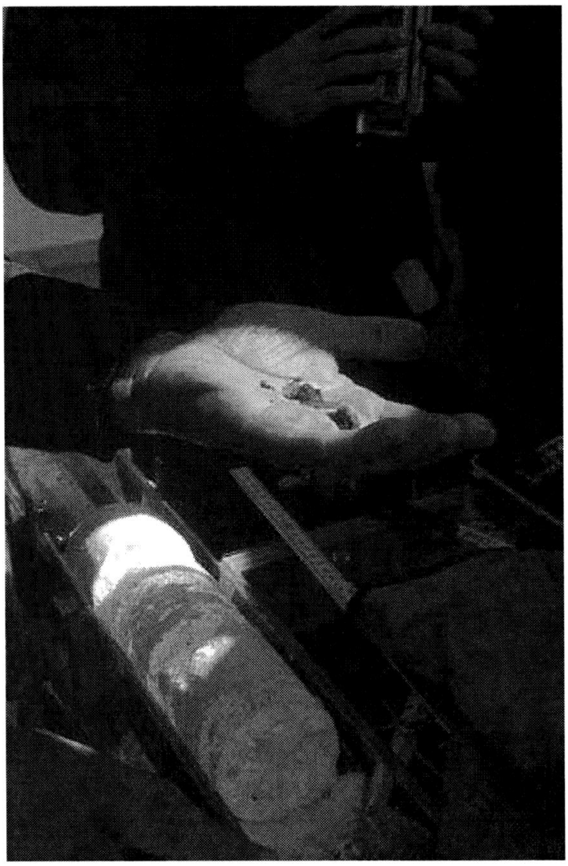

Figure 2. Illimani Ice core. The hand is holding small pieces of bed rock.

Ice cores provide a wealth of environmental proxies, such as the dust content, volcanic ashes, or sulphuric acid, each significant for certain environmental and climatic processes and variations. Here we focus on the water isotopic composition (18O/16O or D/H) of the ice expressed as δ-deviation from a standard (SMOW=Standard Mean Ocean Water). The δ value is classically interpreted as a proxy for the strength of the rainout of an air mass typically travelling from low latitudes to high latitudes (Dansgaard, 1964). However, in the tropics, the importance of convective rain formation weakens the impact of surface temperatures on the rainout and requires a complete analysis of the processes involved. In the global water isotope network, the relationship between local surface temperatures and water isotopes break down at about 15°C mean annual temperature.

In the case of the Andean isotope signal, several studies have highlighted the importance of large-scale transport of water vapour from the tropical Atlantic over the Amazon basin to the final rain/snow deposition (Grootes, et al., 1989; Hoffmann, et al., 2003; Vimeux, et al., 2005). At least on the seasonal to inter-annual time scale, observations and modelling indicate that the variability of vapour transport, convective activity and rainout over South America can be understood as being part of a monsoon-type system. Its intensity is controlled by both mid-tropospheric zonal flow and SST gradients both being in a complex relation to Pacific ENSO variability. In (Hoffmann, 2003; Vuille and Werner, 2005), the authors underline that such circulation variability co-varies with precipitation rather than with temperature variability.

Several efforts have been made to properly calibrate the water isotopes. Seasonally, only a spurious link has been found between the water isotopes and local temperatures or local rainout intensity. However, on a continental scale, significant correlations with Amazon-wide precipitation (and therefore rainout) intensity have been demonstrated (Vimeux, et al., 2005). The seasonal movement of the InterTropical Convergence Zone (ITCZ) controls the moisture flux into the basin. The final isotope signal therefore appears to be the result of possibly several precipitation/evaporation events and of the convective intensity over the basin. Such a large-scale interpretation of the water isotopes was confirmed by numerical experiments with Atmospheric General Circulation Models (AGCMs) fitted with water isotope diagnostics (Vimeux, et al., 2005). On an interannual scale, certain studies (Bradley, et al., 2003; Hoffmann, 2003; Vuille, et al., 2003) cite Pacific Sea Surface Temperatures (SST) as an important factor in the control of atmospheric circulation and moisture flux over South America. In particular the South American Monsoon System reacts sensitively to vertical wind stress anomalies between low level inflow and high level outflows. The latter have also been linked to the atmospheric tele-connection patterns associated with Pacific SSTs and ENSO (Vuille and Werner, 2005). In summary, empirical and model studies point to a region-wide "amount effect" (i.e. a link between the precipitation amount and the water isotopes) controlling the water isotopes on the South American Altiplano both on the seasonal and the inter-annual scale. However, one should bear in mind that a proper calibration (in the sense of a linear relation between the water isotopes and regional precipitation) is difficult. The heterogeneity of the water isotope signals and of the rainfall observations hamper such a classical "simple" calibration. Furthermore, discussion is still underway as to whether on longer time scales (centennial to glacial/interglacial cycles) other factors might influence the water isotopes. In particular, variations in high altitude temperatures might have an important impact on the overall tropical convective intensity (Thompson, et al., 2000).

On a centennial time scale, no calibration or even qualitative interpretation of the water isotope signal has been undertaken, but, based on the interpretation of the common 20th century decadal signal, one would associate more depleted isotope values with mean wet/cold conditions over tropical South America and more enriched water isotope values with dry/warm conditions. The isotope signals in different cores show relatively high coherence over the 20th century on a decadal scale (Hoffmann, et al., 2003). However on shorter time scales, this is not the case and the question remains if the greater heterogeneity of the isotopes is due to faulty dating of the different cores or to real differences in the atmospheric processes involved.

2.2. Tree Rings

Tree rings are a very famous method that provides high-resolution climate reconstructions (Fritts, 1976). However, before applying this method in the Tropical Andes, several problems have to be taken into account. Most tropical trees do not have anatomical rings (Jacoby, 1989; Worbes 1995), which may be due to the reduced temperature variation at low latitudes. However, in high altitude subtropical mountains, during certain periods of the year temperatures reach minima that limit the development of the vegetation, and studies of tree rings are of interest in several species.

Polylepis is a genus that is widely distributed at high altitudes from Venezuela to Argentina. Many ecological studies have been undertaken on this endemic genus of the Rosaceae (Kessler, 1995; Fjeldsa and Kessler, 1996; Kessler and Schmidt-Lebuhn, 2006) and more than thirty species have been identified to date. *Polylepis* are small trees or shrubs that grow in locations with dry atmospheres to high mountain areas with humid atmospheres and cover slopes from 3000 m asl up to 5500 m. This wide distribution makes it possible to document the different climatic conditions and the altitudinal gradients. The first research in dendroclimatology on the *Polylepis* species began approximately 15 years ago starting from samples of *P. tarapacana* collected in Bolivia.

Figure 3. Cross section of *P. tarapacana.*

One of the particularities of the *Polylepis* is the irregularity of radial growth of some species which has an impact on climate reconstructions (figure 3). In certain cases, the trunk is almost circular. In other species (especially *P. tarapacana*) growth is not circular and the pith is not located at the centre of the tree but exhibits a lobed shape. This means that in most cases chronologies are carried out with cross sections of trunks or branches. Another consequence of the lobed growth is that the traditional approaches used in dendroclimatology have to be adapted, as chronologies are generally based on 2-3 radii per tree. In the case of *Polylepis*, the number of rings can differ from one lobe to another. If the number of rings does not differ too much, a mean is calculated over a common time period. However, the largest radius of each tree is used to construct the chronology.

When observed with a magnifying glass, the wood of *P. tarapacana* displays clearly distinguishable growth rings (Argollo et al., 2004). The boundary between annual rings of *P. tarapacana* is defined by an adjustment of the ligneous elements characterized by the presence of more abundant cells with a large diameter at the beginning of the earlywood that contrast with a ligneous band with thicker cellular walls that constitute the latewood (figure 4). In general, the cells have a small diameter and a semicircular pattern. This specific pattern of growth layers makes the identification of rings and the establishment of chronologies based on densitometry more difficult. Hence, the lobed growth, combined with the difficulty in identifying rings -especially false rings, which are common in mountain species- explain why correlations between samples taken from the same tree or from different trees are sometimes low. In addition, genetic analyses revealed that difficulties in identifying species can be partly due to hybridising between species in the same area (Schmidt-Lebuhn et al., 2006). Consequently species are difficult to identify in the field and the classical keys used for the determination of species, for example the number of leaves per branch, are not always relevant. Nevertheless different tree ring chronologies have been established in recent years (figure 5). Most of these chronologies are located in Cordillera Occidental on the slope of high volcanoes. The longest chronology extends over seven centuries. New sites were recently discovered in Cordillera Real (Bolivia) and in Peru (Cusco region) (Jomelli et al., in preparation). Based on other tree species samples have been partly analysed, but results are not published.

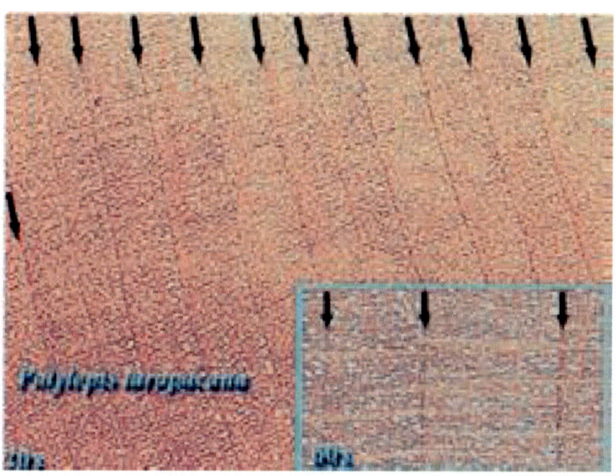

Figure 4. Rings of *tarapacana (arrows)* (magnification x 20)

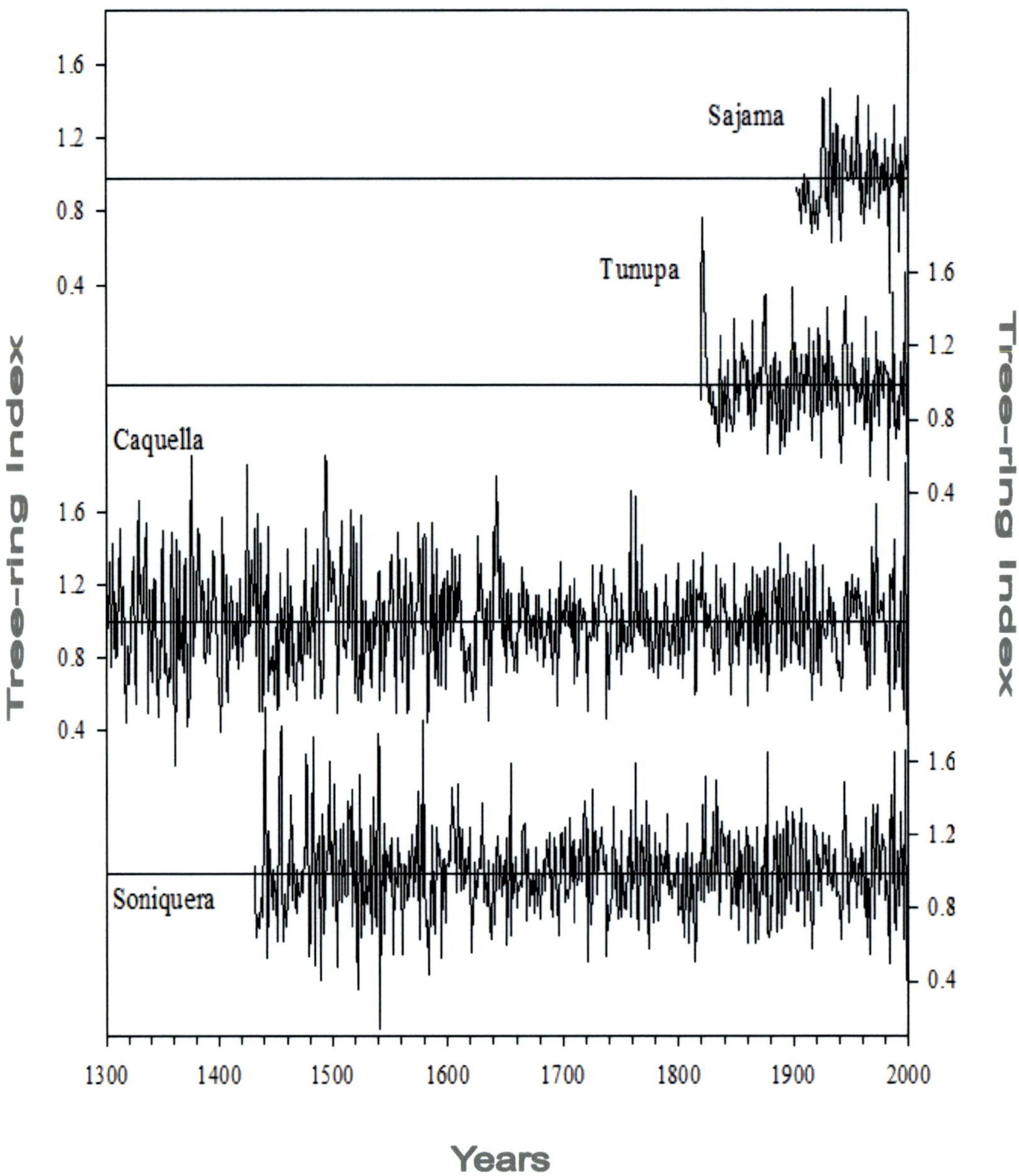

Figure 5. Chronologies of *P. tarapacana* in Western Cordillera (Bolivia).

Due to the lack of data, the impacts of climatic conditions on growth processes are not perfectly understood, although progress is expected in the next few years. Indeed, sensors have been fixed in different trees to better characterize the influence of meteorological conditions on tree growth. For example it has been demonstrated that *P. tarapacana* is a frost tolerant species during the cold unfavourable season which is able to avoid freezing during the more favourable season when minimum night-time temperatures are not as extreme (Rada et al., 2001).

In contrast, statistical analyses of different tree ring series collected in Bolivia revealed clear relationships between tree ring variations and climate (figure 6). In general, there is a closer link with the climatic conditions of the previous year. Moreover, tree ring series located on Amazonian slopes are more sensitive to temperature variations during the growing season, while sites located on Pacific slopes are more sensitive to variations in precipitation during the rainy season. However, relationships between either temperature or precipitation and tree ring indices are complex. The tree ring index is generally positively correlated with precipitation in December and January of the previous growth year (figure 6). Relationships between precipitation that occurred during the cycle of formation of the ring and ring width are even more complex. They can be negatively correlated or there may be no significant correlation with growth. The role of temperature in the tree ring index is also complex. Both negative and positive (figure 6) relationships can be observed during the growing season (November-March) depending on the location. Generally an inverse relationship is observed during the previous growth season.

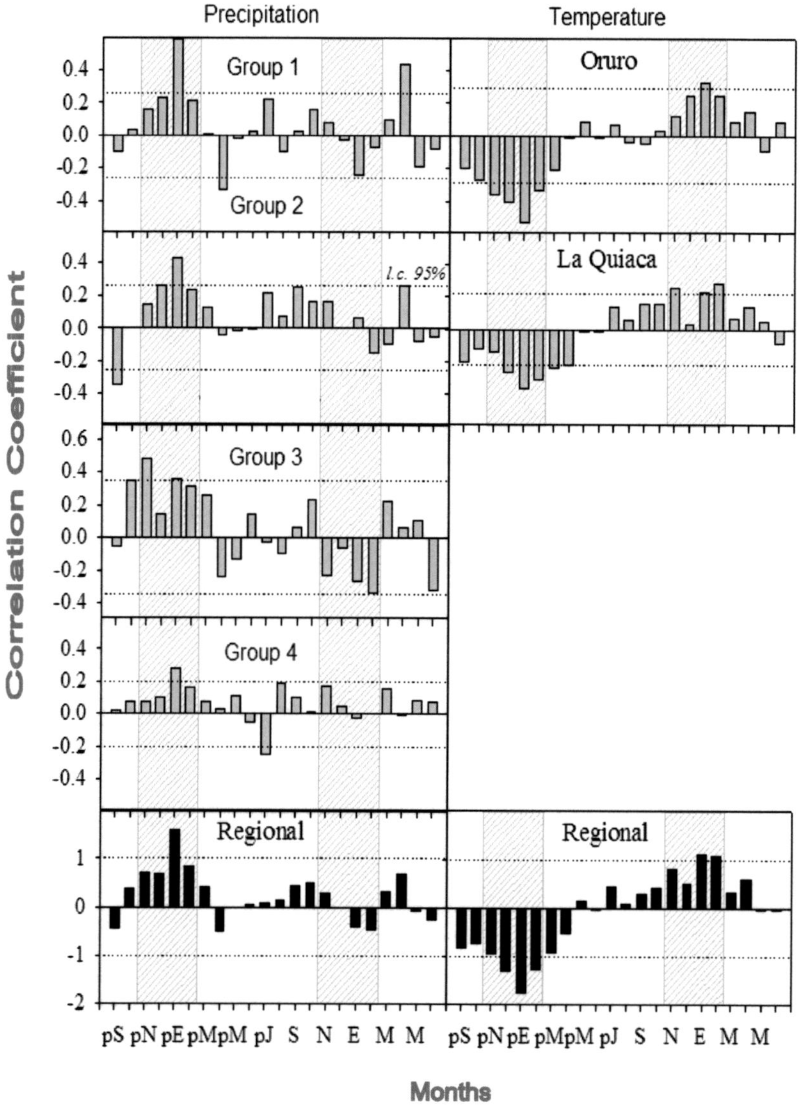

Figure 6. Correlation functions between precipitation and tree ring variability of *Polylepis tarapacana*.

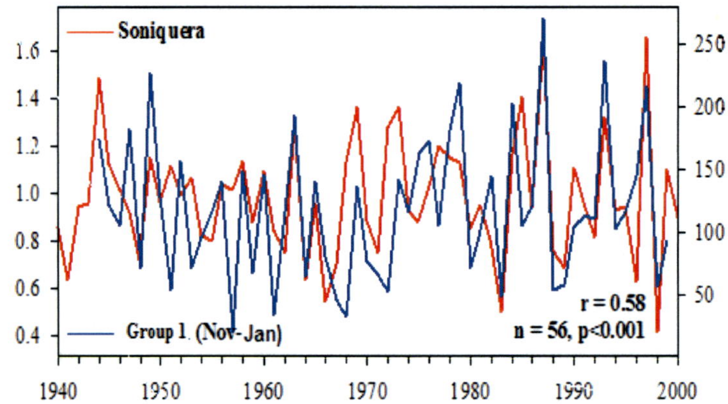

Figure 7. Relationship between winter precipitation and tree ring index in Soniquera.

3-Low Resolution Proxies

3.1. Moraine Landscapes

Of the different proxies available in the tropical Andes, moraine landscapes appear to be the most appropriate thanks to the number of study sites, and even if questions still remain, we have quite a good knowledge of the relationship between glacier fluctuations and climate.

Moraine landscapes testify to the former positions reached by the glacier and are the most obvious evidence of climatic changes in the tropical Andes. However, using a glacial moraine as an indicator of climatic change means the time-lag between mass balance fluctuations and the movements of the snout of the glacier has to be taken into account. Moreover, the glacial moraine landscape is by nature discontinuous, and more recent larger advances of a glacier destroy the moraines deposited during older, less-extensive advances, leaving an incomplete geomorphic record.

These geomorphic features can be used to determine glaciological parameters like Equilibrium Line Altitude (ELA), or mass balance, which are used to infer climate hypotheses that could explain the evolution of the glacier. However, the temporal resolution is lower than that obtained from ice core records or tree rings. It depends on the number of moraines preserved in the glacier foreland and the method of dating used. Uncertainties associated with dates are generally between one and five decades. Fortunately, in the Andes, the moraines are well preserved (figure 8) because of reduced activity of the periglacial processes and of a superficial frost wave (Francou et al., 2001), as well as the specific glacial evolution described in this paragraph. However, the number of climatic hypotheses inferred from glacier moraines for the last millennium is limited compared with the substantial number of works focused on the Last Glacial Maximum or the neo-glacial period (Clapperton, 1983; Seltzer, 1990; Schubert and Clapperton, 1990; Seltzer et al, 1995). The first evidence of climatic changes inferred from glacial moraines created during the last millennium in the tropical Andes was discovered in Peru, which concentrates 71% of the total surface area of tropical glaciers (Mercer and Palacios 1977; Clapperton, 1981; Röthlisberger, 1987; Rodbell, 1992). These investigations revealed higher precipitation amounts. The same climatic changes were considered to be responsible for glacier fluctuations in Bolivia (which contains 20% of tropical glaciers, Kaser et al., 1996) and in Ecuador (Hastenrath, 1981; Gouze et al., 1986; Seltzer, 1992).

Figure 8. The moraines are well preserved because of reduced activity of the periglacial processes and of a superficial frost wave in the Andes.

In recent years, progress in lichenometry (Cooley et al., 2006; Naveau et al., 2007; Jomelli et al., 2007a) has made it possible to improve the chronology of glacier fluctuations in the last century. As a result, published papers have improved and are usually based on a large number of glaciers. New investigations have been carried out in Peru (Solomina et al., 2007; Jomelli et al., 2007b) and in Ecuador (Jomelli et al., 2007c). Using this new approach in lichenometry, Rabatel (2005) published a chronology of glacier fluctuations over recent centuries in Bolivia (Rabatel et al., 2005, 2006, 2007). Recent studies based on lake sediment records combined with estimations of ELA from moraine records enabled palaeoclimatic conditions inferred from glacier fluctuations (Polissar et al., 2006) to be extended.

In addition, substantial work has been undertaken to link climatic change to glacier fluctuations. First, researchers identified the significant climatic parameters that control the behaviour of the glaciers, i.e. they established a transfer function between current climate and glacier variations. Second, they used modelling to infer paleoclimatic conditions from past glacier fluctuations. These two aspects are presented in the followings paragraphs.

Before using glacier fluctuations estimated from glacial landforms as a paleoclimatic proxy, i.e. to reconstruct past climate condition, the role of the climatic parameters that control the ablation and accumulation processes of tropical glaciers must be understood.

Due to a long ablation period (Kuhn, 1980), the vertical budget gradient is steep in the ablation area of tropical glaciers, around 2 m water equivalent (w.e.) per year and per 100 m (figure 9). For example, the net balance close to the tongue of Zongo glacier located in Cordillera Real (Bolivia) is around –7 m w.e. per year, whereas in the accumulation area, the balance is around +1 m w.e. per year. On this glacier, the altitude of the equilibrium line is around 5250 m asl. when the balance is equilibrated (e.g. 1999-2000). The ablation was very high in 1997-98 due to a marked El Niño event (Wagnon et al., 2001). In the last ten years, the mass balances of the Zongo Glacier were generally negative varying between -2 to +1 m w.e. per year.

Climate controls the mass balance of glaciers through the energy and mass fluxes at the ice or snow interface. Consequently, to analyze the causes of changes in mass balance, energy fluxes have to be measured, but measurements are scarce on tropical glaciers and are also generally short-term (e.g., Platt, 1966; Hastenrath, 1978; Hardy et al., 1998; Mölg et al., 2003; Wagnon et al., 2001; Favier et al., 2004; Sicart et al., 2005).

The energy balance equation is an application of the law of conservation of energy on the vertical components (or the components perpendicular to the surface if the surface is tilted) of the energy fluxes at the glacier surface:

$$R + H + LE + P = \Delta Q_M + \Delta Q_S \tag{1}$$

where R is the net radiation, H and LE are the turbulent fluxes of sensible and latent heat, respectively, and P is the heat advected by precipitation. The term ΔQ_S represents the change in heat content in a control volume and ΔQ_M is the energy used for melting (positive) or freezing (negative). The fluxes are expressed in W m^{-2} and are counted as positive if they provide energy to the control volume. P remains very low on the tropical glaciers and is neglected.

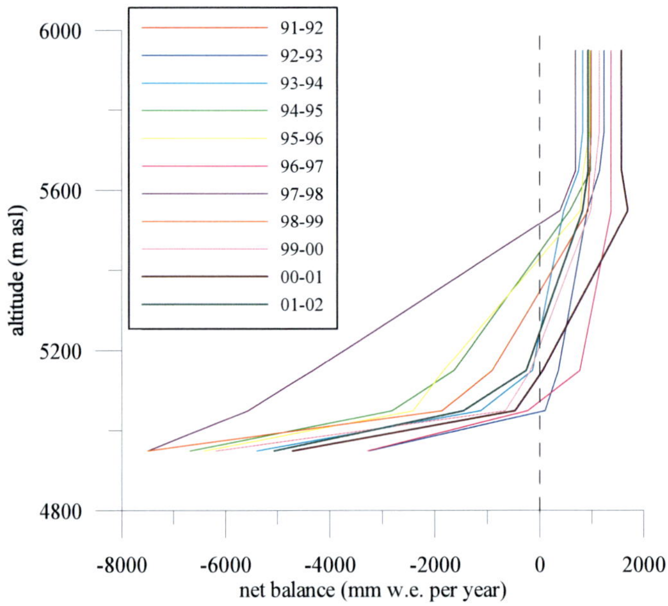

Figure 9. Annual mass balance of Zongo glacier (Cordillera Real, Bolivia) according to elevation from 1991 to 2002. The net balance close to the tongue is around –7 m w.e. per year, whereas in the accumulation area, the balance is around +1 m w.e. per year. The altitude of the equilibrium line is around 5250 m asl when the balance is equilibrated (e.g. 1999-2000).

The net radiation can be written as:

$$R = S{\downarrow} - S{\uparrow} + L{\downarrow} - L{\uparrow} = S{\downarrow}(1-\alpha) + L{\downarrow}\varepsilon - \varepsilon\sigma T_s^4 \tag{2}$$

where $S{\downarrow}$ is the short-wave irradiance, $S{\uparrow}$ is the reflected short-wave radiation ($S = S{\downarrow} - S{\uparrow}$ is the net short-wave radiation), $L{\downarrow}$ and $L{\uparrow}$ are the long-wave irradiance and emittance, respectively ($L = L\downarrow - L\uparrow$ is the net long-wave radiation). α is the surface albedo, ε is the long-wave emissivity of ice ($\varepsilon \approx 1$), $\sigma = 5.67\ 10^{-8}$ W m^{-2} K^{-4} is the Stefan-Boltzmann constant, and T_s is the surface temperature.

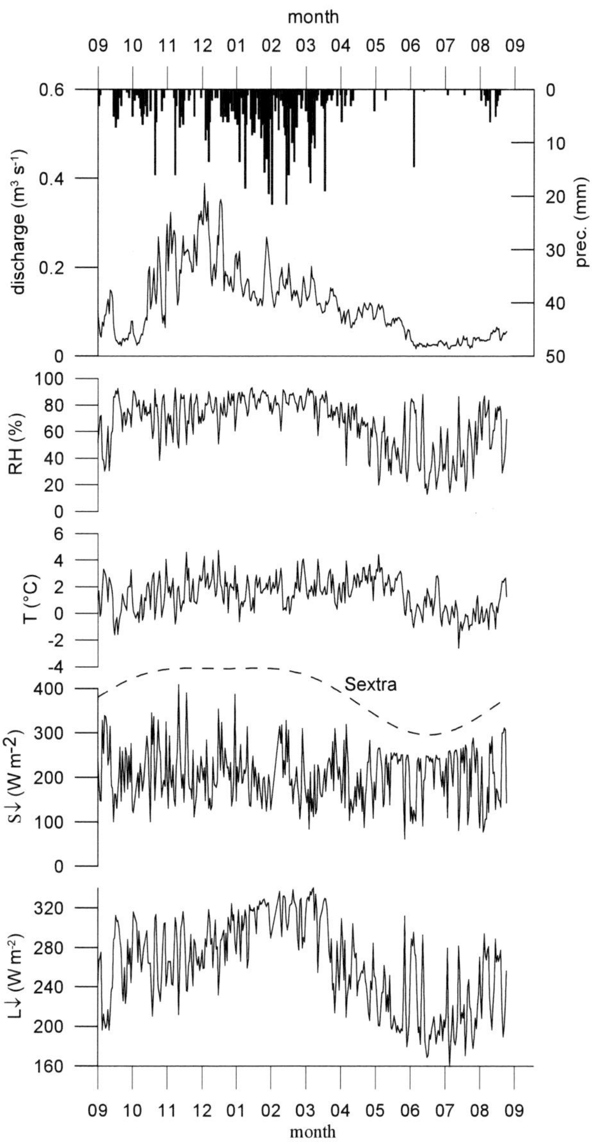

Figure 10. From the upper to the lower panel; daily values of proglacial discharge, precipitation, air temperature (*T*) and relative humidity (*RH*), recorded outside the glacier, and of incoming short-wave ($S\downarrow$) and long-wave ($L\downarrow$) radiation recorded at the surface of glacier at 5050 m asl from September 1, 1999 to August 31, 2000. The dashed line shows the theoretical extraterrestrial solar irradiance (S_{extra}).

Figure 10 shows changes in precipitation, melting discharge, air relative humidity and temperature, and shortwave (solar) and longwave incident radiation on and around the Zongo Glacier (Cordillere Real, Bolivia), which is located in the outer tropics (Kaser, 1999) and which, during the hydrological year 1999-2000, was characterized by marked seasonality of precipitation and cloud cover with a single wet season in the austral summer and a pronounced dry season in winter. The proglacial discharge is maximal in the wet season (austral summer) and minimum in the dry season (winter). At the outlet of the basin, around 85% of the precipitation and 70% of the discharge occur between October and March (Sicart et al., 2003). The annual precipitation is roughly 900 mm w.e. A gradual build-up to the wet season from September to November is observed in the Andes from Bolivia to Peru (Schwerdtfeger, 1976) and can be considered as a transition period. The wet season, which is characterized by frequent precipitations, lasts roughly from December to April. In April, the change to the dry season is rather sudden. A few snowfall events may occur but the dry season is generally a period of moderate ablation until the end of August. The annual cycle of air humidity clearly reflects the alternation of the wet and dry seasons (figure 10). In contrast, thermal seasonality is low -less than 10°C using daily averages- due to the low latitude.

Extraterrestrial solar radiation is maximal from November to February, but exhibits the small seasonal variations that are typical for low latitudes (figure 10). If short-term variations (mainly due to clouds) are disregarded, solar irradiance at the glacier surface is fairly constant throughout the year; the high extraterrestrial irradiance during the austral summer is attenuated by frequent clouds (wet season). In the wet season, long-wave irradiance is enhanced by the warm humid atmosphere and frequent clouds (figure 10). In the dry season, the long-wave irradiance is very low due to the thin (high altitude), dry (generally cloudless) and cold atmosphere.

Figure 11 shows monthly variations in the energy fluxes at 5060 m asl on Zongo Glacier during the hydrological year 1999-2000. Most of the year, the turbulent sensible (H) and latent heat fluxes (LE) tended to cancel each other out, so that the sum of the turbulent fluxes remained small, except in the dry season when $H+LE$ was a significant loss of energy due to strong wind and dry air. The net short-wave radiation varied with the cloud cover and the albedo, without following any marked annual cycle. In contrast, net long-wave radiation varied from very negative values in the dry season, due to a cold and dry atmosphere, to values close to nil in the wet season, due to cloud emissions.

Melting energy mainly depends on the balance between solar energy input and long-wave radiation output. The sum of the atmospheric energy fluxes at 5060 m asl was positive in the wet season (September–March), and was slightly negative in the dry season (April–August). Energy loss by turbulent fluxes was significant only in the dry season when the glacier wind prevailed for most of the day. The energy deficit in long-wave radiation was larger and reduced melting every time the atmosphere was cloudless and solar radiation high in the dry season, but also when the energy inputs were high at the beginning of the hydrological year (figures 10 and 11). In the core of the wet season, from January to March or April, cloud emissions offset most of the glacier emission, so that melting continued despite the high albedo of fresh snow.

Related to the cloud cover, snowfalls reduce the melting energy through a rapid increase in the surface albedo. Due to high solar radiation and low surface albedo, the period of substantial melting from September to December was interrupted by the frequent snowfalls that characterize the core of the wet season (figure 10). The arrival of frequent clouds and snowfalls around January is a key period for snow accumulation, but also for the ablation of the tropical glaciers.

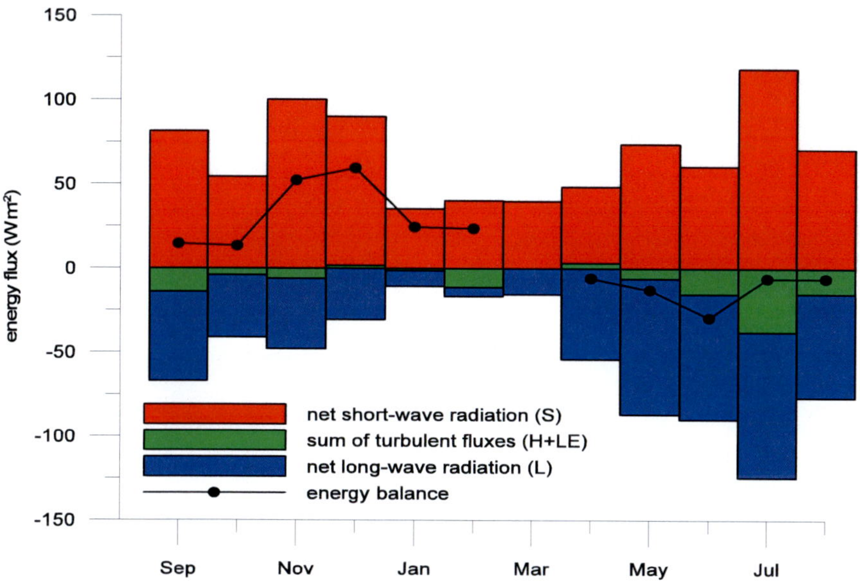

Figure 11. Monthly averages of energetic fluxes at 5060 m asl for the hydrological year 1999–2000. No turbulent data are available for March due to sensor breakdown.

The mean atmospheric temperature and humidity are not clearly linked to the mass balance of tropical glaciers, which are surrounded by a very dry and thin atmosphere. The El Niño Southern Oscillation (ENSO) warm phase prevailed during the last two decades of the 20th century, leading to frequent El Niño events, but several authors reported a likely shift to the cold phase regime in the late 1990s (Trenberth and Hoar, 1996; Fedorov and Philander, 2000). During an El Niño event, the wet season is delayed in Bolivia, and as a consequence, the period of strong melting by solar radiation around the summer solstice lasts longer (figure 10), and the mass balance of glaciers is highly negative. El Niño events also cause high ablation on Equatorial glaciers, probably due to high altitude liquid precipitations (Favier et al., 2004).

In summary, high-altitude tropical glaciers are characterized by the marked seasonality of net long-wave radiation because cloud emissions drastically enhance the low long-wave emittance of the thin atmosphere. The deficit in long-wave radiation is maximal during clear-sky days, counteracting maximal energy inputs. Directly linked to cloud cover and humidity - the main seasonal factors that control low latitude climates- long-wave radiation is a key variable of the energy balance of tropical glaciers.

On tropical glaciers, ablation is closely linked to accumulation: the high albedo of fresh snow in the wet season interrupts the period of high solar energy. Any delay in the wet season, such as during an El Niño event, causes a very negative mass balance due to low accumulation and high ablation.

Based on this knowledge, researchers have inferred palaeoclimatic conditions from glacier fluctuations. Many models have been developed, ranging from black boxes, such as empirical relationships (Oerlemans, 1997; Vincent et al., 2005) or purely statistical approaches (e.g. neural networks), to physically based models (Gerbaux et al., 2005). All these methods are based on the assumption of the stationarity principle (e.g. Kull and Grosjean, 2000), which supposes that modern processes were also active in the past. These different modeling approaches can be separated into two main groups: one in which the mass balance of the glacier is computed, and the other in which only a characteristic of the mass balance (e.g. the ELA) is required which is then linked to climatic conditions. Both types of modeling are used in the tropical Andes.

The mass balance modeling approach uses mass balance data derived from volume variations between different moraine stages, assuming steady state conditions for each glacial stage. The assumption of steady state is generally used to reconstruct the paleoclimate and glacier characteristics from one-time moraine deposits and supposes that the glacier mass balance (B_n) is in equilibrium ($B_n = 0$), and that the computed mass balance is in accordance with dynamical aspects (Kull and Grosjean, 2000). However, due to the lack of data, the study of ablation with a physical model is currently impossible in the tropics and semi-empirical or empirical models are consequently used. Semi-empirical relationships (e.g. Kuhn, 1989; Kaser, 2001) are based on a simplification of the physical equations to reduce the number of input variables, whereas empirical approaches are generally based on degree-day models (Kull and Grosejan, 2000; Hostetler and Clark, 2002). However, these simple models should include changes in temperature and precipitation (e.g. Kaser and Osmaston, 2002; Benn et al., 2005; Stansell et al., 2007). Thus, taking into account conclusions derived from figures 10 and 11, quantitative paleo-temperature records estimated from glacier fluctuations (e.g. Oerlemans, 2005) may be crude if an independent record of precipitation at the time of glaciations is not available (e.g. Harrison, 2005; Benn and Ballantyne, 2005).

The other modeling approach based on palaeo-ELA variations over recent centuries is also regularly used. Climatic information is then inferred from ELA altitude through simple models. Several simple methods are commonly used, such as the accumulation area ratio (AAR) and the area altitude balance ratio (AABR) which are described in detail in Benn et al. (2005). Most of these methods (AAR, AABR, THAR) consider that the glacier mass balance characteristics can be summarized in simple coefficients. The approaches assume that these coefficients are stationary over time, however, this is questionable. Indeed, important changes may have occurred due to variations in the climatic regime, debris cover, or glacier hypsometry. Moreover, these methods do not take into account the effect of specific local accumulation or ablation (e.g. avalanches from valley sides or debris cover). Where good topographic maps and air photograph coverage are available, the AABR method is the most rigorous (Benn et al., 2005).

The models developed to link ELA values to temperature and precipitation values are particularly sensitive to ELA error estimations. For instance, according to the equation of Greene et al. (2002), an uncertainty of 100 meters in the ELA corresponds to a change of 196 mm y^{-1} in precipitation and a change of 0.64°C in temperature or any combination of these parameters. However, according to Kaser's model (2001), the same uncertainty in ELA variations corresponds to a much bigger difference in precipitation (1250 mm y^{-1}). As a consequence, the choice of a model is fundamental and can strongly influence the final result.

3.2. Pollen Records

Material available to enable identification of environmental changes during the LIA in the Andes through pollen analysis is rather scarce. Pollen grains used for paleoenvironmental reconstruction are deposited in lakes or peat bogs and preserved for thousands of years. The sediment core is sampled and radiocarbon dated before proceeding to the extraction of the pollens. During this first phase, the presence of hiatus in sedimentation and the sampling resolution of the core can already be discussed. A hiatus in sedimentation is characterized by a « jump » between two radiocarbon dates e.g. between 25 000 and 17 000 yr B.P. throughout the South American lowlands (Ledru, et al., 1998). To be able to detect such an event, one needs to obtain regular dating throughout the core. Strong erosive conditions are inferred to explain the lack of deposition and when such a gap is detected, the pollen analyst has no data to define the missing time period and its related environment. In the Bolivian Andes, sedimentologists showed that erosive phases prevented sediment deposition at 15 000 years B.P., 7000 years B.P. and in the last 1000 years. The absence of deposition in the last 1000 years due to the installation of a strong seasonal precipitation regime, is the main explanation for how difficult it is to find appropriate sedimentological material covering the LIA in the Andes. (Servant, et al., 1987; Servant and Fontes, 1984). The sampling resolution represents the number of years covered by one sample. In the early stages of palynology, the sampling resolution was often 20 cm per sample which mostly corresponded to a 1000-year time-interval (Ledru and Mourguiart, 2001). Since Quaternarists are aware that the mechanisms that enhance abrupt and rapid environmental changes need to be studied in detail, core sampling resolution has greatly improved. Today most analyses use 1 cm sampling, or when possible, 0.25 cm. Depending on the location of the record and on the sedimentation rate, one sample can represent a time interval of between 2 years (mostly at high elevations), to 50 years, (at low elevations). Ice records show a year or even a seasonal resolution for each sample. To be statistically representative of the regional and local landscape, the total number a minimum of 400 pollen grains per sample need to be counted. This is easy to achieve in peat bogs or organic lakes where a high concentration of pollen grains is deposited but ice cores show a maximum of 100 grains per litre. As a consequence, such a large quantity of ice would be required, that this type of analysis remains limited.

To realistically interpret the data, calibration is required. Relations between pollen, vegetation and climate are based on surface samples collected under a vegetation cover (quantified or not) and related to local climatic data (Kuentz et al., 2007; Reese and Liu, 2005a; Rull, 2006). A single taxon may have several different ecological implications depending on where it grows. For instance Cyperaceae may be associated with a dry or a wet environment. At Cuzco (3355m) arid phases are detected through the presence of Cyperaceae and interpreted as a lowering of the lake level which was replaced by marshy vegetation. In the Mérida Andes range (Venezuela) (4080m) Cyperaceae are found at high elevations, in inundated areas associated with isolated peat bogs within a dry superparamo environment. In one case, an increase in evaporation was inferred and in the other, a decrease. Only the definition of a rigorous regional transfer function would allow the researcher to distinguish between the two interpretations and identify the regional significance of the detected pollen signal. However this transfer function is only available for the immediate vicinity of the study area and cannot be expanded to a general interpretation of the data. The transfer function in the Venezuelan Andes resulted in a relation between the elevation of an ecosystem and the associated precipitation and temperature (Rull, 2006). On the Altiplano of Bolivia, calibration was based on the ratio between Poaceae and Asteraceae pollen frequencies (Reese and Liu, 2005a; Reese and Liu, 2005b). A decrease in Poaceae frequencies was associated with a dry environment while an increase in Asteraceae frequencies was interpreted as a wet environment. Changes in precipitation are easily detected, as the tropical ecosystems react rapidly to changes in moisture rates such as the length of the dry season, whereas changes in temperature are more difficult to characterize in the tropics. A transfer function links the pollen abundance identified with a given altitude. This function is then applied on quaternary pollen records for quantitative reconstructions of vertical displacements and consequently paleotemperature change (Rull, 2006). In Venezuela such a function gave an accuracy of 256 m of vertical displacement, which is equivalent to ~1.5°C. This value falls within the estimation of a 2°C decrease in temperature for the LIA but temperature shifts of the order of 1°C remain undetectable. In any case, such a procedure is only possible when a suitable network of climatic data is available and this is still too rare in the Tropics. Another problem is the fact that today no pollen data set has been used to calibrate traditional agriculture crops. Consequently, in historically densely populated regions such as Cuzco in Peru, it may be difficult to distinguish between climatic and human impacts on the landscape (Chepstow-Lusty and Winfield, 2000).

Comparisons of different pollen records should take all this information into account before a regional synthesis is established. Today, only three pollen records have a high enough resolution to reconstruct LIA environmental changes based on pollen analysis, although the calibration of the ecological data differs with the analyst.

4. Climatic Changes Over the Last Millennium Inferred from the Different Proxies

The analysis of the climate of the last millennium in the tropical Andes based on the different proxies reveals moisture variations over several centuries with a " W " shape made up of three dry periods, one between AD 1000 and 1270, another between 1360 and 1600, and another between 1730 and 1880, which were interrupted by two wet periods, the first between 1270 and 1360, and the second between 1600 and 1730. These distinct phases were identified by analysing the synchronous behaviour of the different proxies discussed above in different regions of the Altiplano.

Although the tropical climate is characterized by slight variations in temperature on an interannual scale, various climate reconstructions revealed that changes in this parameter during the last millennium could be rather close to those observed in the middle latitudes during the same period. In the Cordillera de Merida (10°N) for instance, climatic changes during the last millennium are said to have been characterized by a decrease of 3.2 ± 1.4°C in temperature and by an increase of 20% in precipitation (Polissar et al., 2006). In the southern part of the tropical Andes, climatic reconstructions based on glacial moraines suggest variations in temperatures of about 1 to 2°C combined with variations in precipitation of more than 20% (Rabatel et al., 2005). However, these values can change appreciably depending on the proxy, and this aspect is discussed in the followed paragraphs. The different studies also demonstrated that, as for middle and high latitudes, the climate of the tropical Andes was not stable, and abrupt climatic variations occurred at the decennial or centennial time scale.

1000-1270 AD - a dry period

The only proxy data that make it possible to document the climate at the beginning of the last millennium are lake-sediment records in Venezuela and isotopic variations in ice core records in Peru and Bolivia. However the chronology of most ice core records remains inaccurate for the first part of the last millennium. Between 1000 and 1200 AD, isotopic variations in the ice core of Quelccaya show relatively high values suggesting that rather dry conditions close to those observed during the 20th century prevailed at 13°S at that time. Here we refrain from attempting a quantitative interpretation of the water isotope record because of the considerable uncertainties concerning which region the isotope signal is really representative of, and the exact sensitivity of the isotopes in terms of temperature and precipitation. However, it should be noted that high (more enriched) isotope values are consistent with dry/warm conditions and low (more depleted) isotope levels correspond to wet/cold conditions. In fact, the variation between these two modes (wet/cold vs. dry/warm) seems to be a comparably robust feature of the tropical South American climate. However even over the relatively short period of the last millennium, we cannot exclude different modes of variability (like warm and wet) which are never or only very rarely represented in modern observations. In summary, our preliminary interpretation of the slightly enriched isotope signal during the late Medieval up to the year 1200, agrees with the evidence cited above of a warm/dry climate.

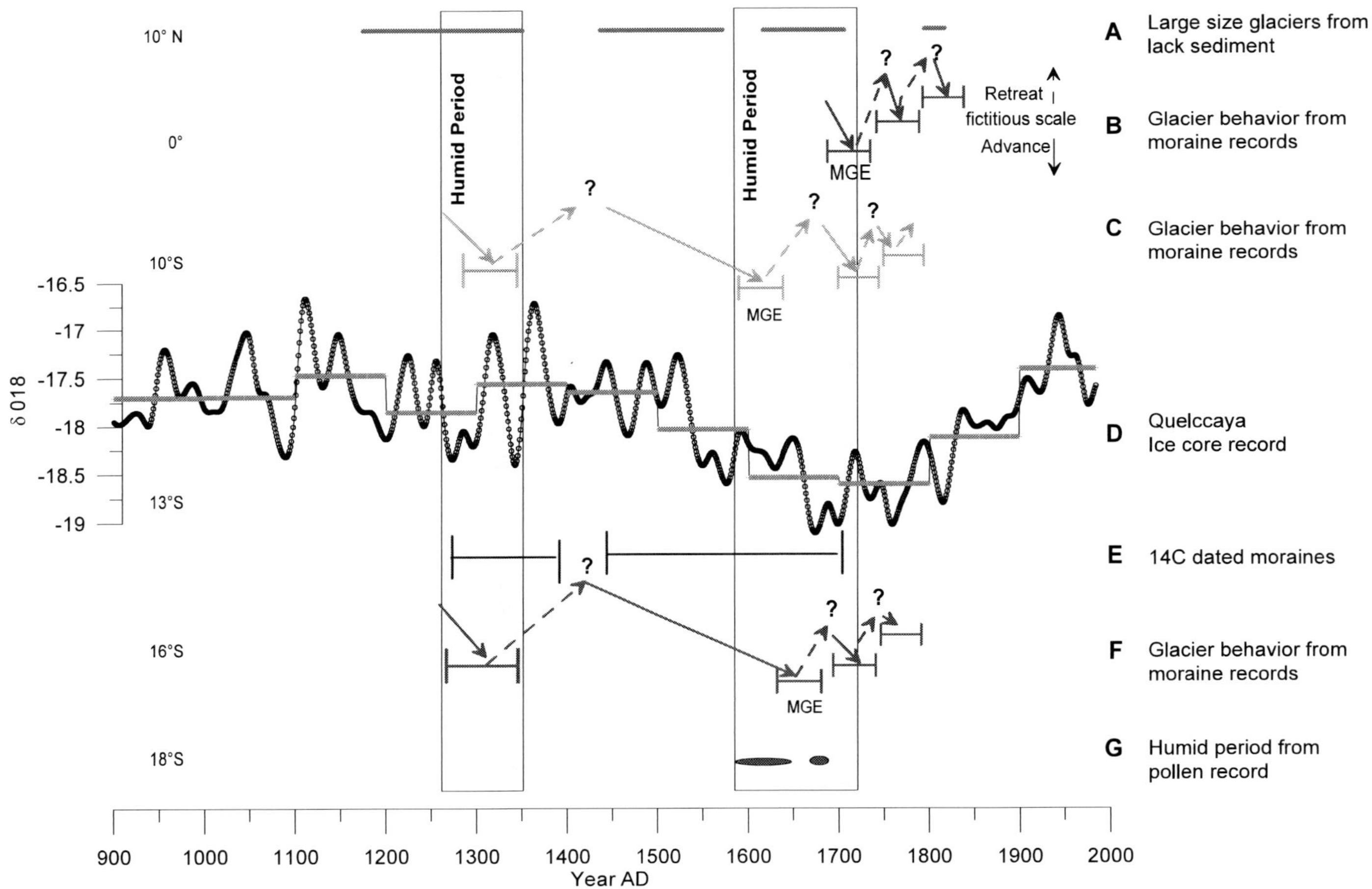

Figure 12. Climatic changes from the different climate proxies.

In addition, the lack of dated moraines for this time period, suggesting that -in these areas- the snout of the glaciers was located at higher altitudes than during the later centuries, is in agreement with the interpretation of the isotopic signal of Quelccaya. At this period, low concentrations of fine-grained magnetic minerals are present in sediment cores and can be identified visually by colour changes and quantified by magnetic susceptibility measurements suggesting glaciers in Venezuela were small due to dry climate conditions in the Cordillera de Merida. Palynological data support this trend with low concentrations of *Cyperaceae* pollen, whereas pollen from this species is abundant in wet environments.

1270-1360 AD - a humid period

Between approximately 1270 and 1360 AD, several proxies show the return of wetter and colder conditions. In Venezuela, data from lake-sediment cores suggest increased glacierization of the catchment with the highest value reached around 1300 AD combined with an increase in plant developed in humid conditions (increase of 25% in *Cyperaceae*) from around 1180 until around 1350 AD. ELA depression suggests a decrease in temperature as well, which unfortunately cannot be accurately quantified from the paper of Polissar et al., (2006). Further to the south, the geomorphological observations revealed an early glacial advance in Peru and in Bolivia that was more or less synchronous with enhanced glacierization in Venezuela. Indeed lichen measurements in the Cordillera Blanca (Peru) enabled Jomelli et al., (2007b) to date remnants of moraines to around 1330±29 suggesting the end of a glacial advance around 1300 AD. The elevation of the remnants of moraines corresponding to this glacial advance indicates an ELA of about 250-300 m lower than today, suggesting cooler and wetter conditions than before. However the small number of moraine deposits analyzed and their poor conservation do not allow a palaeoenvironmental condition to be inferred. The rather good agreement between these dates and those obtained in other cordilleras should be mentioned. In the Cordillera Vilcanota (Peru), Mercer and Palacios (1977) dated glacial moraines around 630±65 years BP, i.e. 1290–1400 AD (radiocarbon date). In the Cordillera Real (Bolivia), if the lichen size-age relationship is extrapolated, the moraines at the base of the snout of a few glaciers could correspond to this 14th century glacial stage. However, as is true for the Cordillera Blanca (Peru), this 14th century glacial stage is absent in most valley glaciers, suggesting that in most cases, younger glacial advances extended further than the 14th century glacial stage. Glacierization may have occurred in Ecuador but its extent was necessarily shorter than advances in more recent centuries and moraines deposited during this glacial stage were destroyed by a more recent glacial advance.

This hypothesis of a wet and cold period being responsible for the glacial advance in Peru and in Venezuela is less clear in isotopic variations of the Quelccaya ice core (13°S). The water isotopes only show a very slight variation in the sense of a wet/cold climate. Even if values recorded between 1200 and 1300 AD correspond to the lowest value for the first half part of the last millennium and suggest an increase in moisture, changes are rather limited compared to the previous period. Nevertheless, the relatively short duration of this phase (around 100 years) and the altogether higher values than those observed during second half part of the last millennium support the geomorphological observations revealing a glacial advance of reduced amplitude.

1360 and 1600 AD - a dry period

Between 1360 and 1600 glaciers on most countries were probably shorter than previously as suggested by the lack of dated moraines between 1350 and 1590 in Cordillera Blanca (Peru), in Cordillera Real (Bolivia) and in Ecuador. In addition, isotopic variations reach the same values as those observed at the beginning of the last millennium which also suggests drier conditions. In Venezuela according to the concentration of fine grained minerals, the size of the glaciers was reduced until 1450. Pollen records show a reduced concentration of *Cyperaceae* attesting to drier conditions than before.

1600-1730 AD - a humid period interrupted by dry events

From the end of the 16th century onward, climatic conditions seem to have changed again with an increase in moisture which lasted until the middle of the 18th century. The increase in the number of proxies in the second half part of the millennium makes it possible to give a better description of the environmental changes that occurred. Pollen records in Venezuela and the glacial advance in Cordillera Villcanota suggest that the humid period already started around 1450. Nevertheless considering the other proxies and the limited accuracy of radiocarbon dating in Villcanota, the date around 1580 is preferred. Indeed pollen records in Bolivia (Altipano) attest to a wet period characterized by a high *Cyperaceae* and Poaceae/Asteraceae ratio. Nevertheless, a drier event is recorded between 1660 and 1680 AD as indicated by the decrease in the Poaceae/Asteraceaea ratio (synchronously with a glacial retreat in Cordillera Real). Pollen data do not make possible to draw conclusions about the temperature values or the location and the size of the Altiplano, far from the upper Andean

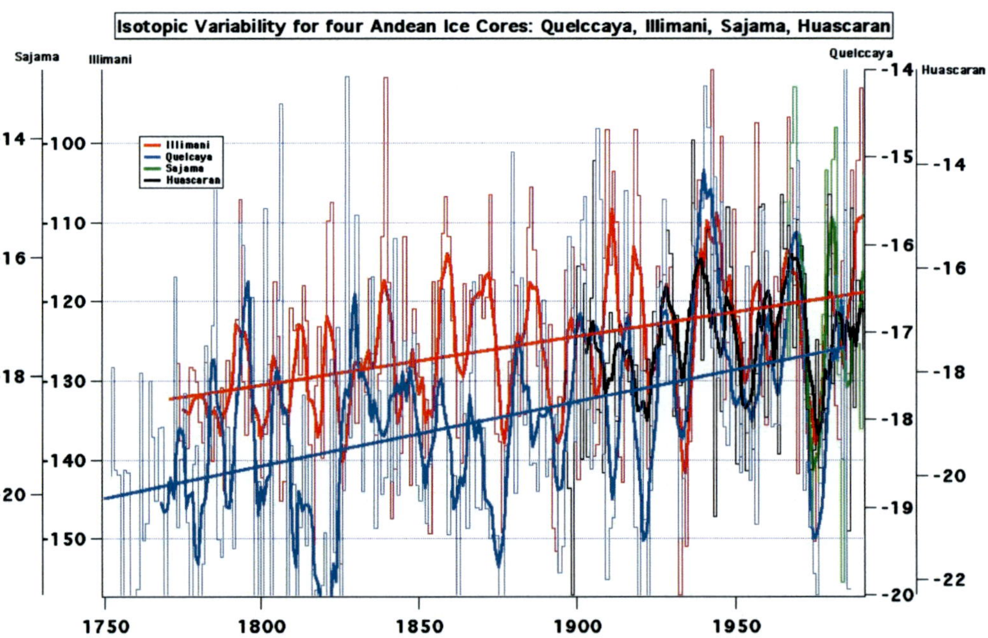

Figure 13. Isotopic records from Huascaran (d18O), Quelccaya(d18O), Illimani (dD) and Sajama(d18O).

forest limit. This humid period is in agreement with considerably reduced values in the Quelccaya ice core.

Logically a glacial advance occurred during this period. In Venezuela, Polissar et al., (2006) deduced two glacial advances from variations in the concentrations of fine grained minerals. The first one occurred between 1450 and 1590 AD and the second one between 1640 and 1730 AD. In Ecuador, lichens measured on high altitude volcanoes (Chimborazo for instance) revealed that glaciers located at the base of the highest summits reached their furthest down-valley extent of the last millennium (defined in this paper as the Maximum Glacial Extent - MGE) at the beginning of the 18th century (1730±16), in agreement with analysis of historical documents (Hastenrath, 1981).

In the Cordillera Blanca (Peru), Röthlisberger (1987) dated a buried soil horizon at 440±185 years BP (i.e. 1300–1660 AD) (Ocshapalca Glacier) and suggested that the advance occurred some time later. A more systematic analysis of moraines revealed that the glaciers reached their maximum extent around 1630±27 (Jomelli et al., 2007b; Solomina et al., 2007) in this cordillera. From then on, the glaciers stated to decrease until a second but minor glacial advance around 1730. At 15°S in the Cordillera Vilcanota (southern Peru), Mercer and Palacios (1977) dated glacial moraines around 270±80 years BP (1447-1697 AD). Finally, in Bolivia, the glacial pattern was analogous. In the Cordillera Real and Cordillera Quimsa Cruz (Bolivia), the glacial advance recorded on 13 glaciers using lichenometry ended around 1657±20 or around 1686±20 depending on the glaciers (Rabatel, 2005; Rabatel et al., 2005, 2006, 2007). After the MGE, glaciers started to retreat until a minor glacial advance occurred that ended in the 1730s.

Climatic conditions estimated from glacier fluctuations indicate a cooler and wetter period than today. Indeed, in order to explain the formation of the MGE moraines in the cordilleras in Peru, Bolivia and Ecuador, the specific mass balance in the accumulation area had to be highly positive to cause ice transfer towards the base of the glaciers. Current physical ablation processes suggest that stronger solid precipitations due to wetter conditions than today could explain the glacial advance. Between 1650 and 1730 in Bolivia for instance, modelling suggests temperatures were about 0.5°C colder than today and precipitation was 25% higher than today (Rabatel et al., 2006). In addition, colder temperatures may have contributed to the LIA advance phase. In Ecuador, modelling suggests that if the mean temperature was 0.5°C to 1.5°C lower at the beginning of the 18th century than today, precipitation at the summit was about 45 to 85% higher than today. If precipitation was the same as today, the ELA altitude would require a difference in temperature of 3.8°C (Jomelli et al., 2007c). In Venezuela, Polissar et al., (2006) concluded that a decrease of 3.2 ± 1.4°C in temperature and an increase of 20% in precipitation would be required to produce the glacial responses observed. In Bolivia, Hastenrath and Ames' (1995) mass-balance sensitivity analysis and Kaser's model (2001) suggest that to increase mass balance and induce the pronounced glacier advance, precipitation and cloudiness would have had to increase by 20-30% and 1-2/10 respectively compared to present conditions. Independently, temperature would have had to decrease by 1.1-1.2°C (Rabatel, 2005; Rabatel et al., 2006, 2007).

1730-1880 AD - a dry period

From 1730 until around 1880 AD, other climatic conditions appear to have prevailed. Pollen records in both Venezuela and Bolivia indicate drier conditions. In Venezuela, the percentage of Cyperaceae decreases abruptly from 1730 to around 1850, i.e. the same values as those observed between 1000 and 1200 AD. Unfortunately palynologic results do not make it possible to associate these drier conditions with a decrease in temperature in this area. In the Sajama region (Bolivia), the period between 1700 and 1880 AD is characterized by the expansion of xerophytic shrubs to replace the puna grasses, together with a sharp decrease in the Poaceae-Asteraceae ratio. This environmental change is attributed to a dry period when the regional vegetation was desert-like shrublands.

Moraine records also attest to progressively drier conditions. In the different cordilleras, glaciers began to recede after 1740 (Jomelli et al., 2007b, c; Rabatel et al., 2005). The glacier retreat was moderate but continuous until about 1870 except for synchronous minor glacial advances like those in 1760 and 1820 recorded in the different countries. Moreover, glacier modeling revealed that the retreat of the glaciers after 1740 was mainly a consequence of continuously drier conditions. In Bolivia, using Kaser's model and Hastenrath and Ames' mass-balance sensitivity analysis, Rabatel et al., (2005), calculated that such dry conditions resulted in a decrease of about 20% in current mean accumulation rates on glaciers.

In Ecuador, modeling revealed that precipitation must have been less than in the 18th century (about 35% less) resulting in a decrease in the ELA value (Jomelli et al., 2007c). However, ELA modeling revealed that a decrease in temperature must also have occurred, with a short very cold period (about -1.4°C lasting for one or two decades) just before 1830.

In the 18th century, dating of different ice cores became more and more precise and the overlapping of several records allows a more robust analysis. Figure 13 shows four high-altitude Andean ice core records that were used in a previous publication (Hoffmann, et al., 2003) to define a stack record (the Andean Isotope Index, AII) for the 20th century. Obviously the high coherency between these records argues for their large-scale climatic significance. Though the two isotope records, the Quelccaya and the Illimani, which together cover the 19th and 18th century, apparently diverge more and more with time and are both still dominated by primarily decadal variability. Within the dating uncertainties, in particular of the Illimani record, it would be an easy exercise to line up the different decadal scale variations in both records. In addition to similar variability, Illimani and Quelccaya isotope records are also similar in their overall centennial trends: from a LIA isotopic minimum (i.e. presumably relatively wet/cold conditions) Figure 13 shows a nearly identical long-term trend to more enriched values in both isotope records over the 19th century. In the 20th century, all records appear to level off pointing to the long-term influence of the outgoing LIA, while changes in the tropical hydrological cycle due to the recent greenhouse gas induced warming trend are not yet clearly visible.

From 1880 onward

A general glacial recession throughout the 20th century is observed in the Andes (e.g., Hastenrath and Ames, 1995; Kaser, 1999; Francou et al., 2000; Ramirez et al., 2001), and even in other tropical regions in Africa (e.g., Mölg et al., 2003). There was a marked general glacial retreat in the Cordillera Blanca and Cordillera Real at the end of the 19th century (data not shown) but this retreat slowed down in the first half of the 20th century. Another significant retreat took place in the 1930s-1940s only interrupted by an advance in the 1920s (Kaser and Georges, 1997; Georges, 2004). Finally, the glacier retreat accelerated during the second half of the 20th century compared to the LIA period. Though the first phase of glacier retreat was in fact accompanied by a corresponding water isotope signal, the second retreat phase is hardly discernible in the Andean isotope records. In addition, direct meteorological observations from the middle of the 20th century indicate that changes in precipitation amount or cloud cover lasting recent decades are minor in most regions (Vuille et al., 2003b). The only exceptions are southern Peru and western Bolivia where there is a general tendency toward slightly drier conditions. On the other hand, near-surface temperature has increased significantly throughout most of the tropical Andes (Vuille et al., 2003b). The increase in temperature varies considerably in the eastern and western Andean slopes with a much larger increase in the west. Gaffen et al. (2000) reported a cooling of the tropical low troposphere after 1979. Satellite images revealed a strengthening of the tropical general circulation in the 1990s, leading to an increase (decrease) in cloud cover at the Equator (in the sub-tropical latitudes) (Chen et al., 2002; Wielicki et al., 2002).

Conclusion

The comparison of the different proxies in the tropical Andes revealed a rather synchronous climatic "W" pattern from 10°N to 18°S for the last millennium at a centennial time scale. Two humid periods around 1180-1360 and 1600-1730 AD were separated by three drier and probably warmer periods. The dates of the start and the end of these humid periods can differ within a few decades depending on the region and the proxy. As in the tropical Andes, a W-shape was initially proposed to describe glacier fluctuations during the last millennium in Europe. However, multiproxy climate reconstruction from high altitudes revealed a much more complex pattern (Jones et al., 2001). Further work based on high-resolution proxies may reveal the same complexity in the tropical Andes. However, for the time being, the correspondence between the qualitatively very different climate proxies presented here certainly supports the description of the climate. Though this correspondence holds for most of the record, it should be noted that the continuous glacier retreat in the 19th century is apparently not accompanied by a corresponding isotope trend. This points to certain mechanisms in the physics of the corresponding tracer which might level off or become less sensitive to recent Pacific warming and convective activity. In addition, although this qualitative documentation of climatic change is interesting, it needs to be improved by a more quantitative interpretation of these proxies. Moreover, none of the proxies analysed in this paper is able to document a change in seasonality of precipitation. This aspect is important for a better understanding of climate during the last millennium in this region that is strongly influenced by the ENSO phenomenon, which is known to modify the climate at the seasonal time scale.

References

Aceituno, P. (1988). On the functioning of the southern oscillation in the South American Sector: Part I. Surface climate. *Monthly Weather Review* **116**, 505– 524.

Argollo, J. Soliz, C. Villalba, R. (2004). Potencialidad dendrocronológica de Polylepis tarapacana en los andes de Bolivia. *Ecologia de Bolivia*, 4-21.

Auer, I., (2007). HISTALP - historical instrumental climatological surface time series of the Greater Alpine Region. *International Journal of Climatology*, **27**, 17-46.

Benn, D.I., Owen, L.A., Osmaston, H.A., Seltzer, G.O., Porter, S.C., Mark, B. (2005). Reconstruction of equilibrium-line altitudes for tropical and subtropical glaciers. *Quaternary International*, **138**, 8-21.

Benn, D.I., Ballantyne, C.K. (2005). Palaeoclimatic reconstruction from Loch Lomond Readvance glaciers in the West Drumochter Hills, Scotland. *Journal of Quaternary Science* **20**, 577-592.

Binford, M, Kolata, A. L., Brenner, M., J. Janusek,W., Seddon, M.T., Abbot, M., Curtis, J.H. (1997). Climate variation and the rise and fall of an Andean civilisation, *Quaternary Research* **47**, 235-248.

Böhm, R., Auer, I., Brunetti, M., Maugeri, M., Nanni, and Schöner, W. (2001). Regional temperature variability in the European Alps: 1760-1998 from homogenized instrumental time series. *International Journal of Climatology* **21**, 1779-1801.

Bradley, R., R.S. Vuille, M. Hardy, D., L.G. Thomson (2003). Low latitude ice core record Pacific sea surface temperatures, *Geophysical Research Letters*, **30**, 10.1029/2002GL016549.

Broecker WS, Denton GH. (1989). The role of ocean–atmosphere reorganizations in glacial cycles. *Geochimica et Cosmochimica Acta* **53**, 2465–2501.

Chepstow-Lusty, A., Frogley, M. R., Bauer, B. S., Bush, M. B., and Herrera, A. T. (2003). A late Holocene record of arid events from the Cuzco region, Peru. *Journal of Quaternary Science* **18**, 491-502.

Chepstow-Lusty, A., and M. Winfield (2000), Inca agroforestry: Lessons from the past, *Ambio*, **29**, 322-328.

Clapperton, C.M. (1981). Quaternary glaciations in the Cordillera Blanca, Peru and the Cordillera Real, Bolivia, *Revista CIAF (Bogota)* **6**, 93-111.

Clapperton, C.M. (1983). The glaciation of the Andes. *Quaternary Science Review*, **2,** 85-155.

Cooley, D., Naveau, P., Jomelli, V. Rabatel, A., Grancher, D. (2006). A bayesian Hierarchical Extreme Value Model for Lichenometry. *Environmetrics* **17**, 555-574.

Craig, H. (1961), Standard for reporting concentrations of deuterium and oxygen-18 in natural water, *Science*, **133**, 1833-1834.

Dansgaard, W. (1964). Stable isotopes in precipitation, *Tellus*, **16**, 436-468.

Favier, V., P. Wagnon, J. P. Chazarin, L. Maisincho, and A. Coudrain (2004). One-year measurements of surface heat budget on the ablation zone of Antizana glacier 15, Ecuadorian Andes, *Journal of Geophysical Research*, **109**, doi: 10.1029/2003JD004359.

Francou, B., Ramírez, E., Cáceres, B., Mendoza, J. (2000). Glacier Evolution in the Tropical Andes During the Last Decades of the 20th Century: Chacaltaya, Bolivia, and Antizana, Ecuador. *Ambio*, **29**, 416-422.

Francou, B., Le Mehauté, N., Jomelli, V. (2001). Factors controlling spacing distances of sorted stripes in a low-latitude, Alpine Environment (Cordillera Real, Bolivia, 16°S). *Permafrost and Periglacial Processes*, **12**, 367-377.

Fedorov, A. V., and Philander S. G. (2000). Is El Nino Changing? *Science*, 1997-2002.

Fjedsa, J. and Kessler. M. (1996). Conserving the biological diversity of *Polylepis* woodlands of the Highland of Peru and Bolivia. NORDECO. Copenhagen, Dinamarca.

Fritts, H.C. (1976). Tree rings and climate. Academic Press, New York. 567 pp.

Garreaud, R. Vuille, M., Clement. A. (2003). The climate of the Altiplano: observed current conditions and mechanisms of past changes. *Paleo. Paleo. Paleo.* **194**, 5-22.

Gerbaux, M., Genthon, C., Etchevers, P., Vincent, C., Dedieu, J.P. (2005). Surface mass balance of glaciers in the French Alps: distributed modeling and sensitivity to climate change. *Journal of Glaciology*, vol. 51(175), 561-572.

Georges, C. (2004). The 20th century glacier fluctuations in the tropical Cordillera Blanca (Peru). *Arctic Antarctic and Alpine Research*, **36**, 100-107.

Gouze, P., Argollo, J., Saliège, J.F. Servant, M. (1986). Interprétation paléoclimatique des oscillations des glaciers au cours des 20 derniers millénaires dans les régions tropicales, exemple des Andes boliviennes. *Comptes rendus de l'Académie des Sciences* **303**, 219-224.

Greene, A.M., Seager, R., Broecker, W.S. (2002). Tropical snowline depression at the last glacial maximum: comparison with proxy records using a single-cell tropical climate model. *Journal of Geophysical Research* **107** (ACL 4-1 to 17.).

Grootes, P. M., Stuiver, M. Thompson L. C. and Mosley-Thompson. E. (1989). Oxygen isotope changes in tropical ice, Quelccaya, Peru, *Journal of Geophysical Research*, **94**, 1187--1194.

Grove, J. (2004). *Little Ice Ages ancient and modern.* Routledge.

Hardy, J. P., R. E. Davis, R. Jordan, W. Ni, and C. E. Woodcock (1998). Snow ablation modelling in a mature aspen stand of the boreal forest, *Hydrological Processes*, 1763-1778.

Harrison, S.P. (2005). Snowlines at the last glacial maximum and tropical cooling. *Quaternary International* 138-139, 5-7.

Hastenrath, S. (1978). Heat-budget measurements on the Quelccaya Ice Cap, Peruvian Andes, *Journal of Glaciology*, 85-97.

Hastenrath, S. (1981). The Glaciation of the Equadorian Andes. Rotterdam: Balkema (160 p).

Hastenrath, S., Ames. A. (1995). Recession of Yanamarey Glacier in Cordillera Blanca, Peru, during the 20th century. *J. Glaciol.*, **41**, 191-196.

Hoffmann, G. (2003). Taking the Pulse of the tropical water cycle, *Science*, **301**, 776-777.

Hoffmann, G. Ramirez, E. Taupin J.D. Francou, B. Ribstein, P. Delmas, R. Dürr, H. Gallaire, R. Simoes, J. Schotterer, U. Stievenard, M. and Werner, M. (2003). Coherent isotope history of Andean ice cores over the last century. *Geophys. Res. Lett.* **30**, 10.1029/2002GL014870, 2178-2181.

Hostetler, S. W., Clark, P.U. (2000). Inferred from Glacier Mass-Balance Modeling Tropical Climate at the Last Glacial Maximum, *Science*, **290**, 1747-1750.

IPCC (Ed.) (2007), Climate Change 2007 - The Physical Science Basis.

Jacoby G.C. (1989). Overview of tree-ring analysis in tropical regions. *IAWA Bulletin* **10**: 99-100.

Jomelli, V., Grancher, D., Naveau, P., Cooley, D. (2007a). Assessment study of lichenometric methods for dating surface. *Geomorphology*, **86**, 131-143.

Jomelli, V., Grancher, D., Brunstein, D. Solomina, O. (2007b). Recalibration of the Rhizocarpon growth curve in Cordillera Blanca (Peru) and LIA chronology implication. Geomorphology (in press).

Jomelli, V., Grancher, D., Brunstein, D., Favier, V, Francou, B. (2007c). Little Ice Age glacial extension in Ecuador from lichenometry and historical observations. *Quaternary Research (under review).*

Jones, P. D., Osborn, T. J., and Briffa, K. R. (2001). The evolution of climate over the last Millenium. *Science* **292**, 662-667.

Kaser, G., Georges. C., Ames A. (1996). Modern glacier fluctuations in the Huascaran-Chopicalqui massif of the Cordillera Blanca, Peru. Zeitschrift für Glestsberkunde und *Glazialgeologie*, **32**, 91-99.

Kaser, G., Georges. C. (1997). Changes of the equilibrium line altitude in the tropical Cordillera Blanca, Peru, 1930-50, and their spatial variations. *Ann. Glaciol*., **24**, 344-348.

Kaser, G. (1999). A review of the modern fluctuations of tropical glaciers, *Global Planetary Change*, 93-103.

Kaser, G. (2001). Glacier-climate interaction at low latitudes. *Journal of Glaciology*, **47**, 195-204.

Kaser, G., Osmaston, H.A. (2002). *Tropical Glaciers*, Cambridge University Press, New York.

Kessler, M. 1995. The genus Polylepis (Rosaceae) in Bolivia. Candollea 50. Conservatore et Jardin Botaniques de Geneve. 172 pp.

Kaser, G., Juen, I., Georges, C., Gomez, J., Tamayo, W. (2003). The impact of glaciers on the runoff and the reconstruction of mass balance history from hydrological data in the tropical Cordillera Blanca, Perú. *Journal of Hydrology*, **282**, 130-144.

Kessler, M., Schmidt-Lebuhn, A. (2006). Taxonomical and distributional notes on Polylepis (Rosaceae). Organisms, *Diversity and Evolution* **6**: 67–69.

Kuentz, A., Galan de Mera, A., Ledru, M.-P., and Thouret, J. C. (2007). Phytogeographical data and modern pollen rain of the puna belt in southern Peru (Nevado Coropuna, Western Cordillera). *Journal of Biogeography*, in press.

Kuhn, M. (1980). Climate and glaciers. Sea Level, Ice and Climatic Change (Proceedings of the Canberra Symposium, Dec. 1979), *IAHS*, 3-20.

Kuhn, M. (1989). The response of the equilibrium line altitude to climate fluctuations. Theroy and observations. In Oerlemans, J. eds. *Glacier fluctuations and climatic change*. Kluwer, Dordrecht, 407-417.

Kull, C. and Grosjean, M. (2000). Late Pleistocene climate conditions in the north Chilean Andes drawn from a climate-glacier model, *Journal of Glaciology*, **46** (155), 622-632.

Ledru, M.-P., Bertaux, J., Sifeddine, A., and Suguio, K. (1998). Absence of Last Glacial Maximum records in lowland tropical forests. *Quaternary Research*, **49**, 233-237.

Ledru, M.-P., and Mourguiart, P. (2001). Late Glacial vegetation records in the Americas and climatic implications. In "Interhemispheric Climate Linkages." (V. Markgraf, Ed.), pp. 371-390. Academic Press, San Diego.

Le Roy Ladurie, E., (1986). *Histoire du climat depuis l'an mil*. Flammarion, Paris, 853 pp.

Mark, B. G., Seltzer, G. O. (2003). Tropical glacial meltwater contribution to stream discharge: a case study in the Cordillera Blanca, Perú. *Journal of Glaciology* **49** (165), 271-281.

Mercer, J.H. and Palacios, M.O. (1977). Radiocarbon dating of the last glaciation in Peru. *Geology*, **5**, 600-604.

McGregor, G. R., and Nieuwolt, S. (1998). "Tropical climatology." Wiley, New York.

Mölg, T., C. Georges, and Kaser, G. (2003). The contribution of increased incoming shortwave radiation to the retreat of the Rwenzori Glaciers, East Africa, during the 20th century, *International Journal of Climatology*, 291-303.

Naveau, P., Jomelli, V., Cooley, D., Grancher, D. Rabatel, A. (2007). Modeling uncertainties in lichenometry studies with an application: The Tropical Andes (Charquini Glacier in Bolivia). *Arctic, Antarctic and Alpine Research*,**39**, 277-285.

Oerlemans., J. (1997). A flow-line model for Nigardsbreen: projection of future glacier length based on dynamic calibration with the historic record. Annals of Glaciology 24, 382-389.

Oerlemans, J. (2005). Extracting a climate signal from 169 glacier records. *Science* **308**, 675-677; 10.1126/science.1107046.

Ortloff, C.R. Kolata, A.L. (1993). Climate and collapse: agro-ecological perspectives on the decline of the Tiwanaku State. *Journal of Archaeological Science*, **20**, 195-221.

Platt, C. M. (1966). Some observations of the climate of Lewis Glacier, Mount Kenya, during the rainy season, *Journal of Glaciology*, 267-287.

Polissar, P.J. Abbott, M.B., Wolfe, A.P., Bezada, M., Rull, V., Bradley, R.S. (2006). Solar modulation of Little Ice Age climate in the tropical Andes. *PNAS*, vol. 103, 8937–8942.

Pouyaud, B., Zapata, M., Yerren, J., Gomez, J., Rosas, G., Suarez, W., Ribstein, P. (2005) Avenir des ressources en eau glaciaire de la Cordillère Blanche. *Hydrol. Sci. J.* **50**, 999–1022

Rabatel, A., Jomelli, V., Francou, B., Naveau, P., Grancher, D. (2005). Dating Little Ice Age in the Tropics from the moraines of Charquini glaciers (Andes of Bolivia, 16°S). *CRAS, Géosciences*, **337**, 1311-1322.

Rabatel, A., Machaca, A., Francou, B., Jomelli, V. (2006). Glacier recession on Cerro Charquini (Bolivia, 16°S) since the maximum of the Little Ice Age (17th Century). *Journal of Glaciology*, vol. 52, 110-118.

Rabatel, A., Francou, B., Jomelli, V., Naveau, P., Grancher, D. (2007). A chronology of the Little Ice Age in the tropical Andes of Bolivia (16°S) based on moraine dating by lichenometry: implications for climate reconstruction. *Quaternary Research*, (under review).

Rada, F., Gacia-Nunez, C., Borero, C., Gallardo, M., Hidal, M., Gonzales, J., Prado, M., Liberman-Cruz, M. and Azucar, A. (2001). Low temperature resistance in Polylepis tarapacana, a tree growing at the highest altitudes in the world. *Plant Cell, and Environment,* **24**, 377-381.

Ramirez, E., Francou, B., Ribstein, P., Descloîtres, M., Guérin, R., Mendoza, J., Gallaire, R., Pouyaud, B., Jordan, E. (2001). Small glaciers disappearing in the tropical Andes. A case study in Bolivia: Glaciar Chacaltaya (16°S), *Journal of Glaciology*, **47**, 187-194.

Ramirez, E., Hoffmann, G. Taupin, J.D. Francou, B. Ribstein, P Caillon, P. Ferron, F.A.Landais, A.Petit, J.R. Pouyaud, B. Schotterer, U. Simoes, J.C. Stievenard, M. (2003). A new Andean deep ice core from the Illimani (6350 m), Bolivia, *EPSL*, **6674**, 1-14.

Reese, C. A., and Liu, K.-B. (2005a). A modern pollen rain study from the central Andes region of South America. *Journal of Biogeography* **32**, 709-718.

Reese, C. A., and Liu, K.-b. (2005b). Interannual variability in pollen dispersal and deposition on the Tropical Quelccaya ice cap. *Professional Geographer* **57**, 185-197.

Rodbell, D.T., (1992). Lichenometric and radiocarbon dating of Holocene glaciation, Cordillera Blanca, Peru. *The Holocene*, **2**, 19-29.

Rothlisberger, F. (1987). 10000 Jahre Gletschergeschichte der Erde. Aarau, Verlag Sauerlander (348 p.).

Rull, V. (2006). A high mountain pollen-altitude calibration set for palaeoclimatic use in the tropical Andes. *The Holocene* **16**, 105-117.

Schmidt-Lebuhn, A., Kessler, M. and Kumar, M. (2006). Promiscuity in the Andes: Species Relationships in Polylepis (Rosaceae, Sanguisorbeae) Based on AFLP and Morphology. *Systematic Botany*, **31**, 547–559

Schubert, C., Claperton, C., (1990). Quaternary glaciations in the Northern Andes (Venezuela, Colombia and Ecuador. *Quaternary Science Reviews*, **9,** 123-135.

Schwerdtfeger, W. (1976). Climates of Central and South America, Elsevier Scientific, Amsterdam, 532 p.

Seltzer, G.O. (1990). Glacial history and palaeoclimate of the Peruvian-Bolivian Andes. *Quaternary Sciences Review*, **9**, 137-152.

Seltzer, G.O. (1992). Late Quaternary glaciation of the Cordillera Real, Bolivia. *Journal of Quaternary Science*, **7**, 87-98.

Seltzer, G.O., Rodbell, D.T., Abbott, M.B. (1995). Andean glacial lakes and climate variability since the last glacial maximum. *Bulletin de l'Institut Français d'Etudes Andines*, **24**, 539-550.

Servant, M., Fontes, J. C. (1984). Les basses terrasses fluviatiles du quaternaire récent des Andes boliviennes: datation par le 14 C, interprétation paléoclimatique. Cahiers ORSTOM. *Série Géologie* **14**, 15-28.

Servant, M., Argollo, J., and Fournier, M. (1987). Dépôts fluviatiles en Bolivie : variations du régime des écoulements au cours du Quaternaire récent. *Géodynamique* **2**, 105-107.

Sicart, J. E., P. Ribstein, B. Francou, and Gallaire, R. (2003). Etude des precipitations et de la fonte sur un glacier tropical: le glacier du Zongo, Bolivie, 16°S, *Hydrological Sciences Journal*, 799-808.

Sicart, J. E., P. Wagnon, and Ribstein, P. (2005). Atmospheric controls of the heat balance of Zongo Glacier (16°S, Bolivia), *Journal of Geophysical Research*, **110**, doi:10.1029/2004JD005732.

Solomina, O., V. Jomelli, G. Kaser, A. Ames and B. Pouyaud. (2007). Little Ice Age moraines in the Cordillera Blanca: lichenometric data replication. *Global Planet. Change*, in press.

Stansell, N.D., Polissar, P.J., and Abbott, M.B. (2007). Last Glacial Maximum Equilibrium-Line Altitude and Paleo-Temperature Reconstructions for the Cordillera de Mérida, Venezuelan Andes. *Quaternary Research*, **67**, 115–127.

Stichler, W.,Shotterer, U., Frölich, K., Ginot, P., Kull, C., Gäggeler, H., and Pouyaud, B. (2000). Influence of sublimation on stable isotope records recovered from high glaciers in the tropical Andes, *J. Geophys. Res.*, **106**, 22,613-622,620.

Thompson, L.G., Mosley-Thompson, E., Dansgaard, W., Grootes, P.M. (1986). The Little Ice Age as Recorded in the Stratigraphy of the Tropical Quelccaya Ice Cap. *Science*, **234**: 361-364.

Thompson, L. G., Davis, M.E. Mosley-Thompson, E. Liu, K.-B. (1988). Pre-Incan agricultural activity recorded in dust layers in two tropical ice cores, *Nature*, **336**, 763-765.

Thompson, L. G., Davis, M.E. Mosley-Thompson, E. Sowers, T.A. Henderson, K.A. Zagorodnov, V.S. Lin, P.N. Mikhalenko, V.N. Campen, R.K. Bolzan, J.F. Cole-Dai, J.A. (1998). A 25,000-Year tropical climate history from Bolivian ice cores, *Science*, **282**, 1858-1864.

Thompson, L., Mosley-Thompson, E., Henderson, K., (2000). Ice-core palaeoclimate records in tropical South America since the Last Glacial Maximum, Journal of Quaternary *Science*, **15**, 377-394.

Thompson, L. G., and Davis, M. (2006). Ice Core Records from South America. In Encyclopedia of Quaternary Science, Elsevier *Science*, (E. S. Scott A. Elias, Ed.), 1233-1242.

Thompson, L. G., Mosley-Thompson, E., Brecher, H., Davis, M., Leon, B., Les_, D.,Lin, P-N., Mashiotta, T.,Mountain, K., (2006). Abrupt tropical climate change: Past and present. *PNAS*, **103**, 10536–10543.

Trenberth, K. E., and T. J. Hoar (1996). The 1990-1995 El Nino-Southern Oscillation event: Longest on record, *Geophysical Research Letters*, 57-60.

Tschumi, J., and Stauffer B. (2000). Reconstructing past atmospheric CO2 concentration based on ice-core analyses: open questions due to in situ production of CO2 in the ice, *Journal of Glaciology*, **46**, 45-53.

Vimeux, F. Gallaire, R. Bony, S. Hoffmann, G. and Chiang, J.C.H. (2005). What are the climate controls on δD in precipitation in the Zongo Valley (Bolivia)? Implications for the Illimani ice core interpretation, *EPSL*, **240**, 205-220.

Vincent, C., Le Meur, E., Six, D., Funk, M. (2005). Solving the paradox of the end of the Little Ice Age in the Alps, *Geophysical Research Letters*, **32**, L09706, doi: 10.1029/2005GL022552.

Vuille, M., and R. S. Bradley, 2000: Mean annual temperature trends and their vertical structure in the tropical Andes. *Geophys. Res. Lett.*, **27**, 3885-3888.

Vuille, M., Bradley, R. Werner, M. Healy, R.Keimig, F. (2003a). Modelling $\delta^{18}O$ in precipitation over the tropical Americas: 2. Simulation of the stable isotope signal in Andean ice cores, *J. Geophys. Res.*, **108**, 4175, doi:4110.1029/2001JD002039.

Vuille, M. Bradley, R.S. Werner, M., Keimig, F., (2003b). 20th Century climate change in the tropical Andes: observations and model results. *Climatic Change*, **59**, 75-99.

Vuille, M., and Werner, M. (2005). Stable isotopes in precipitation recording South American summer monsoon and ENSO variability: observations and model results, *Climate Dynamics*, **25**, 401-413.

Wagnon, P., P. Ribstein, B. Francou, and Sicart J. E. (2001). Anomalous heat and mass budget of Glaciar Zongo, Bolivia, during the 1997/98 El Nino year, *Journal of Glaciology*, 21-28.

Wagnon, P., Sicart, J.E Berthier E. and Chazarin J.P. (2003). Wintertime high-altitude surface energy balance of a Bolivian glacier, Illimani, 6340 m above sea level, *Journal of Geophysical Research-Atmospheres*, **108**, 14.

Wang X, Auler AS, Edwards RL, Cheng H, Cristalli PS, Smart PL, Richards DA, Shen C-C. (2004). Wet periods in northeastern Brazil over the past 210 kyr linked to distant climate anomalies. Na*t*ure **432**: 740–743.

Worbes, M. (1995). How to measure growth dynamics in tropical trees – A review. *IAWA Journal*, **16**: 337-351.

In: Climate Change Research Progress
Editor: Lawrence N. Peretz, pp.161-186
ISBN: 978-1-60021-998-6

Chapter 4

INORGANIC GEOCHEMISTRY IN PALEOCLIMATE: TOOLS AND TRENDS IN THE ANALYSIS OF PEAT AND LAKE SEDIMENTS

***Malin E. Kylander*[1] *and Joanne Muller*[2]**
[1]Department of Geology and Geochemistry, Stockholm University, Stockholm, Sweden
[2]Geology and Geophysics, Woods Hole Oceanographic Institution,
Woods Hole, MA, USA

Abstract

This chapter presents an overview of the application of inorganic geochemistry in peat and lake sediments as a tool for paleoclimate studies. A brief look at the important dynamics occurring in lakes and mires highlights the fact that we are examining small-scale processes to deduce large-scale changes. A summary of analytical techniques, including sample preparation and instrumentation are presented and important statistical and interpretive tools are explored. Specific examples from recent publications demonstrate the practical application of these tools and the paleoclimatic interpretations that can then be drawn.

Introduction

As public and scientific interest in climate change intensifies so does the need to develop new tools to better comprehend the Earth's climate system. One way to increase our understanding of present and future climate dynamics is to look back to times when anthropogenic activities were absent. This picture of the past may be constructed with environmental archives that vary from more classical media like ice (e.g., Barnola et al., 1987; Petit et al., 1999; Johnsen et al., 2001) and marine sediments (e.g., Hughen et al., 1996; Haug et al., 2001) to more unconventional media, such as emu eggshells (Johnson et al., 1999). The tools used to build and interpret these records are likewise diverse and include biological, physical and geochemical indices. A multi-proxy approach is desirable in climate reconstructions as the strength of any one index is greater when there is agreement between several types of proxies see Birks and Birks (2006) for a discussion.

Among terrestrial archives peat and lake sediments have been used extensively to decipher changes in paleoenvironments, with common focus on biological proxies such as diatoms, pollen, plant macrofossils, chironomids, charcoal, protozoa and other microfossils. Similarly, physical proxies like 'loss on ignition', accumulation rates, grain size distribution and magnetic susceptibility have been important. To date geochemical proxies remain largely confined to stable isotopes (H, O, C and N) and carbonate content analyses (Boyle, 2002; Charman, 2002). That is not to say that geochemical analyses have not been made on these archives, but that the focus has been on environmental pollution rather than on paleoclimate. As such, the studied time periods and analytical design are not tailored to divulge past climate signals. This is particularly true in peat studies where reconstruction of heavy metal deposition has been the primary focus.

The purpose of this contribution is to look at some of the analytical and interpretive tools we use in geochemistry to reconstruct paleoclimatic changes recorded in peat and lake sediments. Geochemical data can be divided into four major categories: major elements, trace elements, radiogenic isotopes and stable isotopes. To narrow this broad field our emphasis here is on inorganic major and trace elemental geochemistry, the discussion will center on analytical techniques and tools for interpretation (statistical and interpretive). Specific examples of techniques and tools applied to peat and lake deposits are also given. The aim of this chapter is not to provide an exhaustive review on geochemistry and paleoclimate, but rather to present an overview of the possible paths that one can take when working with peat and lake sediment archives, highlighting new and innovative approaches in this discipline.

Major and Trace Elements

The major elements make up the bulk of the Earth's crust and as such are expressed as a weight percent. These elements include: Si, Ti, Al, Fe, Mn, Mg, Ca, Na, K and P. Major elements are usually determined based on cation concentrations with the understanding that each is accompanied by an appropriate amount of oxygen. Conversely, trace elements are those elements found in crustal concentrations below 0.1% and are often expressed as parts per million (ppm) or parts per billion (ppb). The elements that fall into this group are numerous and include the transition metals (Sc to Zn, excluding Fe and Mn), the rare earth elements (REE), the platinum group elements (PGE) as well as Rb, Sr, Y, Zr, Pb, U and Th see Rollinson (1993) for detailed information.

Dynamics to Consider

Inorganic geochemistry, as applied to lake and peat deposits in a paleoclimatic context involves the measurement of small scale geochemical changes and properties which are then extrapolated to reflect large scale processes. A wide range of processes occur both internally and externally to the deposit, which must be considered when interpreting geochemical data (Figure 1).

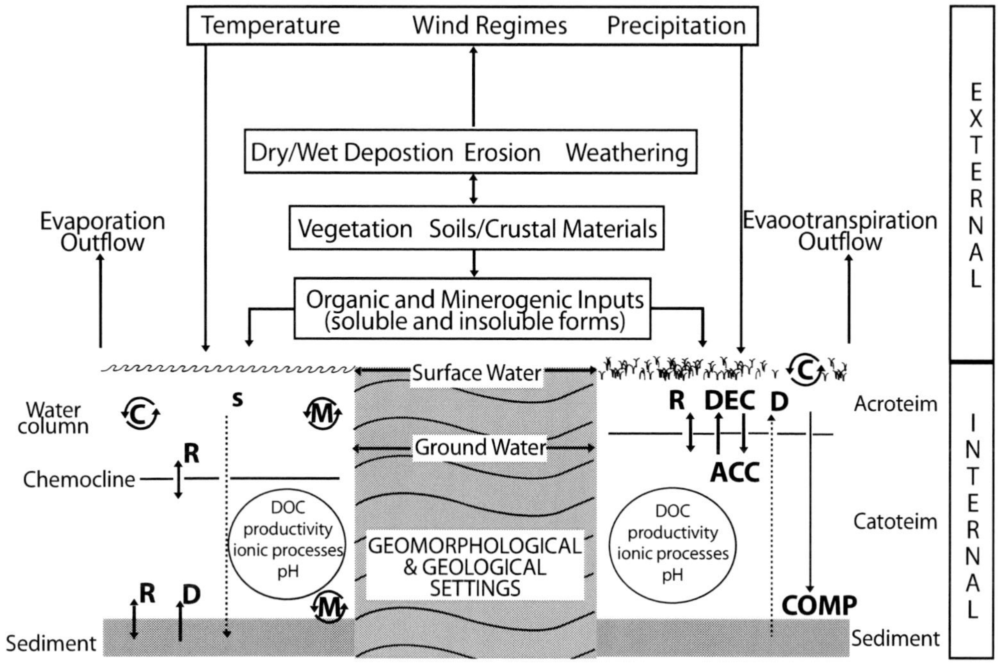

Figure 1. A simplified box diagram showing some of the major dynamics of lake and peat deposits. Physical, biological, geochemical and hydrological dynamics are important. External factors affect both the deposit and the processes occurring within. These processes include elemental cycling (within the water column and vegetation) (C), physical mixing (M), redox processes (R), diffusion (D), sedimentation (S), decomposition (DEC), accumulation (ACC) and compression (COMP) of organic material.

Elements archived in lake sediments must be viewed in light of the dynamics within the water column and the interaction between the water column and the sediment. Elements may enter a lake environment in a variety of forms. Once present in the lake, soluble species may not be captured by particles within the water column before they are deposited and buried. In deep lakes elements may be held for longer periods of time within the water column and thus short-term changes may not be recorded due to a lag in signal capture. Surface enrichments in trace elements can also occur, most likely the result of trace element cycles within the water column. After burial a number of processes may affect elemental concentrations such as sediment mixing (e.g., Gerhardt and Schink, 2005; Engstrom et al., 2006), diffusion of trace elements from the sediment back to the water column (e.g., Carignan and Tessier, 1985; Tessier et al., 1989) and redox processes related to the mobilization of Fe and Mn oxides. Changes in climate can affect lake processes such as biological activity and the depth of water mixing (Johnson et al., 2002); the redox conditions in the sediments and within the water column (Brown et al., 2000); and the concentration of elements in the solution (Moreno et al., 2007). For a more in-depth discussion of these issues the reader is directed to Boyle (2002).

Peat deposits are also subject to complex internal processes but unlike lakes these systems are often acidic and organic rich (>70%). The peat bog is anoxic with the exception of the uppermost acrotelm where the water table fluctuates and oxic conditions dominate. Water table fluctuations themselves can differ in ombrotrophic (hydrologically isolated) and minerotrophic (receives both ground and surface water inputs) peatlands. Under such extreme conditions it is important to examine post depositional remobilization of elements by biological incorporation, redox/precipitation and dissolution processes considering how these processes are influenced by climate shifts e.g., variations in dissolved organic carbon (DOC) (Freeman et al., 2001), pH (Clark et al., 2005) or accumulation and decomposition (Page et al., 2004). The latter is important for organically bound elements such as mercury and the halogens (Biester et al., 2003). As much as 50% of the original peat mass is lost in the acrotelm until later burial in deeper layers and as much as 90% is lost over a period of several thousands years (Kuhry and Vitt, 1996; Malmer and Wallen, 1999); this undoubtedly affects organically bound elements. The microtopography of the peat bog is also important, hummocks and lawn environments are preferred sampling sites over hollows since hollows can influence small scale deposition patterns (Bindler et al., 2004).

Given the highly variable dynamics of lake and peatland systems a good understanding of the small-scale dynamics within the lake or mire is paramount. This can be achieved via information on the trophic status and stratigraphy and physical/chemical data such as LOI, grain size distribution (which plays a role in elemental concentrations), pH, eH and dissolved oxygen concentration (DOC). It is also important to sample from several sites within the deposit to avoid over interpretation of small scale variations that may be found in single core analyses (Rowan et al., 1995; Rowan et al., 1995; Bindler et al., 2004).

Analytical Techniques

It is imperative in any geochemical study to acquire trustworthy data where one can quote the accuracy (difference against a standard reference material or using different instrumentation) and precision (i.e., the repeatability of the measurements) of all measurements, in addition one must assess for contamination (i.e., quantify blanks). Laboratory conditions are important in this respect, most crucially when samples with low trace element concentrations are analysed. Pioneering steps in this area came with the first ice and snow cores taken from Greenland by Patterson and his research team in 1965. Their results suggested that previous work suffered contamination from sampling apparatus, ambient dust and/or in-laboratory conditions. New standards for laboratory work have since been established for trace metal analysis. 'Clean laboratories' have HEPA-filtered air, ultra-pure chemicals are used in sample preparation, implements are cleaned via acid bathing, and workers wear particular clean room footwear and clothing (Vallelonga et al., 2002).

Sample Preparation

Depending on the study objectives and target elements, different sample preparation and analytical steps are required. Sample treatment can be beneficial as it reduces some of the interferences (e.g., matrix effects) that occur during analysis. It should not however contaminate the sample, result in elemental loss, or concentrate bulk elements, thereby increasing background signals during instrumental steps. Sample preparation can be varied to target total elemental concentrations, acid-extractable elements or involve a series of sequential extractions (Walsh et al., 1997). Sequential extractions have been criticized in the past due to difficulties associated with the selectivity of the extractants for a given phase and trace metal redistribution during the extraction procedure (e.g., Tipping et al., 1985; Kheboian and Bauer, 1987; Martin et al., 1987; Belzile et al., 1989; Nirel and Morel, 1990). However, the information derived from sequential extractions is still valuable, and in the absence of any other technique provides a means to assess trace metal partitioning in geological materials (e.g. Tessier et al., 1979; Tessier and Campbell, 1991; Teutsch et al., 2001; Emmanue and Erel, 2002).

Analytical Instrumentation

Analytical instrumentation is constantly being refined to improve on achievable detection limits, precision and accuracy. Non-destructive methods for analysis of solids include instrumental neutron activation analysis (INAA), X-ray fluorescence (XRF), energy dispersive X-ray spectroscopy (EDS) and Microprobe. INAA can achieve low detection limits through the measurement of gamma emissions from a sample that has been irradiated by neutrons from a nuclear reactor. XRF, EDS and Microprobe are all X-ray based methods suitable for the analysis of a wide range of elements. These methods differ in the resolution of the stimulating beam (XRF analyses bulk material vs. EDS and Microprobe which can measure individual minerals) and type of beam used (XRF uses X-rays while the other two methods use electron beams to generate X-rays) (Williams, 1987; Lachance, 1994; Gill, 1997).

The instrumentation used in the analysis of solutions includes atomic emission spectrometry (AES), atomic absorption spectrometry (AAS) and inductively coupled plasma-mass spectrometry (ICP-MS). AES is based on the thermal excitation of atoms, which results in the emission of narrow radiation lines of characteristic wavelength within the visible spectrum. AES is sometimes coupled to ICP sources where the plasma (a highly inonized gas) is used to generate radiation emission (Walsh, 1997). AAS measures concentrations of metals based on the absorption of light by gas-phase atoms in either a flame (F-AAS) or a graphite furnace (GF-AAS)(Gill, 1997). ICP-MS pairs the ICP ionization source with a mass analyser which separates ions based on their mass to charge ratio; these ions are then counted by a single ion collector (De Laeter, 2001; Herbert and Johnstone, 2002).

For most elements ICP-MS offers the best accuracy and detection limits on a wide array of elements and uses minimal sample amounts. It is however more expensive than F-AAS methods and requires more sample preparation than XRF methods. In general XRF and ICP-AES are suitable for the analysis of major elements while INAA, ICP-AES, ICP-MS and F-AAS are used for trace element analysis.

New Techniques for Climate Reconstructions Based on Lake Sediments

Trends and events can be overlooked if sampling resolution is too low, yet one wants to maximise the number and quality of analyses. Recent developments in instrumental arrangement can reduce these problems, as analyses are high-resolution and non-destructive, leaving material for further analyses. As such, major and trace element changes can be identified and this information used to plan sub-sampling if further analyses are required. In

Laser Ablation (LA)-ICP-MS a laser is used to vaporize the sample surface and material is carried via a gas stream to the plasma for ionization. As the laser beam scans the surface of the sample continuous measurements are taken, providing intensity data that can be calibrated to provide concentrations (for reviews of LA-ICP-MS principles and applications see Durrant, 1992; Durrant, 1999; Durrant and Ward, 2005). This well-established technique has only recently been applied to the analysis of sediment cores for climate change work. Rauch et al. (2006) used LA-ICP-MS analysis on sediment cores, collected using freeze coring, to identify sub-annual changes in elemental concentrations. Detection limits of 0.09% for Al, 0.006% for Fe, 0.07% for S, 0.75% for Si, 0.70 $\mu g\ g^{-1}$ for As, 68 $\mu g\ g^{-1}$ for Cu, and 32 $\mu g\ g^{-1}$ for Zn allow for the analysis of both major and trace elements.

A new variation on classic XRF techniques offers a similar but complementary method to LA-ICP-MS. The Itrax XRF core scanner is equipped with flat beam optics, this allows for the acquisition of high-resolution elemental XRF data as well as two-dimensional radiographic images. The radiographic image, as well as an optical image may then be lined up with elemental profiles to identify changes in stratigraphy and material influx for e.g. Detection limits reported for the Itrax XRF core scanner are low at 2.2% for Al, 0.9% for Si, 150 $\mu g\ g^{-1}$ for F, 100 $\mu g\ g^{-1}$ for Ca, 60 $\mu g\ g^{-1}$ for Ti, 25 $\mu g\ g^{-1}$ for Mn and Fe and 5 $\mu g\ g^{-1}$ for Rb and Sr (Croudace et al., 2006).

The major difference between LA-ICP-MS and core scanning techniques is that the XRF integrates data from an 8 mm by >200 μm area as opposed to the laser which is a single point The single point varies in size according to which laser wavelength/system is used. As such the heterogeneity expressed during laser analysis is greater since it may reflect single grain composition. Further development and application of both techniques has the potential to identify climate change events down to sub-annual scales and to previously over looked abrupt shifts.

Tools for Interpretation

The nature of geochemical data is such that a large amount of information is generated during analysis and it can be difficult to decide what constitutes a natural variation and what can be considered 'noise' in a record. Geochemists have both statistical (e.g., average, variance, regression) and interpretive tools (e.g., enrichment factors, normalized element diagrams) at their disposal to help simplify the task. Some of the more commonly used tools are described and discussed below.

Statistical Tools

Correlation and Regression

A correlation is the strength of association between two variables (x, y) measured on a number of individuals. This relationship is quantified using the Pearson product-moment coefficient of linear correlation, which is more commonly termed the correlation co-efficient (r). This value ranges from -1 to 1 where zero indicates no relationship between x and y and ± 1 indicates a perfect relationship. A major assumption of this correlation is that x and y follow a normal distribution; if they do not, then a Spearman rank correlation should be used.

The quoting of an r^2 value is valuable as this is a measure of the fraction of the total variance of x and y explained by the linear relationship.

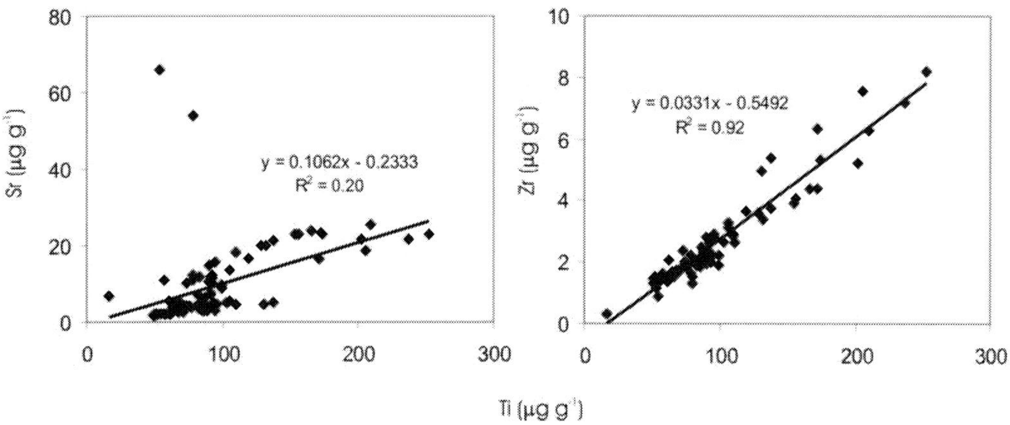

Figure 2. Scatter diagrams of Sr and Zr *vs*. Ti for 5000 years of peat accumulation at PVO in northwestern Spain. The correlation between Zr and Ti is much stronger than that between Sr and Ti (Kylander, unpublished data).

As an example Ti, Zr and Sr data from a peat core from Penido Vello (PVO) in northwestern Spain dating back 5000 years is presented (Kylander, unpublished data) (Figure 2). The correlation between Ti and Sr has an r-value of 0.44 while for Ti and Zr it is 0.96. We know that Ti and Zr are added to the mire via mineral matter input only (versus processes like sediment dissolution) and that these elements are immobile after deposition. Conversely, we know that Sr is subject to post depositional diagenesis associated with fluctuations in the water table. Thus we expect the behaviour of Zr and Ti to differ from Sr, which is reflected in the r-value. The r^2 value tells us that 92% of the total variance in Zr and Ti is explained by this linear relationship.

Table 1. Correlation matrix presenting peat core data from PVO in northwestern Spain (Kylander, unpublished data)

	K	Fe	Y	Zr	Sc	Ti	Sr	Rb	Cu	Zn
K	1.000									
Fe	0.755	1.000								
Y	0.856	0.652	1.000							
Zr	0.740	0.569	0.931	1.000						
Sc	0.799	0.624	0.931	0.959	1.000					
Ti	0.682	0.552	0.872	0.959	0.962	1.000				
Sr	0.277	0.239	0.293	0.355	0.458	0.443	1.000			
Rb	0.820	0.571	0.732	0.662	0.678	0.563	0.362	1.000		
Cu	0.580	0.464	0.385	0.327	0.398	0.291	0.528	0.751	1.000	
Zn	0.876	0.770	0.851	0.826	0.873	0.813	0.437	0.756	0.593	1.000

When there are many variables to consider a correlation matrix can be made that presents the r-value for each pair of variables. Expanding on the Spanish data above it is apparent from the correlation matrix that the elements are strongly associated and might be worth exploring further (Table 1). For example, those elements with the strongest correlations ($r > 0.872$), namely Y, Zr and Sc, are known to be immobile. Zinc also shows a high correlation with Y and Sc and even Zr. Not much work has been done on Zn in peat bogs but the evidence here suggests that Zn may well be immobile.

Regression analysis provides a means to quantify the association between x and y. This is done by fitting a straight line through the data and finding the equation of this line in the form $y = a + bx$. The value for a is the y intercept when $x = 0$ while b is the slope of the line used to fit the data. Most data do not however fall on a straight line and several approaches to estimating this line can be taken. Ordinary least square regression is the most commonly used form because it is often available in statistical packages. The least squares regression that quantifies the relationship between Ti and Sr and Ti and Zr is given in Figure 2 above. This fit minimizes the sum of the squares of the deviations about the line of the random variable. The problems associated with this approach lay in the fact that two regression lines are possible for a data set and it is assumed that errors on either x or y are small. Given that these variables are often determined by the same analytical technique, it is difficult to assign which is the random and which is the non-random variable. Three other types of regression can be used that may better suit geochemical data sets, namely: reduced major axis, weighted least squares and robust regressions. For a more detailed description of these techniques and those described above the reader is directed to Rollinson (1993) and Davis (2002).

Principle Component Analysis (PCA)

When a large geochemical dataset is generated a geochemist may use a Principal Component Analysis (PCA) to reduce the number of observed variables to a small number of factors, thus providing insight into the structure of the variance of the parameters (Jolliffe, 2002). This statistical technique is applied to a single set of variables (e.g., individual elements) to discover which set of variables form coherent subsets that are relatively independent of one another. Variables that are correlated with one another, which are largely independent of other subsets of variables, are combined into factors. Here factors should represent the underlying process that created the correlations among these variables. Loadings relate to the specific association between the factors (processes) and original variables (elements). Therefore it is necessary to find the loadings, then solve for the factors, which approximate the relationship between the original variable and underlying factors. If we apply PCA to measured and calculated variables from a lake sediment or peat core the PCA should relate to the main processes or mechanisms accounting for the actual distribution of the elements in the deposit. Various statistical software packages provide a means for making a PCA (e.g. SPSS).

A New Statistical Approach: Change Point Modeling

When dealing with noisy time or depth series it is difficult to objectively identify the timing and location of distinct changes in a record. To solve this, the general Bayesian approach as described by Denison et al. (2002) for dealing with such ‘change point’ problems may be applied. The primary assumption is that the underlying trend in the data between two change points is either constant or varies linearly with depth. Trends are estimated for a given number and location of change points by least squares regression for the data between each pair of change points. Of primary interest is where the data imply change, not the underlying regression function. In this way, different regression functions for each data type are allowed for, but require that the change points be in the same location for each simulation considered. For example, if we modeled the Spanish Zr and Ti concentrations together, change points in these two data sets would occur at the same place. During modeling numerous simulations are run with 200,000 to 500,000 samples in each simulation and the regression function within each partition is allowed to be either a constant value (i.e., one variable between each change point) or a linear function (i.e., two variables between each change point). An advantage with this type of modeling is that it tends to favour simpler solutions and thus identifies few change points.

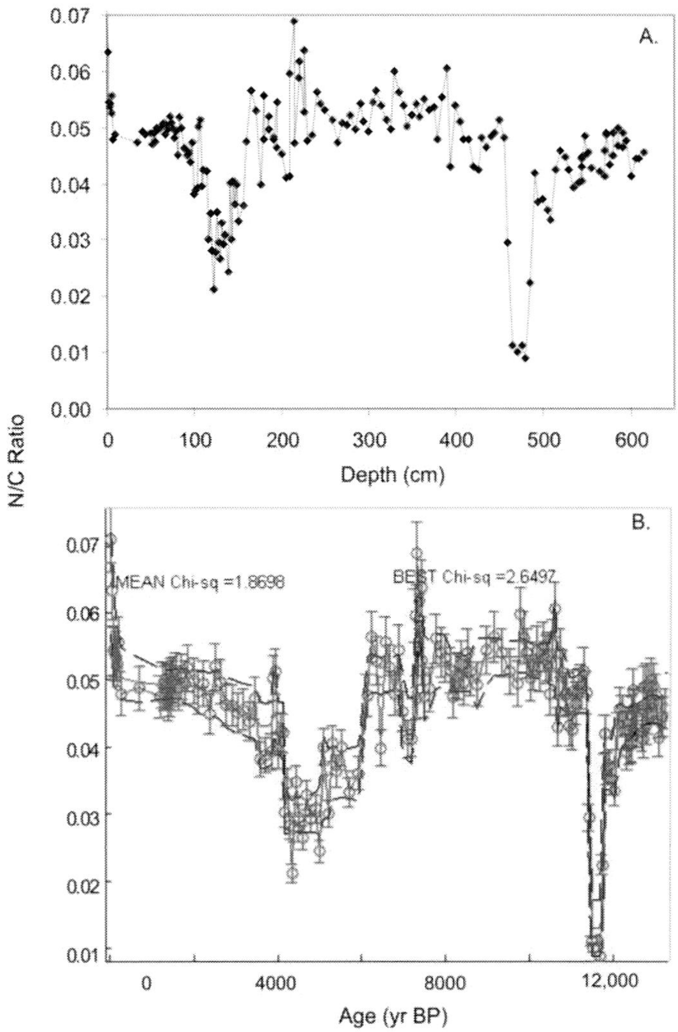

Figure 3. An example of change point modeling taken from a peat core in northwestern Sichuan, China (from Shopland, 2007). Both the original N/C data *vs.* depth (A) and the modeled data *vs.* yr BP show the posterior mean estimate (solid line) as well as the 95% confidence intervals (dashed lines) (B). Note that in this case N/C ratios were modeled together with $\delta^{13}C$ data, the latter are not shown.

Change point modeling has been successfully applied to Eu anomaly and Pb isotope records from Lynch's Crater, Australia (Kylander et al., 2007) see below and to $\delta^{13}C$ and N/C records from northwest Sichuan, China (Shopland, 2007). An example of change point modeling from the latter study is given in Figure 3, which shows the original peat core N/C data as well as the modeled posterior mean estimate of the constant regression function model with 95% credible intervals (note $\delta^{13}C$ and N/C data were modeled together but only the N/C data is shown). This posterior mean estimate is effectively the average of all change point models and is therefore smoother than any individual simulation result. The most significant change points are still captured however, as these will appear in many of the simulations. A full manuscript detailing this data modeling approach is in preparation by Ghallagher et al.

Geochemical Tools

Concentration Profiles and Net Accumulation Rates

In lake sediments and peats elements are measured in concentrations. This is a problem because concentrations are measured relative to the sample mass thus the concentration of one element is dependant on that of all other constituent elements. An apparent increase in an element may be the result of either a real increase in supply or a decrease in the supply of other elements. One way to nullify this effect is to calculate elemental net accumulation rates (I in $\mu g\ cm^{-2}\ y^{-1}$) which for the element X is calculated as:

$$I = [X]_{Total} * (\rho S_{sample}) / \Delta t \qquad (1)$$

where [X] is the total concentration in the sample ($\mu g\ g^{-1}$); ρ is the bulk density ($g\ cm^{-3}$); S_{sample} is the thickness of the sample slice (cm) and Δt is the time span represented in the slice (y). Elemental net accumulation rates are not without their drawbacks, as they rely heavily on the constructed age model, which is less precise than the concentration data. The accumulation rates of sediment and peat can vary significantly within one deposit and local conditions at a sampling site might act to change elemental signals. Additionally, elements that are transported to the lake in soluble forms may not be completely captured. Those elements supplied independently of sediment (e.g., atmospherically deposited materials, in catchment solutions) are susceptible to variable dilution. A review by Boyle (2002) provides a good overview of some of these issues and how one can use e.g., modeling to overcome them. The best starting point is to examine both concentration profiles and elemental net accumulation rates and to decide which of these data representations is more appropriate, if not both.

Elemental Ratios and Enrichment Factors (EF)

Elemental ratios provide a way of assessing changes in elemental inputs to a system. The ratio between a given element "A" and another element "B", to acquire an "A/B" ratio, will remain constant without perturbation of the system under examination. If the system is disturbed through climate induced changes in redox conditions, organic matter or minerological input for example, the ratio will change. Often the element in the denominator position is a conservative, lithogenic element (e.g., Ti, Zr, Sc, Y or Al) and is used for gauging changes in less conservative elements, depending on the intended purpose of normalization. Weiss et al.

(2002) normalize all elements in their peat core record to Ti to take into account changes in the bulk density (connected to accumulation/decomposition compaction processes) of the peat. On the other hand, Moreno et al. (2007) normalize elements to Ti to tease out signals that would have otherwise been overwritten by the dominant volcanic signatures of the surrounding the site.

Enrichment factors (EF) provide a means for gauging changes in elemental ratios relative to a reference material. This can in turn provide a way to compare changes within a record or between different sites using a reference material as a common denominator. To calculate EF for the element "A" for example, the following equation is used:

$$EF_{Reference,B} = (A/B)_{Sample}/(A/B)_{Reference} \quad (2)$$

where $EF_{Reference,B}$ is used to communicate the reference material and the element used for normalization in the equation, $(A/B)_{Sample}$ is the ratio of "A" to the selected (lithogenic) element in a given sample and $(A/B)_{Reference}$ is the ratio of "A" to the same selected lithogenic element in the reference material. The "A/B" values used for reference can be external (from the continental crust, a given geological material, soil, etc.) or internal (the average of a set of samples from a baseline portion of the same record).

Elemental ratios and EF should not be applied indiscriminantly. A main assumption associated with these tools is that the normalizing element/ratios in the reference material remains constant with time, which is rarely the case. Elements differ in their representation among minerals and are subject to different levels of fractionation during transport and deposition which can create enrichements and depletions in some elements. Post deposition processes will also affect elemental ratios and EF. Kylander et al. (2006) demostrate how the selection of the normalizing element and reference material in EF calculations can lead to widely different results at a single site and even more so when comparing several sites.

Mixing Equations

In geochemistry the term mixing is used to describe the process in which several mineral phases or components are brought together in a multi-phase or multi-component system to form an array of mixtures. A 'phase' retains its chemical character within the mixture like quartz in sediment for example, while a 'component' looses its physical identity after mixing, as in the mixing of lake and river waters. These relationships can be described using a simple mass balance equation where the sum of the contribution from *n* end members expressed as a fraction is equal to 1:

$$f_1 + f_2 + \ldots\ldots f_n = 1 \quad (3)$$

To solve for each member additional equations are required depending on the number of end-members. Here we will describe binary (or two-component) mixing and provide a simple graphical method to solve for ternary (or three-component) mixing.

In a conservative system binary mixing can easily be calculated. This is demonstrated here using data from the Lynch's Crater peat deposit in northeastern Australia (Kylander et al., 2007). In this case Eu anomalies are used to identify mineral dust inputs from local rocks and soils where two distinct source groups, I and II, are defined. Eu anomalies are a parameter often used in REE work. It is calculated by dividing Eu concentrations by the geometric mean of Sm and Gd to get Eu/Eu^*_{PAAS} (in this case these values have all been normalized to Post-Archaean Australian Shale (PAAS)). To simplify things we will use the average Eu anomaly (instead of the range) of Groups I and II such that they have values of 1.245 and 1.025, respectively. Samples from 145 to 595 cm of depth have Eu anomalies ranging from 1.09-1.22. We can calculate the relative contributions of Group I and II to these samples as follows:

$$f_{\text{Group I}} + f_{\text{Group II}} = 1 = \text{Samples}_{145\text{-}595\text{ cm}} \tag{4}$$

To solve for $f_{\text{Group I}}$ and $f_{\text{Group II}}$ the equation must be re-arranged and a conservative parameter introduced which in this case is the Eu anomaly signature of each group. In such a mass balance equation the units will cancel out during calculation (note that Eu anomalies do not have a unit in any case).

$$\text{Eu/Eu*}_{145\text{-}595\text{ cm}} = \text{Eu/Eu*}_{\text{Group I}} \cdot f_{\text{Group I}} + \text{Eu/Eu*}_{\text{Group II}} \cdot f_{\text{Group II}} \tag{5}$$

Re-arranging Equation 4 and substituting in Equation 5 we have:

$$f_{\text{Group I}} = 1 - f_{\text{Group II}} \tag{6}$$

$$\text{Eu/Eu*}_{145\text{-}595\text{ cm}} = \text{Eu/Eu*}_{\text{Group I}} \cdot (1 - f_{\text{Group II}}) + \text{Eu/Eu*}_{\text{Group II}} \cdot f_{\text{Group II}} \tag{7}$$

Substituting in real values we can now determine the contributions of Group I and II at the lower range limit of Eu anomalies in the peat samples (1.09):

$$1.09 = 1.245 \cdot (1 - f_{\text{Group II}}) + 1.0245 \cdot f_{\text{Group II}} \tag{8}$$

with a final result of 0.70 for $f_{\text{Group II}}$ and using Equation 6, 0.30 for $f_{\text{Group I}}$ contributions. If we solve the same set of equations using the upper range limit of the peat sample Eu anomalies (1.22) we calculate that Group I and II contribute 11% and 89%, respectively. Thus at these depth contributions these two source groups vary significantly.

A similar set of equations can be used for three component mixing, but with some risk of complication. A simpler way to calculate mixing between three components when they are presented graphically is to use the lever rule. If we have contributions from three different end-members (A, B and C in Figure 4) to a sample (S) it is possible to use the measured distances between these points to calculate the relative contribution of each end-member. The contribution of end-member A to the sample S is calculated by measuring the distance S*m* divided by the distance A*m* (1.5 cm/2.8 cm = 0.54). Thus A contributes 54% of the signal at S and end-members B and C together contribute 46%. Dividing the distances C*m*/CB gives the proportion of that 46% contributed by end-member B (1.5 cm/3.4 cm = 0.44), resulting in overall proportions of 54A: 20B: 26C. Worked examples of this method as well as other mixing applications (e.g., isotopes) can be found in Albarède (1995).

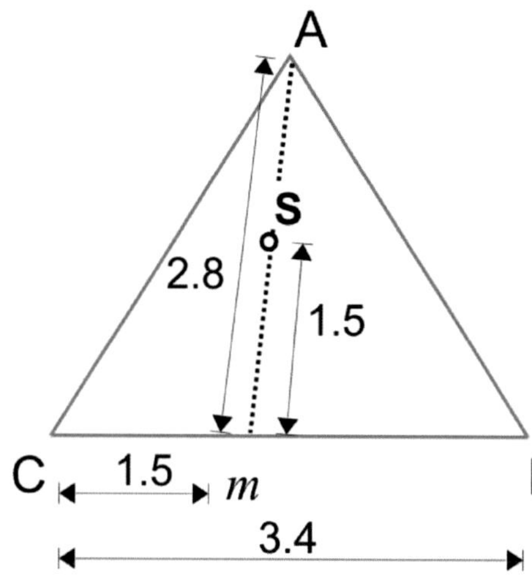

Figure 4. Diagramatic example of how to calculate the relative mixing between three components (A, B and C) to a sample (S) using measured distances when data are presented graphically.

Bivariate and Triangular Plots

Bivariate and Triangular plots are a simple way to display chemical differences between a suite of related samples. Bivariate plots show two selected variables plotted against each other. By convention that variable showing the greatest variability should be plotted on the x-axis. A triangular plot has the same function but shows three variables. The spatial distribution of the samples can help to establish geochemical relationships within a data set.

Practical Examples

It is unusual in geochemistry to rely on a single interpretive tool; the complexity of geochemical data sets requires that they be approached from a variety of angles to extract the most important information. Advantages lie in considering parallel biological and physical measurements even though these indices are often driven by separate processes they are not unconnected processes relating paleoenvironmental changes. In this section practical examples from recent publications are given to demonstrate how various geochemical tools have been applied in paleoclimatic reconstructions.

Concentration Profiles, Net Accumulation Rates and PCA

Muller et al. (in press) provide an example of how PCA can be used on a geochemical data set to obtain paleoclimatic information. The study site, Lynch's Crater, is an in-filled volcanic crater containing ~ 60 m of peat and lake sediments. The crater is located in northeastern Australia in the tropical Southern Hemisphere, an underrepresented region in terms of paleoclimate records. Past climate variations in the Earth's tropical regions are important to understand, as they help us to determine if, and how, the Northern and Southern Hemispheres are teleconnected.

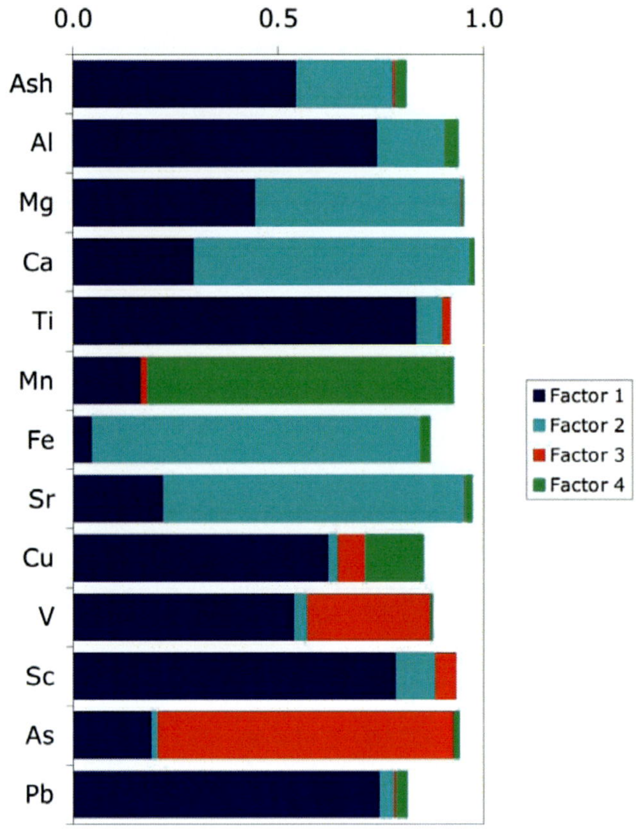

Figure 5. Communalities (total length of the bars) of the elements and their association with the main factors for concentrations identified using PCA at Lynch's Crater (Muller et al., submitted).

Using data from the top 13 m of peat, the SPSS statistical software package was used to perform a PCA on the following trace and major elements: Al, Mg, Ca, Ti, Mn, Fe, Sr, Cu, V, Sc, As and Pb, with ash yield as an additional variable. Four factors explained 88% of the total geochemical variance in Lynch's Crater. The observed loadings of each of the variables to these four factors are shown in Figure 5. Factor 1 (F1) accounts for a large proportion of the variance (45% of total variance) and is dominated by ash yield, Al, Sc, Ti, Pb, Cu and V. Factor 2 (F2) (27% of total variance) is mainly associated with Mg, Ca, Sr and Fe. Factor 3 (F3) (9% of total variance) is largely associated with As while V also has 30% variation associated to F3. Factor 4 (8%) is almost exclusively related to Mn. From these factors, four main processes were identified including; detrital influx (catchment and mineral dusts) (F1); mobilization within the peat due to chemical changes (pH, Eh, DOC, etc.) and biocycling of plant essential elements post-depositionally (F2); long-range atmospheric dust inputs (F3); and diagenetic alteration within the profile, most likely associated with groundwater flux along with aerobic/anaerobic changes (F4).

A PCA was also performed using elemental net accumulation rates but in this case, the concentration data actually provided more significant factors and that variance, not explained by these factors (the residual variance), was lower than with elemental net accumulation rates.

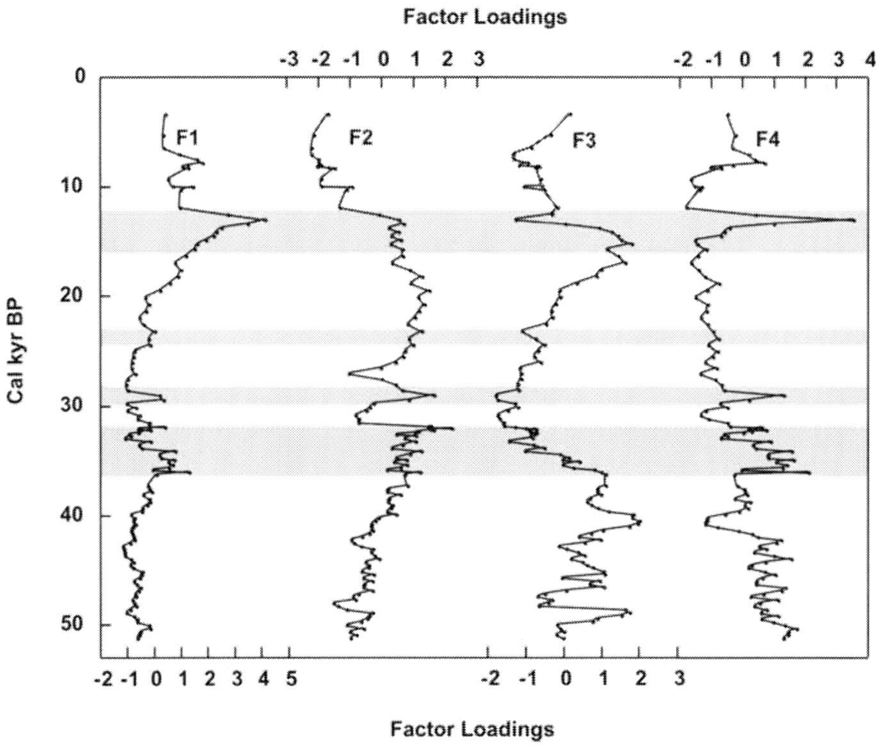

Figure 6. Variations of factor scores (F1-F4) through time of the Lynch's Crater peat deposit (Muller et al., submitted).

To derive paleoclimatic information from these four factors each was plotted against time (Figure 6). Consistent low F1 scores between 55 and 36 kyr BP correspond with palynological evidence for extensive vine forest cover (Kershaw, 1994), implying reduced soil erosion and mineral matter influx to the mire. In contrast, where F1 loadings were high (32.2-29.4, 25, 22.5, 12.2 and 8.3 kyr BP) pollen evidence infers the onset of an arid trend beginning ~36 kyr BP on the surrounding Atherton Tablelands; this supports increased weathering of the catchment, shown here by F1. Factor 2 (Figure 6) also shows agreement with pollen studies where low values correlate with the expansion of rainforest vegetation (Kershaw, 1994), seen here in sections older than 36 kyr BP, at 24.4-22.8 kyr BP, 21.5-20.1 kyr BP and during the Holocene. This implies a relationship between F2 and rainforest vegetation, which is thought to involve greater recycling of soil nutrients (Mg, Ca and Sr) in the catchment area and lower export to the mire. High values of F3 (Figure 6), associated with As and V, during the late Pleistocene (48-46, 37-34 and 15-12 kyr BP) are thought to signal dust inputs to Lynch's Crater from a long-range dust source indicating possible changes in atmospheric circulation patterns and dust sources over the last 55 kyr.

Looking to lake systems, the use of elemental profiles is exemplified by a maar lake record from Lago Grande di Monticchio, in southern Italy. Brauer et al. (2000) used Al_2O_3, TiO_2, MgO as well as dry density and pollen as indicators of detritus additions to the lake. In this case, greater detrital inputs are indicative of higher surface erosion within the catchment and hence, a harsher climate and lower amounts of vegetation. Using the core section from 101,000 to 65,000 cal yr BP that represents the early Weichselian, three peaks in detrital indicators occurred at 87,000 to 84,000 ca yr BP (LM 6); 78,800 to 77,800 cal yr BP (LM 5d); and 76,200 to 75,200 cal yr BP (LM 5b). The authors link these peaks to a major climatic deterioration between interstadial conditions (LM 6) and a period of transition from interstadial to full glacial conditions where several abrupt, short-term oscillations occurred (LM 5b,d).

Concentration Profiles, Correlation Matrices and XRF Core Scanning Techniques

Core scanning techniques are a recent development in the field of paleoclimate with much of the currently published material coming from marine sites (e.g., Jaccard et al., 2005; Boning et al., 2007). A recent publication by Moreno et al. (2007) demonstrates how the high resolution data acquired by XRF core scanners allows for the identification of abrupt climate shifts in a terrestrial environment that may have otherwise been overlooked.

The lake sediment cores from Lago Chungará in the Andes represent 14 kyr of accumulation in an area of intense volcanic activity of the tropical Southern Hemisphere. The authors rely on a correlation matrix to determine the relationships between mineralogical profiles, magnetic susceptibility, total organic and total carbon as well as elemental profiles. The results of this highlighted three main inputs to the Lago Chungará sediments: volcanic minerals, endogenic carbonates and siliceous biological remains. Changes in lake productivity were gauged by variations in lithostratigraphy, TOC, Si/Ti and S/Ti ratios; in turn increased productivity signaled the beginning of the Holocene (11,500 to 7500 cal yr BP). A short and prominent peak in Mn/Ti ratios around 10,000 cal yr BP was associated with an increase in oxygen of the bottom waters caused by a drop in lake levels during a significant drying event; this allowed for mixing and aeration of the water column. Coincident with changes in the Mn/Ti ratios, enhanced carbonate precipitation and distinct changes in Ca/Ti were observed in mid-Holocene portions of the profile. The authors use the variations in Ca/Ti and carbonate formation as evidence for a series of short, abrupt, dry spells (8600 to 6400 cal yr BP); during these dry spells reduced lake levels would have concentrated Ca in the water.

Identifying Changes in Dust Deposition Using Elemental Ratios and EF

Early promise for the use of peat in paleoclimate came from work by Shotyk et al. (2001). Elemental profiles for a wide range of conservative, lithogenic elements (Sc, Ti, Y, Zr and REE) from a 12,370 ^{14}C yr old core from Etang de la Gruère in Switzerland allowed for the assessment of fluxes of soil dust to the site. In this study, Sc was used as an indicator element to calculate the average annual flux of soil dust to the peat bog. In samples dating between 11,440 and 9320 ^{14}C yr BP the calculated annual soil dust flux was at its highest and peaks in Al/Sc, La/Sc, Zr/Sc, Hf/Sc, and REE/Sc occurred. This timeframe coincides with the Younger Dryas (YD) climate event in the Northern Hemisphere, which is characterized by increased storminess, the expansion of dry dusty areas and a reduction in vegetation cover. The elemental variations recorded at Etang de la Gruère confirm not only this increased dustiness but suggest a change in dust source to the mire. Interestingly, Zr, Hf and REE are all hosted by minerals with high specific densities (e.g., zircon) and while this supports a change in dust source, it also invokes a change in particle size distribution and/or increased wind strength.

In a peat core from Kalimantan in Borneo, southeast Asia dating back to 22,120 ^{14}C yr BP, Weiss et al. (2002) similarly demonstrated how trace and major element profiles can be used to examine and quantify soil dust inputs. Below the zone of biological influence (~200 cm) significant correlations were found between Al, Fe (which interestingly in this core was found to be immobile) Si, S, Ca and Ti and Y and Zr. Both Y and Zr are lithophilic suggesting that these elements are all derived from atmospheric dusts. Enrichment factor calculations showed Al/Ti enrichments (relative to the upper continental crust), peaks in Ti, Si and Fe concentrations and increased dust flux (Table 2, calculated here using Al rather than Sc as the indicator element) between ~10,830-9060 ^{14}C yr BP. These changes correspond in time to the YD in the Northern Hemisphere. The actual dust fluxes in Borneo however are lower than those reported for archives from the Northern Hemisphere. The authors hypothesized that the increase in dust deposition at this time must be due to long-range transport of dusts from the Northern Hemisphere, rather than the occurrence of the YD in the Southern Hemisphere.

Table 2. Calculated mineral aerosol flux to a peat bog in Kalimantan, Borneo for different time periods (modified from Weiss et al., 2002)

Time period (^{14}C yr BP)	Period	Mineral Aerosol mg m^{-2} y^{-1}
5790- Present	Later Holocene	1.63
9060-5790	Background	2.75
10,830-9060	Younger Dryas-Like Event	10.88
22,120-10,830	Late Pleistocene	0.38

Lake sediments from Lake Malawi covering the last 25,000 years were recovered to study climate variations in tropical Africa (Johnson et al., 2002). The main analysis was of biogenic silica but, trace elements, total and inorganic phosphorous, diatom concentrations and species assemblages were also examined. The biogenic silica mass accumulation rate (BSi MAR) in the sediments reflects the abundance of diatoms, and hence primary productivity, which is at its maximum in Lake Malawi during the dry, windy winter season. BSi MAR values were low during the Pleistocene compared to Holocene values and the Pleistocene-Holocene transition occurs in two abrupt steps; this transition differs in timing from that of the Greenland ice core (GRIP) δ^{18}O record. In late Pleistocene sediments the Lake Malawi BSi MAR is anti-phased with the GRIP δ^{18}O record suggesting that during this period the African tropical climate was linked to ice sheet and climate dynamics in the Northern Hemisphere. Though this relationship seems to weaken during the Holocene. Analysis of volcanic ash layers in sediments from northern Lake Malawi showed that these are enriched in certain elements, namely Nb. Ratios of Nb/Ti were used by Johnson et al. (2002) to quantify the input of volcanic ash from the sole available source, the Rungwe volcanoes to the north of the lake. The Nb/Ti ratios showed higher average values in the Pleistocene than in the Holocene and are thought to be the result of winds blowing over the cool, dry volcanic landscape to the north of the lake. These winds would have caused upwelling and increased diatom productivity and may be the result of the southward movement of the Intertropical Convergence Zone drawing northeasterly winds into Malawi.

Further work in Lake Malawi related the changes in heat balance, wind stress and lake levels to redox conditions in the bottom waters of this deep lake (Brown et al., 2000). By examining the mobility/immobility of certain elements under oxidizing/reducing conditions (Fe, Mn and U); the formation of insoluble sulphides (Cd, Mo, Pb and Zn); and the susceptibility of certain elements (Co, Mo, Pb, V and Zn) to scavenging by Fe and Mn oxyhydroxide precipitates, changes in the chemocline (the transition layer in a stratified lake between upper mixed waters and lower unmixed waters) were identified. These behavior variations were related back to increased water column mixing associated with stronger winds and cooler temperatures rather than to changing lake levels.

Mixing Equations and Rare Earth Element Diagrams

In a study by Kylander et al. (2007) REE, and more specifically $Eu/Eu*_{PAAS}$, was developed as a new paleoproxy for dust source changes exploiting a 55 kyr old peat core from Lynch's Crater in northeastern Australia. The REE form a coherent group (from La to Lu) their properties show ordered responses across the series during chemical and physical processes. During crustal weathering the chemical composition of source rocks is inherited and REE data can be used as a source tracer in paleoclimate studies (McLennan, 1989). REE are usually normalized to an external material like PAAS and plotted by increasing atomic number to generate REE patterns.

As mentioned above in *Geochemical Tools*, Groups I and II represent possible local mineral dust inputs to Lynch's Crater. REE patterns for these two groups as well as that from the Atherton Basaltic Province where the crater is located are shown in Figure 7. It is evident that Group I signatures overlap with those of the Atherton Basaltic Province and that the peat REE patterns agree with those of Group I and the Atherton Basaltic Province. The use of mixing equations can better clarify the different source contributions however.

If we re-define the range of Group I to reflect inputs from the range of volcanic terrains of eastern Australia, the importance of Group II is shown to increase between 105 to 895 cm (8.5-40.8 kyr BP)(Figure 8). This increase in Group II importance is unlikely to be the result of its preferential weathering. Rather, overlap between regional and Group II $Eu/Eu*_{PAAS}$ source signatures suggest contributions from dust sources beyond the Atherton Basaltic Province. Similarly, $Eu/Eu*_{PAAS}$ signatures at the bottom of the profile are too high to be explained by either Group I or II but good correspondence is found with long-distance source signatures; this is in agreement with findings from the PCA described above (Muller et al., in press).

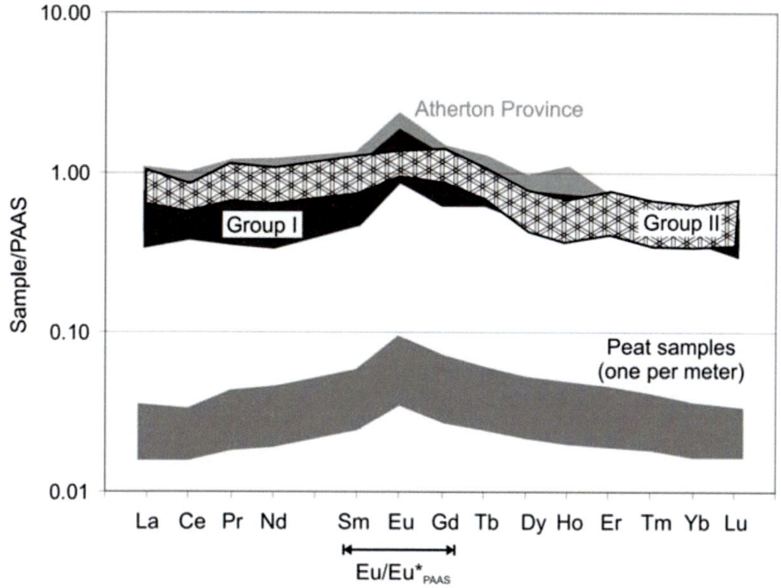

Figure 7. REE patterns normalized to PAAS for local source Groups I and II to the peats accumulated at Lynch's Crater. The REE pattern for the surrounding basaltic province (Atherton Basaltic Province) agrees well with that for Group I and patterns from the peat samples (modified from Kylander et al., 2007).

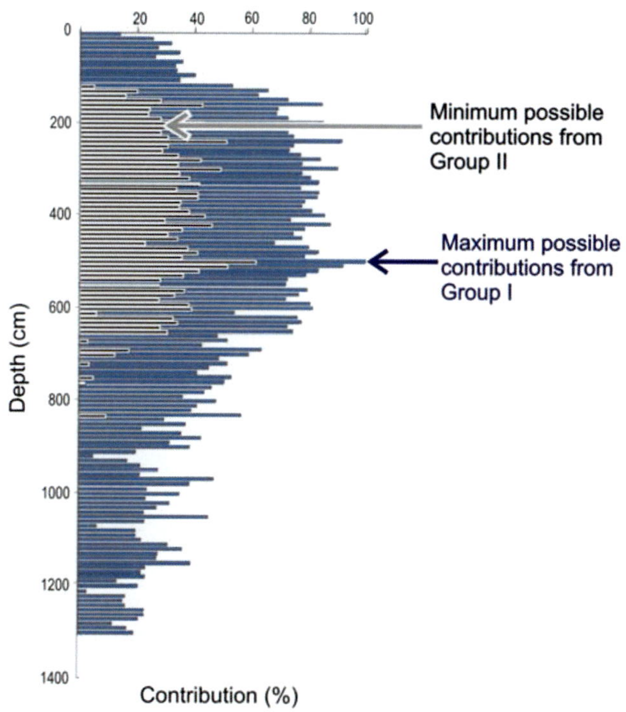

Figure 8. Changing contributions of Group I and II with depth as calculated using mixing equations based on Eu anomaly data from Lynch's Crater (modified from Kylander et al., 2007).

Triangular Plots and XRF Core Scanning

Tephrochronology is a geochronological technique that uses descrete layers of tephra, or volcanic ash from a single eruption, to create a chronological framework. It is thus, an important dating tool for Quaternary sedimentary sequences and allows for the correlation of sites across a region. With the discovery of a new tephra layer geochemical characterization is essential, allowing for its identification (or not) in other sequences. The work of Davies et al. (2003) examines two lake sediment sequences from Skallahult and Hässeldala ports in southern Sweden. At these sites several tephra layers are identified, the most significant being the Borrobol Tephra (12,300 ^{14}C yr BP)(Turney et al., 1997) previously not known to have reached southern Sweden. A new tephra, the Hässeldala Port Tephra (estimated age of 11,500 cal yr BP), was identified based on its unique geochemistry as plotted in Figure 9, which also shows the chemical composition of several other relevant tephras (Boygle, 1999; Langdon and Barber, 2001; Wastegard, 2002). A third tephra was sourced back to Askja caldera in Iceland (10,000 ^{14}C yr BP)(Sigvaldason, 2002) based on its distinct CaO and FeO concentrations which were in excess of those found for either the Borrobol or Hässeldala Port Tephras.

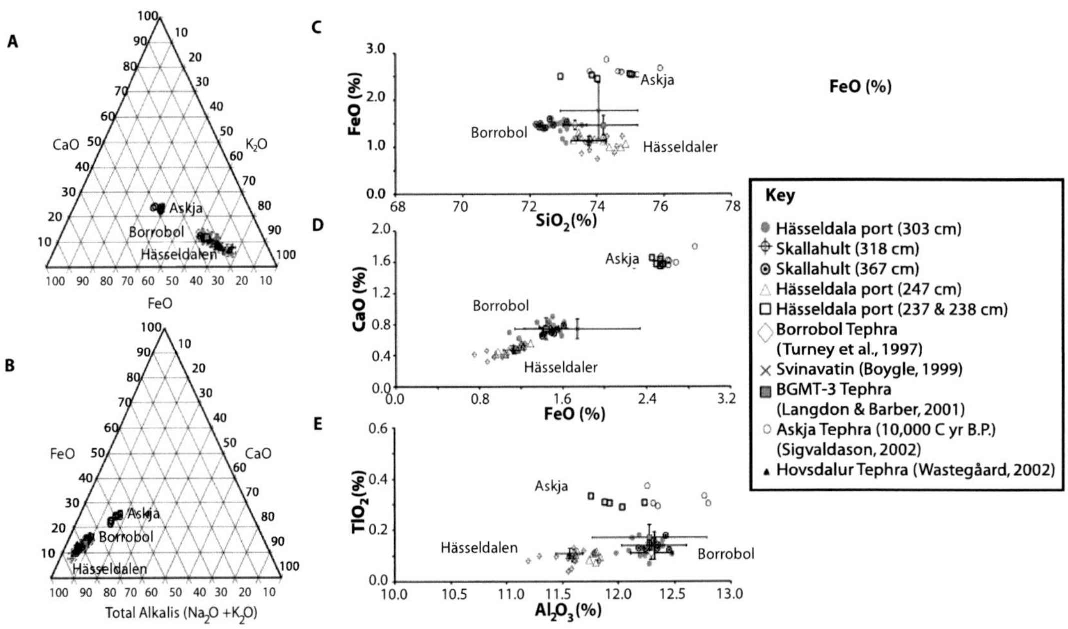

Figure 9. Major oxides data of glass shards discovered at Skallahult and Hässeldala port lakes in Southern Sweden plotted as ternary (FeO_{Tot}, CaO and K_2O (A) and FeO_{Tot}, CaO and total alkalis (B)) and bivariate (FeO vs. SiO_2 (C), CaO vs. FeO (D) and TiO_2 vs. Al_2O_3 (E)) plots. The mean compositions of the Borrobol Tephra (Turney et al., 1997), Askja Tephra (10,000 ^{14}C yr B.P.; Sigvaldason, 2001) as well as several other tephras are shown (Boygle, 1999; Langdon and Barber, 2001; Wastegard, 2002)(from Davies et al., 2003, published by permission from Elsevier).

The Hässeldala Port core has recently been analysed using an Itrax XRF core scanner. While this work is still in its preliminary stages, three tephra layers were identified using this technique. By comparing with previously published data and stratigraphy from Wohlfarth et al. (2006) and Davies et al. (2003) the Hässeldala Port Tephra was identified (Figure 10). The identity of the two other tephras requires further geochemical investigation but this work shows future potential for the application of core scanning techniques in tephrochronology.

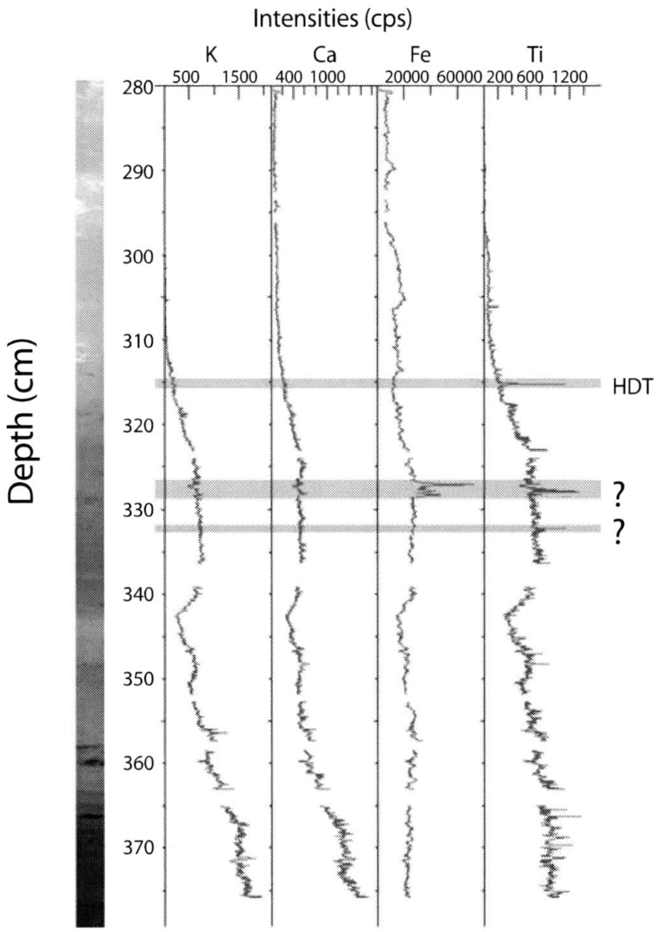

Figure 10. Intensities (in counts per second) for K, Ca, Fe and Ti *vs.* depth for the Hässeldala port lake sediment core taken using an Itrax XRF core scanner. Given also is the radiographic image of the sediments. Three possible tephras have been found of which one has been identified as the Hässeldala Port Tephra (HDT)(Kylander, unpublished data).

Chemical Indices as Paleoproxies

Minyuk et al. (2007) present a study from El'gygytgyn Lake in northeastern Russia from which 12.9 m of sediment, covering the last 250 kyr was collected. The authors endeavour to associate correlation analyses, elemental profiles and the chemical index of alteration (CIA, developed by Nesbitt and Young, 1984; 1989) to four major climate shifts (relatively warm, peak warm, cold and dry, cold and moist) identified previously by other proxies. The CIA provides a measure of the degree of alteration of feldspars to clay minerals during weathering and is calculated as follows:

$$CIA = [Al_2O_3/(Al_2O_3 + CaO + Na_2O + K_2O]*100 \quad (9)$$

During warm periods when the lake was seasonally free of ice and experiencing full mixing, SiO_2, CaO, Na_2O, K_2O and Rb concentrations increased while those of TiO_2, Fe_2O_3, Al_2O_3 and MgO decreased. The CIA values were lower during warm periods than during cold periods. The CIA in this context provides some indication of changes in hydrological settings: during warmer periods runoff is more intense and there is a low residence time for sediments in transport to the lake so the CIA is lower than during cold periods.

In some cases hydrological fluctuations in a basin can cause extreme changes in lake water chemistry limiting the paleoproxies that can be used; this is the case with some of the closed basins of east Africa. Deocampo (2004) broaches this problem by exploiting chemical variability in ultra fine clays (<0.1 μm) where Mg uptake occurs in conditions of elevated salinity and alkalinity. These changes were quantified using an octahedral cation index that is calculated as follows:

$$OCI = (Mg/ (Al + Fe) \quad (10)$$

OCI values were shown to range from <0.1 for upland soils and detrital sources to >4 for saline and alkaline deposits. While the application of this paleoproxy is very specific (it is limited to settings rich in alkaline waters) the author was able to show two periods of lake expansion starting some 1.79 Ma, with an intervening lowstand that lasted some 10 kyr. The initial lake expansion was found to agree with marine records of increased African humidity while the later expansion suggests some high-frequency environmental oscillations may be independent from obliquity and precession control.

Mercury as a Paleoproxy in Lake Sediments and Peats

Mercury has a complicated behavior in the environment where the primary form of Hg emitted from both natural and anthropogenic sources is as metallic gaseous Hg. Its main transport pathway is through the atmosphere and the residence time of Hg can be up to one year. Natural sources include volcanism, hot spring activity and fracturing and faulting (Schroeder and Munthe, 1998). Lindeberg et al. (2006) used lake sediment profiles from Greenland to explore variations in both Pb and Hg. In pre-pollution sections of the records the authors exclude natural sources due to lack of evidence (e.g., ash layers from volcanic eruptions) and interpret aeolian activity and dusts from the nearby sandy outwash plains at the margins of surrounding ice sheets to be the primary natural Hg and Pb sources. This is supported by a shift in Pb isotopes *ca.*1000 BC to 55 AD in several of the lakes, coinciding with lower temperatures recorded in other Greenland ice and lake sediment records.

Martinez-Cortizas et al. (1999) used the variable geochemistry of Hg to an advantage by exploiting its differences in thermal stabilities to reconstruct temperature changes at a site in northwestern Spain over the last 4000 ^{14}C yr BP. Peat samples were either measured wet or dried at 30°C or 105°C before analysis. The main assumption was that each type of treatment would reflect progressively higher thermal stabilities that could then be related to climate conditions at the time of deposition. Indeed, it was found that cold climates promoted and enhanced the storage of Hg with low thermal stability while warmer conditions preserved Hg with higher thermal stabilities. Later work by Beister et al. (2003) suggests that these differences may be a function of the properties of the organic matter and its decomposition rather than any temperature variations, however this has yet to be unequivocally established.

Conclusion

The application of inorganic geochemistry to the analysis of peat and lake sediments in paleoclimate reconstructions shows great potential for the future. To date the analytical techniques and tools available for interpretation of paleoclimate records have received little attention. All paleoclimate records are unique and different tools are appropriate in different settings therefore it is important to consider that geochemistry applications in this field attempt to extract large-scale processes from what may be small-scale variations. A well planned sampling protocol and a sound understanding of deposit dynamics are essential; with this more confidence may be placed in geochemical proxies especially when they are supported by both biological and physical proxies.

References

Albarède, F.; *Introduction to Geochemical Modeling*; Cambridge University Press: Cambridge, 1995; 563.

Barnola, D. R.; Korotkevich, Y. S.; Lorius, C. *Nature* 1987 329, 408–414.

Belzile, N.; Lecomte, P.; Tessier, A. *Environ. Sci. Tech.* 1989 23, 1015-1020.

Biester, H.; Martinez-Cortizas, A.; Birkenstock, S.; Kilian, R. *Environ. Sci. Tech.* 2003 37, 32-39.

Bindler, R.; Klarqvist, M.; Klaminder, J.; Forster, J. *Global Biogeochem. Cycle* 2004 18, art. no.-GB3020.

Birks, H. H.; Birks, H. J. B. *Veget. Hist. Archaeobot.* 2006 15, 235-251.

Boning, P.; Bard, E.; Rose, J. *Geochem. Geophys. Geosys.* 2007 8.

Boygle, J. *Glob. Planet. Change* 1999 21, 129-149.

Boyle, J. F. In *Tracking Environmental Change Using Lake Sediments: Physical and Chemical Techniques*; Last, W. M.; Smol, J.; Ed.; Kluwer Academic: Dordrecht, 2002; 83-141.

Brauer, A.; Mingram, J.; Frank, U.; Gunter, C.; Schettler, G.; Wulf, S.; Zolitschka, B.; Negendank, J. F. W. *Quat. Int.* 2000 73-4, 79-90.

Brown, E. T.; Le Callonnec, L.; German, C. R. *Geochim. Cosmochim. Acta* 2000 64, 3515-3523.

Carignan, R.; Tessier, A. *Science* 1985 228, 1524-1526.

Charman, D. *Peatlands and Environmental Change*; John Wiley and Sons, Ltd.: West Sussex, 2002; 301.

Clark, J. M.; Chapman, P. J.; Adamson, J. K.; Lane, S. N. *Global Change Biol.* 2005 11, 791-809.

Croudace, I. W.; Rindby, A.; Rothwell, R. G. In *New Techniques in Sediment Core Analysis*; Rothwell, R. G.; Ed.; The Geological Society of London: London, 2006; 272.

Davies, S. M.; Wastegard, S.; Wohlfarth, B. *Quat. Res.* 2003 59, 345-352.

Davis, J. C. *Statistics and Data Analysis in Geology*; Wiley: New York, 2002; 638.
De Laeter, J. R. *Applications of Inorganic Mass Spectrometry*; John Wiley and Sons, Ltd: Chichester, 2001; 474.
Denison, D.; Holmes, C.; Mallick, B.; Smith, A. *Bayesian Methods for Nonlinear Classification and Regression*; Wiley: New York, 2002; 277.
Deocampo, D. M. *J. Paleolimnol.* 2004 31, 1-9.
Durrant, S. F. *Analyst* 1992 117, 1585-1592.
Durrant, S. F. *J. Anal. At. Spectrom.* 1999 14, 1385-1403.
Durrant, S. F. Ward, N. I. *J. Anal. At. Spectrom.* 2005 20, 821-829.
Emmanuel, S.; Erel, Y. *Geochim. Cosmochim. Acta* 2002 66, 2517-2527.
Engstrom, D. R.; Schottler, S. P.; Leavitt, P. R.; Havens, K. E. *Ecol. Appl.* 2006 16, 1194-1206.
Freeman, C.; Evans, C. D.; Monteith, D. T. *Nature* 2001 412, 785.
Gerhardt, S.; Schink, B. *Biogeochem.* 2005 74, 341-356.
Gill, R.; Ed. *Modern analytical geochemistry: an introduction to quantitative chemical analysis for earth, environmental and materials scientists*; Longman: Harlow, 1997; 328.
Haug, G. H.; Hughen, K. A.; Sigman, D. M.; Peterson, L. C.; Rohl, U.; *Science* 2001 293, 1304-1308.
Herbert, C. G.; Johnstone, A. W.; *Mass Spectrometry Basics*; CRC Press: Boca Raton, 2002; 474.
Hughen, K. A.; Overpeck, J. T.; Peterson, L. C.; Trumbore, S. *Nature* 1996 380, 51-54.
Jaccard, S. L.; Haug, G. H.; Sigman, D. M.; Pedersen, T. F.; Thierstein, H. R.; Rohl, U. *Science* 2005 308, 1003-1006.
Johnsen, S. J.; DahlJensen, D.; Gundestrup, N.; Steffensen, J. P.; Clausen, H. B.; Miller, H.; Masson-Delmotte, V.; Sveinbjornsdottir, A. E.; White, J. *J. Quat. Sci.* 2001 16, 299-307.
Johnson, B. J.; Miller, G. H.; Fogel, M. L.; Magee, J. W.; Gagan, M. K.; Chivas, A. R. *Science* 1999 284, 1150-1152.
Johnson, T. C.; Brown, E. T.; McManus, J.; Barry, S.; Barker, P.; Gasse, F. *Science* 2002 296, 113-132.
Jolliffe, I. T. *Principal Component Analysis*; Springer Verlag: New York, 2002; 502.
Kershaw, A. P. *Palaeogeogr. Palaeoecol.* 1994 109, 399-412.
Kheboian, C.; Bauer, C. F. *Anal. Chem.* 1987 59, 1417-1423.
Kuhry, P.; Vitt, D. H. *Ecology* 1996 77, 271-275.
Kylander, M.; Weiss, D.; Peiteado Varela, E.; Taboada Rodriguez, T.; Martinez Cortizas, A. In *Peatlands: Evolution and Records of Environmental and Climate Changes*; Martini, I. P.; Martinez Cortizas, A.; Chesworth, W.; Ed.; Elsevier: 2006; 479-497.
Kylander, M. E.; Muller, J.; Wust, R. A. J.; Gallagher, K.; Garcia-Sanchez, R.; Coles, B. J.; Weiss, D. J. *Geochim. Cosmochim. Acta* 2007 71, 942-960.
Lachance, G. R.; *Quantitative X-Ray Fluorescence Analysis: Theory and Application* Wiley: Chichester, 1994; 424.
Langdon, P. G.; Barber, K. E. *J. Quat. Sci.* 2001 16, 753-759.
Lindeberg, C.; Bindler, R.; Renberg, I.; Emteryd, O.; Karlsson, E.; Anderson, N. J. *Environ. Sci. Tech.* 2006 40, 90-95.
Malmer, N.; Wallen, B. *Ecography* 1999 22, 736-750.
Martin, J. M.; Nirel, P.; Thomas, A. J. *Mar. Chem.* 1987 22, 313-341.

Martinez-Cortizas, A.; Pontevedra-Pombal, X.; Garcia-Rodeja, E.; Novoa-Munoz, J. C.; Shotyk, W. *Science* 1999 284, 939-942.

McLennan, S. M. In *Geochemistry and Mineralogy of Rare Earth Elements*; Lipin, B. R.; McKay, G. A.; Ed.; Mineralogical Society of America: Washington, 1989; 169-200.

Minyuk, P. S.; Brigham-Grette, J.; Melles, M.; Borkhodoev, V. Y.; Glushkova, O. Y. *J. Paleolimnol.* 2007 37, 123-133.

Moreno, A.; Giralt, S.; Valero-Garcés, B.; Sáez, A.; Bao, R.; Prego, R.; Pueyo, J. J.; González-Sampériz, P.; Taberner, C. *Quat. Int.* 2007 161, 4-21.

Muller, J.; Kylander, M.; Martinez Cortizas, A.; Wust, R. A. J.; Weiss, D. J.; Blake, K.; Coles, B. J.; Garcia Sanchez, R. *Geochem. Cosmochem. Acta* in press.

Nesbitt, H. W.; Young, G. M. *Geochim. Cosmochim. Acta* 1984 48, 1523-1534.

Nesbitt, H. W.; Young, G. M. *J.Geol.* 1989 97, 129-147.

Nirel, P. M. V.; Morel, F. M. M. *Wat. Res.* 1990 24, 1055-1056.

Page, S. E.; Wust, R. A. J.; Weiss, D.; Rieley, J. O.; Shotyk, W.; Limin, S. H. *J. Quat. Sci.* 2004 19, 625-635.

Petit, J. R.; Jouzel, J.; Raynaud, D.; Barkov, N. I.; Barnola, J.-M.; Basile, I.; Bender, M.; Chappellaz, J.; Davis, M.; Delaygue, G.; Delmotte, M.; Kotlyakov, V. M.; Legrand, M.; Lipenkov, V. Y.; Lorius, C.; Pepin, L.; Ritz, C.; Saltzman, E.; Stievenard, M. *Nature* 1999 399, 429-436.

Rauch, S.; Hemond, H. F.; Brabander, D. J. *Limnol. Oceanogr.* 2006 4, 268-274.

Rollinson, H. R. *Using Geochemical Data: Evaluation, Presentation, Interpretation*; Addison Wesley Longman Limited: Essex, UK, 1993; 352.

Rowan, D. J.; Cornett, R. J.; King, K.; Risto, B. *J. Paleolimnol.* 1995 13, 107-118.

Rowan, D. J.; Rasmussen, J. B.; Kalff, J. *Can. J. Fish. Aquat. Sci.* 1995 52, 2146-2158.

Schroeder, W. H.; Munthe, J. *Atmosph. Environ.* 1998 32, 809-822.

Shopland, M. (2007). Variations of Δ13c and N/C in a Peat Core from Nw Sichuan, China, and Their Climatic Implications. Department of Earth Science and Engineering, Imperial College London. MSci**.**, 24.

Shotyk, W.; Weiss, D.; Kramers, J. D.; Frei, R.; Cheburkin, A. K.; Gloor, M.; Reese, S. *Geochim. Cosmochim. Acta* 2001 65, 2337-2360.

Sigvaldason, G. E. *B. Volcanol.* 2002 64, 192-205.

Tessier, A.; Campbell, P. G. C. *Wat. Res.* 1991 25, 115-117.

Tessier, A.; Campbell, P. G. C.; Bisson, M. *Anal. Chem.* 1979 51, 844-851.

Tessier, A.; Carignan, R.; Dubreuil, B.; Rapin, F. *Geochim. Cosmochim. Acta* 1989 53, 1511-1522.

Teutsch, N.; Erel, Y.; Halicz, L.; Banin, A. *Geochim. Cosmochim. Acta* 2001 65, 2853-2864.

Tipping, E.; Hetherington, N. B.; Hilton, J.; Thompson, D. W.; Bowles, E.; Hamilton-Taylor, J. *Anal. Chem.* 1985 57, 1944-1946.

Turney, C. S. M.; Harkness, D. D.; Lowe, J. J. *J. Quat. Sci.* 1997 12, 525-531.

Vallelonga, P.; Van de Valde, K.; Candelone, J. P.; Ly, C.; Rosman, K. J. R.; Boutron, C. F.; Morgan, V. I.; Mackey, D. J. *Anal. Chim. Acta* 2002 453, 1-12.

Walsh, J. N. In *Modern Analytical Geochemistry*; Gill, R.; Ed.; Addison Wesley Longman Limited: Essex, UK, 1997; 41-66.

Walsh, J. N.; Gill, R.; Thirlwall, M. F. In *Modern Analytical Geochemistry*; Gill, R.; Ed.; Addison Wesley Longman Limited: Essex, UK, 1997; 29-40.

Wastegard, S. *J. Quat. Sci.* 2002 17, 723-730.

Weiss, D.; Shotyk, W.; Rieley, J.; Page, S.; Gloor, M.; Reese, S.; Martinez-Cortizas, A. *Geochim. Cosmochim. Acta* 2002 66, 2307-2323.

Williams, K. L. *An Introduction to X-Ray Spectrometry : X-Ray Fluorescence and Electron Microprobe Analysis*; Allen & Unwin: London, 1987; 370.

Wohlfarth, B.; Blaauw, M.; Davies, S. M.; Andersson, M.; Wastegard, S.; Hormes, A.; Possnert, G. *J. Quat. Sci.* 2006 21, 321-334.

In: Climate Change Research Progress
Editor: Lawrence N. Peretz, pp. 187-215
ISBN: 978-1-60021-998-6

Chapter 5

IMPACTS OF CLIMATE CHANGES ON MARINE ECOSYSTEMS AND FISHERIES PRODUCTIVITY IN MARGINAL SEAS OF THE NORTHWEST PACIFIC

***Chang Ik Zhang*[1]*, Suam Kim*[1*]*, Jai Ho Oh*[1] *and Donald R. Gunderson*[2]**
[1]Pukyong National University, Busan, 608-737 Korea
[2]University of Washington, Seattle, WA 98115 U.S.A.

Abstract

Climate changes such as global warming, decadal climatic regime shifts and interannual variability of the ENSO phenomenon can affect ocean conditions, and thus affect the functioning of marine ecosystems and the distribution and abundance of fisheries resources. There has been an increasing trend in sea surface temperatures in the last 100 years which has accelerated in the last decade in the marginal seas of the Northwest Pacific. The recent warming trend in the Korean marine ecosystem is associated with a decline in cold-water species (e.g. walleye pollock) and an increase in warm-water species (e.g. squids, jellyfish, mackerels). There were also some indications of decadal regime shifts in 1976/77 and in 1988/89 in the region. Thus, it is anticipated that climate change will affect the productivity and distribution of species important to Korean fisheries.

There are more than 300 fish species in the marginal seas of the Northwest Pacific of which about 100 are of commercial value. Most commercially important species can be categorized into three fish communities. A demersal community (small yellow croaker, hairtail) exists in the Yellow Sea, and in the East/Japan Sea and the Okhotsk Sea (walleye pollock, Pacific cod). A pelagic community (mackerels, squids, anchovy) exists in the East China Sea. Overfishing has been a serious threat to productivity, resulting in a higher proportion of smaller, very young individuals. Climate related ocean changes also profoundly affect productivity. There has been a decline in the catch of demersal species and an increase in the catch of pelagic species. Chub mackerel, Pacific saury and common squid abundance increased in periods of weak Aleutian Lows and a negative PDO. In contrast, filefish and Jack mackerel productivity declined during these periods but increased when the regimes shifted to strong Aleutian Lows and a positive PDO. Pacific sardine have a well-known climate related

* E-mail address: suamkim@pknu.ac.kr (Corresponding author)

alternating cycle of abundance. Warm sea surface temperatures reduce the marine survival of chum salmon fry and restrict the southward migration of walleye pollock. Each ecosystem responded differently to climate impacts under different exploitation systems. It is recognized that future climate change will have important impacts on marine productivity that will require developing a holistic ecosystem-based approach to fisheries management. This requirement will trigger integrated multidisciplinary research in climatology, oceanography and fisheries sciences to understand the impacts of climate changes on marine ecosystems and fisheries productivity.

Key words: climate change, marine ecosystem, impacts to fisheries, Northwest Pacific

Introduction

Early ideas of a steady-state ocean have long been replaced by the present view of an ocean varying at periods from less than days to those of years, decades, centuries, and beyond. While seasonal and inter-annual variability is now well known, that at decadal and inter-decadal periods has only become evident as longer time series of observations have become available (Wooster and Zhang, 2004). Climate changes such as global warming, decadal climatic regime shifts and inter-annual variability of the ENSO phenomenon can affect ocean conditions, and thus affect the functioning mechanism of marine ecosystems and the distribution and abundance of fisheries resources. The air-sea interactions, closely connected to large-scale environmental changes, is the main driving force in controlling climate and the fluctuation of marine organisms. Ocean productivity is not stable under these changing environments. Comparing the current ecosystem to the past, it was found that the present conditions of the ecosystem have been modified seriously. Stock abundance and distribution, species composition, and carrying capacity of ecosystems vary from one climate regime to another (Kim et al., 2007).

However, knowledge of physical variability has benefited from the relatively long time series of atmospheric and surface ocean data, while that of ecosystem variability generally depends on much shorter records of less systematic observations. On multi-year time scales, the duration of periods with characteristic environmental conditions, so-called regimes, is of the order of decades, while transitions between them, or regime shifts, usually take place with in a year. These events can be seen in a 100-year record of the Pacific Decadal Oscillation (PDO) index, the leading Principal Component of monthly sea surface temperature (SST) anomalies in the Pacific Ocean of 20°N (Mantua and Hare, 2002). Consequently, the biomass of fishery resources, which usually occupy the upper trophic level in marine ecosystems, has changed from year to year because their growth and survival depends on the availability of food organisms which are sensitive to environmental conditions.

Major changes in species distribution and abundance in North Pacific marine ecosystems are often correlated with climate shifts in the twentieth century (Overland and Wang, 2007). Fish populations have been adapting to oceanic phenomena over evolutionary time scales, so that their survival rate should be high in the resident area during an ordinary period. The episodic changes in climate/environments, however, result in changes in survival, productivity and species composition of the ocean ecosystem. Recruitment failure of fish populations caused by high mortality during the early life stages is detrimental in maintaining healthy stock conditions. Most hypotheses on recruitment processes are closely related to the plankton-based ecosystem, which largely depends upon oceanic variability (Cole and McGlade, 1998). In order to delineate the effect of ecological interactions among fisheries resources, time-series on production of fisheries resources and observation on the environment are required.

The new paradigm of ecosystem response to environmental variability has become the main theme in marine ecology and fishery science. This type of research provides the understanding of cause and effect mechanisms, as well as prediction capabilities for fisheries recruitment related to climate changes. These in turn could eventually be used to establish appropriate fisheries resource management procedures (Kim et al., 2006; Kim and Lo, 2001).

The purpose of this chapter is to review effects of events such as global warming, atmospheric circulation patterns, climate regime shifts in the North Pacific, and El Niño events in the tropical Pacific on Korean marine ecosystems and fisheries, to introduce the projected climate changes and discuss possible impacts to fisheries, and to propose the implication for fisheries management. Finally, we suggest a research direction to support an ecosystem-based fisheries management system capable of responding to changing climates and marine ecosystems.

Climate Effects on Marine Ecosystems

Long-term climatic changes are widely known to lead to larger changes in marine ecosystems, by affecting ocean currents and production patterns. Climate parameters such as air temperature and precipitation, often expressed or reflected by indices, have significant impacts on the ocean, and reconstruct its ecosystem. Pan-Pacific climate events would influence on physical properties of seawater and biological phenomena in the entire north Pacific. The most significant indices in the Pacific are the Pacific Decadal Oscillation (PDO) Index, Southern Oscillation Index (SOI), and Arctic Oscillation (AO) Index. Those indices were developed from different spatial coverage, and indicate different characteristics in periodicity. In the northeastern Pacific, there were two climate regime shifts in 1976/77 and 1988/89. The PDO index, which is the first component from EOF analysis using monthly SST anomalies in the North Pacific, shows such inter-decadal climate fluctuation over the north Pacific (Mantua et al., 1997). In Korean waters, some direct or indirect indications of changes were found around the late 1970s and late 1980s. Details of variability and relationships between indices are well described in Chang et al. (in press) and Otterson et al. (in press).

The Korean Peninsula is located in monsoon areas of the mid latitude where the Siberian High and the subtropical Pacific Low collide producing cold-dry winters and warm-wet summers. North and northwest winds in the autumn and winter are strong, and wind speeds easily reach 10 m s^{-1}. Winds reverse direction and become weaker in the spring and summer. Typhoons developed in the west subtropical Pacific bring heavy rains in the summer and autumn, and about nine typhoons pass through the region every year (Kim and Khang, 2000). East Asia is also influenced by global warming, and the increasing trend in air temperature over the Korean Peninsula is obvious. The rate of increase was estimated as 0.23°C/decade over the Korean Peninsula since 1950s (Fig. 1a). After the mid-1970s, the frequency of strong storms and heavy rain also increased. The precipitation pattern over the Seoul area indicated that 2-4 year wet and dry periods alternated since the 1800s except for a long dry period from

1880-1910. Recently, a wet period appeared in Seoul during the mid 1950s through the early 1970s, a dry period from the mid 1970s to the mid 1980s, and a highly wet period after the late 1980s (Fig. 1b). The AO index changed to positive and Asian monsoons weakened after the late 1980s, resulting in high temperatures and enhanced zooplankton biomass in the Tsushima Warm Current (Tian, in press).

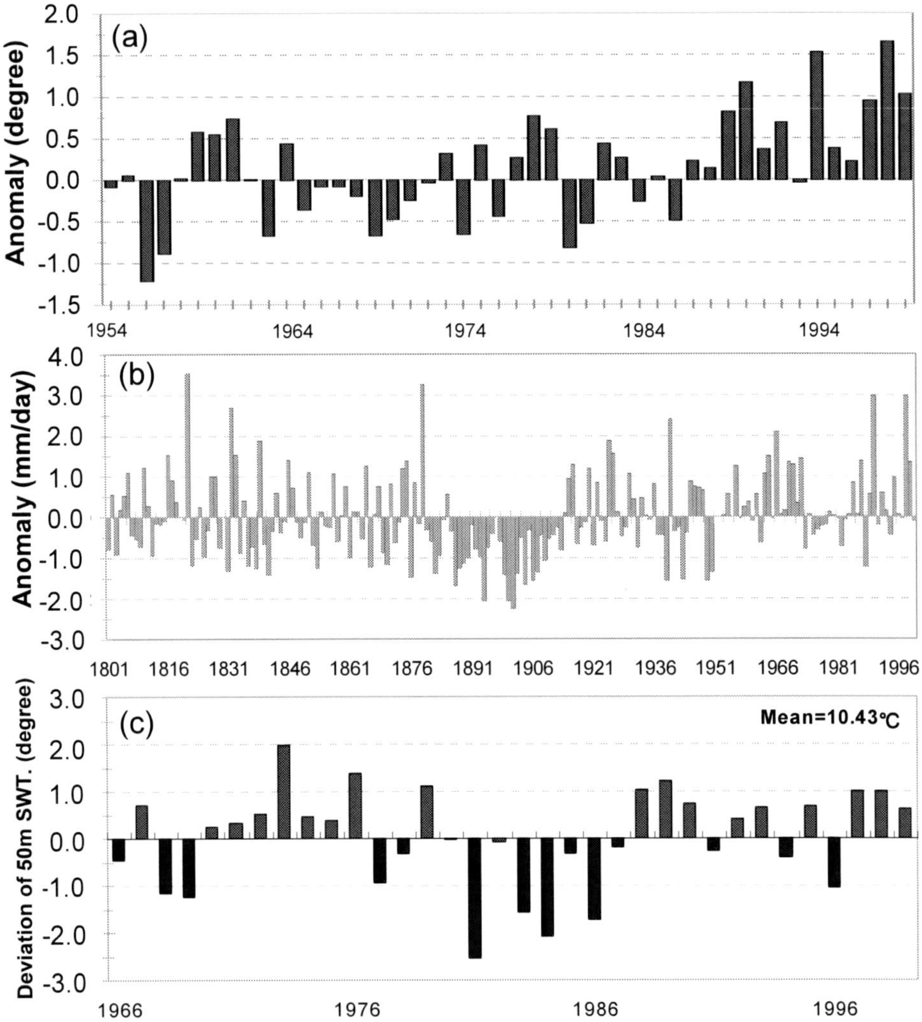

Figure 1. Time series of (a) mean air temperature anomaly over the Korean Peninsula, (b) precipitation in Seoul, and (c) change in seawater temperature in April at 50 m in the East/Japan Sea in last century (Kim et al., 2007).

Changes in environment in spring have an especially profound effect on oceanic productivity. Sea surface temperature (SST) is usually influenced by horizontal advection and direct exchange of heat through air-sea interaction. As indicated by Hahn (1994), there has been an increase of around 2°C in February SST in Korean waters during the past one hundred years. The increase in August water temperature was about 1.0°C. The rate of change has increased during the past decade. Hahn (1994) also showed a northward movement of isotherms during the same period.

Along with the long-term trend of warming, there was an alternate pattern of mean seawater temperature. In the surface layer (above 200 m) of the southwestern East/Japan Sea, warm water masses were prevalent during the early 1970s, cold ones during the late 1970s through mid 1980s, and warm ones again since 1988 (Fig. 1c). This phenomenon was typical in the spring and the autumn. It is also anticipated that the current pattern as well as subarctic front systems could be reformed due to climate variability in the East/Japan Sea. Park and Oh (2000) reported that the SST change over Korean waters also seemed to be connected to the El Niño/Southern Oscillation with phase lags of 5 to 9 months, which meant that a cold summer might occur in the East/Japan Sea after an El Niño winter. Kim and Kang (2000) also showed a significant relationship between SOI and December SST in the East China Sea.

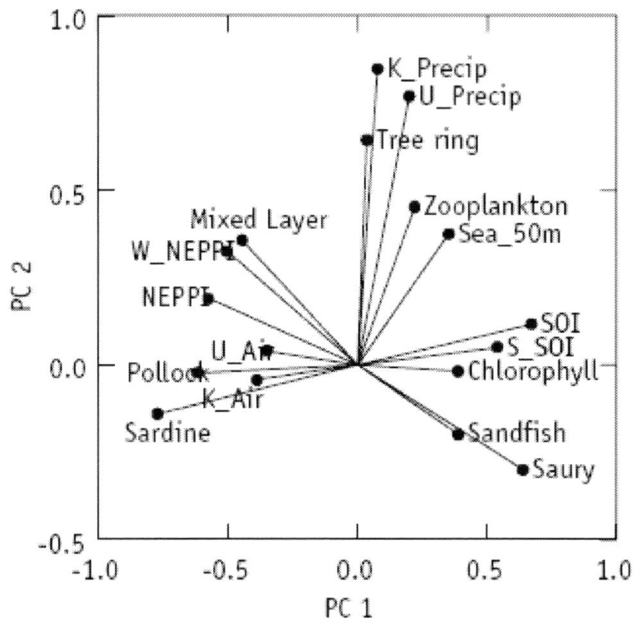

Figure 2. Principal component ordination of correlations between fish catches and environmental variables in Korean waters from 1960-1990. NEPPI (annual North East Pacific atmospheric pressure index), W_NEPPI (winter only), SOI (annual Southern Oscillation Index), S_SOI (spring only), U_Precip, U_Air, K_Precip and K_Air (spring precipitation and air temperature at Ulleung Island and Kangrung, respectively), and catches of different fish species (PICES, 2004).

In the East/Japan Sea ecosystem, differences in ecosystem structure and productivity have been observed during the 1960s through 1990s. Mixed Layer Depth (MLD) increased after 1977 and fluctuated around depths about 40 % deeper than prior to 1976 (Zhang et al., 2000). The mean transparency depth from Secchi disk observations was 11.9 m during the 1960-1975 period, but increased to 14.2 m during the 1976-1990 period. This result implied higher primary productivity (and consequently the zooplankton biomass) during the earlier period (Kang et al., 2000). Correlation studies indicated that large catches of some fish populations (saury and sandfish) occurred when the SOI was positive (i.e., La Nina period).

When spring chlorophyll was high, air temperatures were cooler in coastal cities and catches of sardine and walleye pollock were low (Fig. 2; PICES, 2004). Chiba et al. (2005) argued that planktonic community in surface layer reflected regime shifts in 1976/77 and 1988/89. It was also found that the rapid growth of pine trees in coastal areas appeared after a protracted El Nino event (Kang et al., 2000).

Climate Effects on Fisheries Productivity

Fluctuations in climate conditions can affect the distribution of fish and their growth and survival. A better understanding of how the climate change affects the growth, development and survival of fish during the critical early stages of their life history will allow better predictions of stock abundance, and the development of strategies that promote long-term sustainability of fish stocks.

The Northwest Pacific contains four marginal seas, the Yellow Sea, the East China Sea, the East/Japan Sea (which also is called the Japan/East Sea, the East Sea, or Sea of Japan), and the Okhotsk Sea. In the Northwest Pacific, over 300 marine fishery species are found, of which fishermen commercially utilize about one hundred species. Annual fishery landings in the Northwest Pacific were about 22.1 mmt in 2005, accounting for more than 23% of total global fishery landings of 94.6 mmt (FAO, 2007). Korean fisheries yield over 1 million metric tons (mmt) of marine fish from their waters annually. The increase in fishing activities since the 1970s has reduced a number of fish populations in Korean waters. FAO records indicate that total catches of fish and invertebrates from 1960 to 2004 averaged 1.14 mmt. The largest catch of 1.73 mmt occurred in 1986. Catches were relatively stable from 1981 to 1996 (over 1.4 mmt except the two years of 1991-1992), and have decreased slightly in recent years. Although it is not clear how major climate events influenced ecosystems and fisheries resources in Korean waters, some retrospective analyses with historic data sets have shown correlations between abiotic or biotic factors and fishery production (Kang et al., 2000; Kim and Kang, 2000; Kim et al., 1997; Seo et al., 2006; Zhang and Gong, 2005; Zhang and Lee, 2001; Zhang et al., 2000; Zhang et al., 2004).

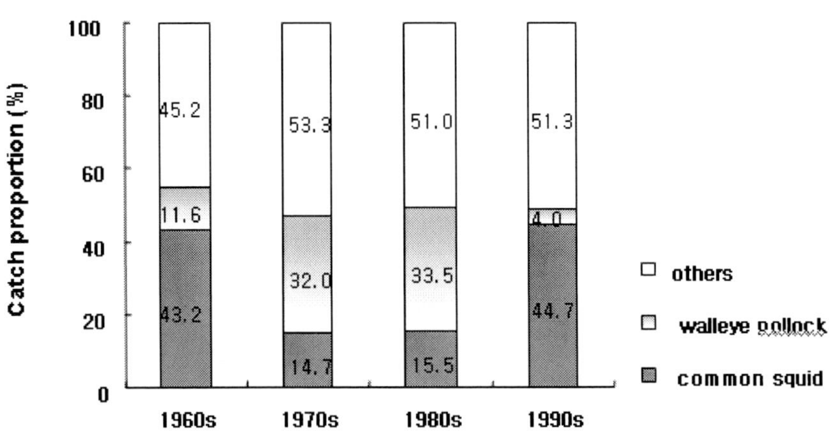

Figure 3. Catch proportions of the East/Japan Sea by species (1961-2000) (modified from Zhang et al. (2004)).

Most species in this region are migratory. Fishes in the Yellow Sea and the East China Sea spawn in coastal areas during the spring, feeding to the north during the summer, and overwintering after southward movement to the East China Sea. The demersal species in the East/Japan Sea seem to have an onshore-offshore or south-north migration seasonally. In addition to these ecosystems, surface layers of the Yellow Sea and the northern part of the East/Japan Sea are important habitats for anchovy and fingerling chum salmon, respectively. Such migratory behavior of species is likely important in distributing biological energy from one area to another in the northwestern Pacific (Kim, 2003).

Commercial catches in the East/Japan Sea have also indicated remarkable changes in species composition (Fig. 3). The ecological regime shift between small pelagic fish and demersal fish was especially evident. In the 1960s, common squid catch accounted for 43 % of the catch in the East/Japan Sea, followed by saury (15 %) and walleye pollock (12 %) (Zhang et al., 2004). Walleye pollock catches subsequently increased, constituting about 33 % of the catch in the 1970s and the 1980s, and concurrently, the proportions of common squid greatly decreased. The dominant fisheries catch also shifted to common squid in the 1990's accounting for about 45 % of the total catch. The abundance of small pelagic fish and invertebrates has increased in the Tsushima Warm Current ecosystem. In recent years, seven major small pelagic species comprised around 57 % of the total catch, and common squid alone around 20 % (Kim et al., 2007).

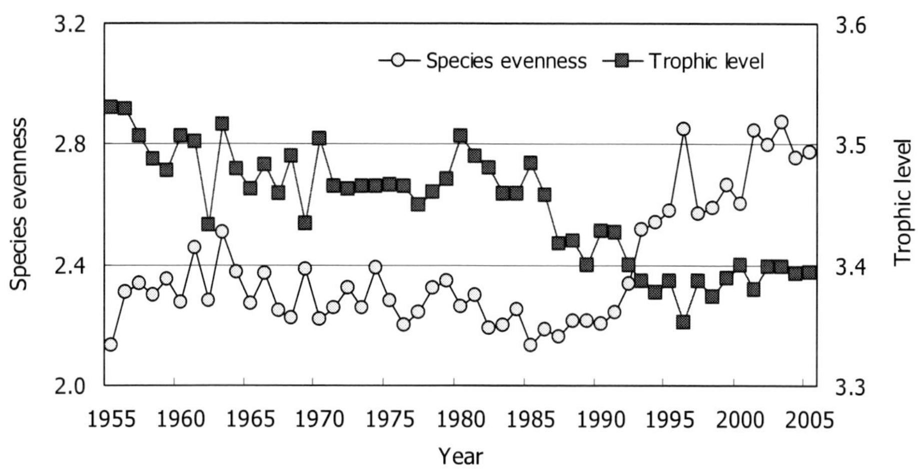

Figure 4. Trophic level and species evenness of fisheries resource in Korean waters (Kim et al., 2007).

The mean trophic levels of resource organisms from Korean waters showed a decreasing trend during the four decades (Fig. 4). Mean trophic level was about 3.46 until the early 1980s, but decreased rapidly during the mid 1980s through early 1990s, and stabilized to about 3.40 after the late 1990s. Demersal higher trophic level fish (e.g., small yellow croaker) gradually decreased, and lower trophic pelagic fish (e.g., anchovy and common squid) increased. More specifically, the decline of the trophic level for the Yellow Sea was steeper than for the East China Sea. In the East/Japan Sea, however, the mean trophic level increased from 3.09 to 3.28 during the 1976 regime shift period (Zhang et al., 2004). The 1988/89 regime shift event changed biomass and production of fisheries resources in the southwestern

Japan/East Sea. Total biomass of all species groups increased by 59 % after this regime shift. The relative contribution of walleye pollock, at trophic level III, to the total flow of energy decreased drastically from 33.0 % to 4.3 %, while that of common squid, at the same trophic level, doubled from 34.2 % to 72.2 % during the period (Zhang et al., in press). An increase in the evenness of fish species in Korean commercial fishery production occurred in the 1990s with species replacement from demersal to pelagics (Fig. 4). Three major small pelagic species, i.e. chub mackerel, anchovy, and common squids occupied about 55% of total commercial catches and averaged about 650 thousand mt during 2003-2005 (Kim et al., 2007).

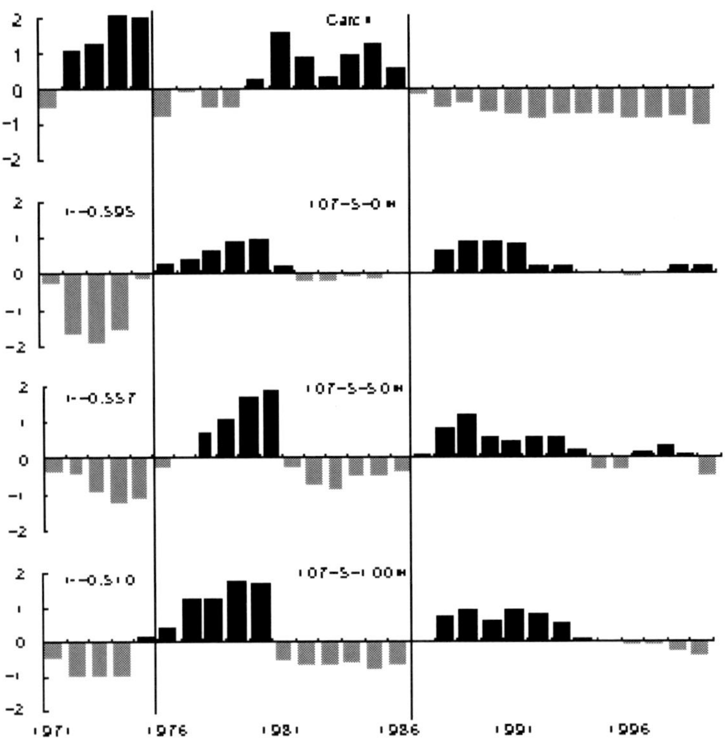

Figure 5. The relationship between walleye pollock catch and seawater temperature (0-100 m) anomalies off the east coast of the Korean Peninsula. Seawater temperatures used were collected from the KODC's oceanographic observation line, 107-5 (modified from Kim et al., 2007).

In the subarctic demersal ecosystem of the East/Japan Sea, walleye pollock are cold-water and demersal species, and their spawning grounds in Korea are at the southern limit of their distribution. The majority of the walleye pollock population resides in North Korean waters. Annual catches in South Korea increased from the early 1970s and peaked in 1981 at 1.7 mmt. Since then, catches declined continuously. The fishing areas in the 1990s were restricted to coastal areas in the north apparently because of the higher SST of the East/Japan Sea. This contrasts to a broader fishing area in the late 1970s when SSTs were cooler. There is a negative correlation between fish catch and local seawater temperatures at a major fishing area off the east coast of the Korean Peninsula (Fig. 5). Negative anomalies in seawater temperature (0~100 m) during the early 1970s and early 1980s were coincident with a positive anomaly of walleye pollock catch, while warm water masses in the late 1970s and late 1990s resulted in low walleye pollock catches. A shift in fishing season was also noticed in the 2000s. The highest catches were recorded in January-March during the 2000s, compared to November-December during the 1980s. Kang et al. (2000) also found that walleye pollock catch is significantly related to the Northeastern Pacific Pressure Index (Fig. 2).

In the arctic demersal ecosystem of the Okhotsk Sea, walleye Pollock biomass increased and reached its highest historical level in the far-eastern seas after the 1976/77 regime shift. In the first half of the 1970s, there were not strong year-classes of walleye Pollock in the Sea of Okhotsk. Abundant year-classes appeared after 1976, especially in the northern part of the Sea of Okhotsk. Thus, the 1976/77 climate regime shift was also observed in the Sea of Okhotsk and was characterized by a general sea surface warming, intensive cyclogenesis in the overlying atmospheric domain, and the intensification of water exchange with the Pacific Ocean (Zhang et al., 2006).

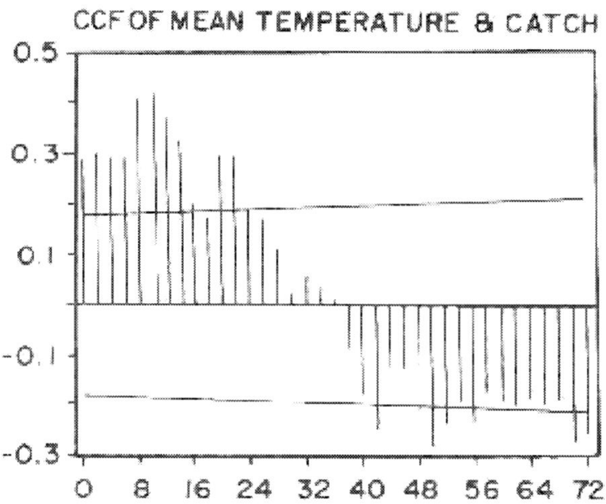

Figure 6. Cross-correlation coefficient between seasonal anomalies of small yellow croaker catch and seawater temperature at 75 m (modified from Kim et al., 1997).

In the subtropical demersal ecosystem of the Yellow Sea and the East China Sea, small yellow croaker is one of the most important commercial fish. This fish migrates to the East China Sea in the winter and returns to the Yellow Sea to spawn in the spring. The recruitment success of small yellow croaker is related to seawater temperature in the deep layer during the early life history. It was found that cold and variable seawater temperatures at 75 m during the early life periods resulted in poor yields or weak year-classes when they recruited to the fisheries (Kim et al., 1997). The correlation between the seasonal anomaly of mean temperature at 75 m depth and the seasonal anomaly of fish catch was highly positive and significant, with a time lag of 8-14 months (Fig. 6). This phenomenon reflected recruitment processes, since the catches of small yellow croaker are made up primarily of fish younger than age two.

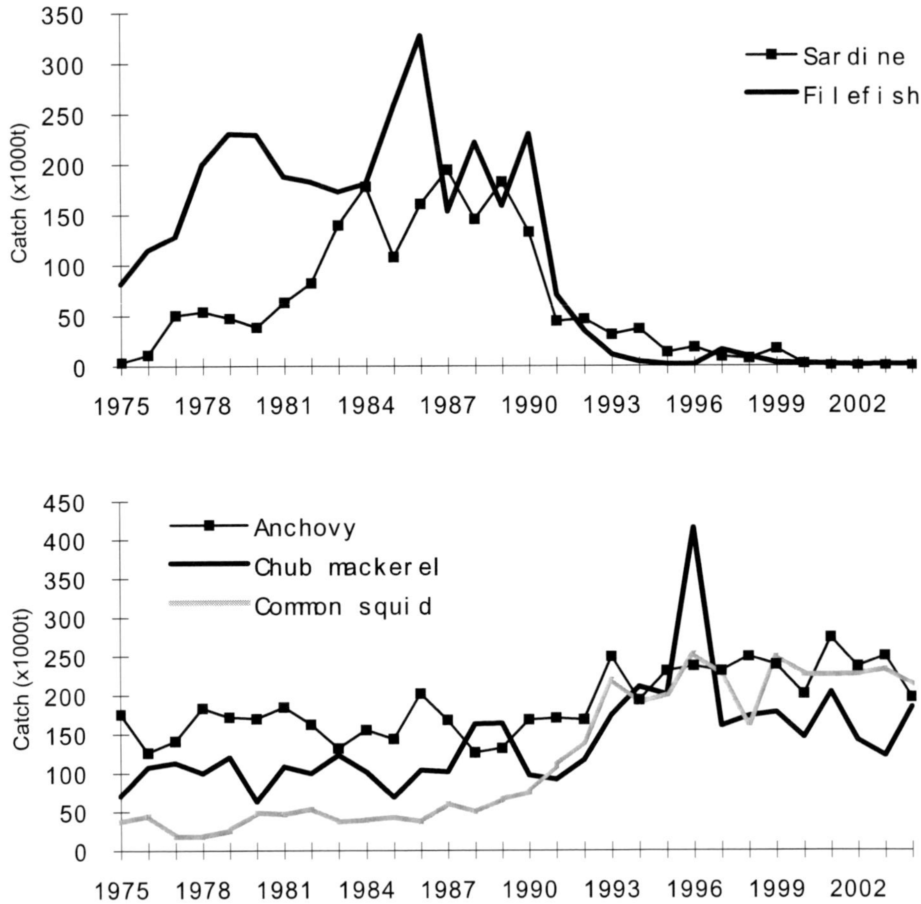

Figure 7. Two types of catch pattern for small pelagic fish in the Tsushima Warm Current ecosystem during the last three decades (Kim et al., 2007).

In the subtropical pelagic ecosystem of the Tsushima Warm Current region, two patterns of catches have characterized during the last three decades (Fig. 7): sporadic increase/decrease and continuous increase with time. Sardine and filefish were not abundance before the early 1970s, but were very abundant in the 1980s, then decreased in the early 1990s. On the other hand, the catches of chub mackerel, anchovy, and common squid showed an increasing pattern since the 1970s. Chub mackerel and common squid increases were particularly abrupt in the early 1990s, and stable throughout the 1990s. Most species share the same spawning and nursery grounds, but peak spawning periods are separated to avoid competition for food.

Filefish occupy a relatively broad range of habitats both horizontally and vertically in Korean waters. They migrate and spawn in the East/Japan Sea and the Yellow Sea within the warm current from April or May. Some schools reach the North Korean coastal areas and spend the summer there. Another group then migrates to the Yellow Sea occupying the whole Yellow Sea in June, and migrates south in October. The production of filefish increased after the late 1970s, and the fishery collapsed in the early 1990s due partially to overfishing (Fig. 7). Zhang et al. (2000) argued that the increased volume transport of the Kuroshio Current and MLD in the East/Japan Sea during 1976-77 might be responsible for the increase in the filefish catch (Fig. 8).

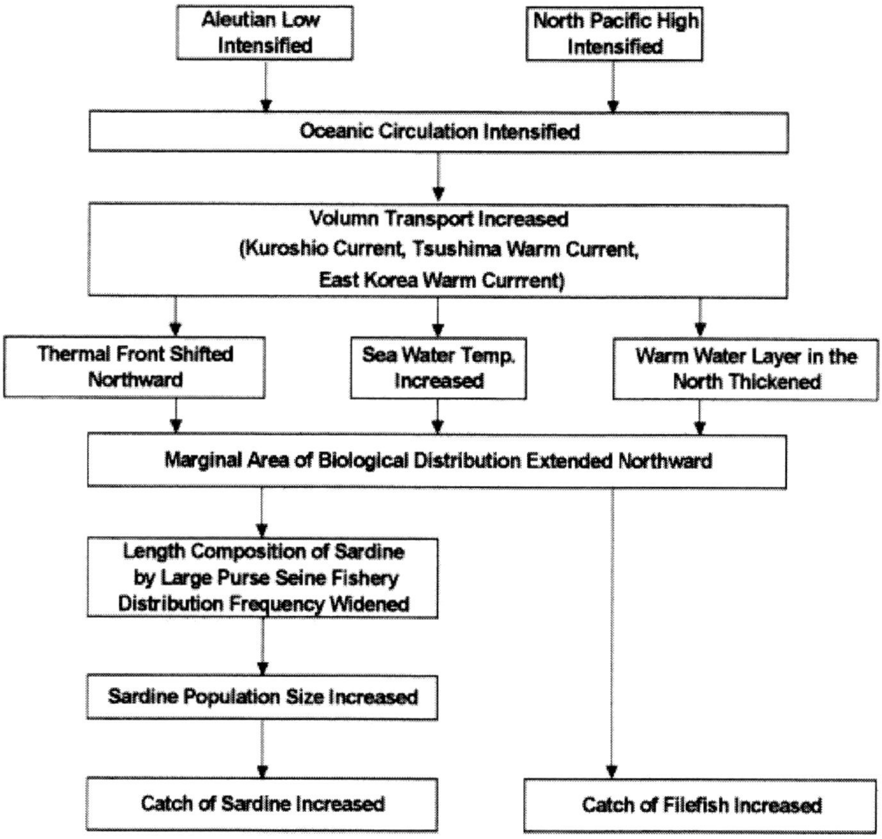

Figure 8. Diagram showing the effects of the 1976 climate-driven regime shift on filefish population in Korean waters (modified from Zhang et al., 2000).

Sardine spawn off the southeastern coast of the Korean Peninsula from December through June with a peak in February-April. They migrate north in the summer to feed near 45° N, along the east coast of the Korean Peninsula. In November, they move to the southern coast of the Korean Peninsula. Variability in the Kuroshio current, which is related to the transport processes for sardine eggs and larvae in Japanese waters, plays a critical role in the fluctuation of sardine populations (Kasai et al., 1995; Yatsu et al., 2005), in addition to the effect of exploitation. In Korean waters, the abundance and geographical coverage of sardine eggs and larvae were high in the 1980s when spawning biomass was high, and vice versa in the 1970s and the 1990s (Kim et al., 2006). Decadal-scale changes in seawater temperature at 50 m during the spring indicated cool temperatures during the early to mid 1980s followed by warm waters since the late 1980s. Also, feeding and spawning areas based on fisheries information exhibited the same pattern of extension/contraction as seen in ichthyoplankton surveys, onforming to

MacCall (1979)'s 'basin' model. The sardine catch had a high positive correlation with chlorophyll-a concentration in April, and negative correlations with chaetognaths in April and euphausiids in June (Table 1).

Table 1. Selected correlation coefficients, which were statistically significant for fish catch vs. their biotic and abiotic environmental factors. Sampling month in parenthesis; * indicates that the correlation is significant at the 0.05 level, and ** at the 0.01 level (Kim and Kang, 2000).

Variable	Anchovy	Chub mackerel	Sardine
Anchovy	1.0	0.790**	0.453*
Chub mackerel	0.790**	1.0	0.602**
Sardine	0.453*	0.602**	1.0
SST (Dec.)	0.419*	0.436*	0.327
Chl. a (Apr.)	0.186	0.019	0.561**
Chl. a (Jun.)	0.635**	0.523**	0.264
Chl. a (Aug.)	0.442*	0.377	0.276
Zooplankton (Feb.)	-0.559**	-0.406*	-0.339
Zooplankton (Apr.)	-0.304	0.408*	-0.291
Copepods (Apr.)	0.563*	0.434	-0.398
Copepods (Jun.)	0.121	0.571*	-0.042
Copepods (Dec.)	0.635*	0.477	-0.277
Chaetognaths (Apr.)	0.647**	0.307	-0.499*
Chaetognaths (Oct.)	0.728**	0.512*	-0.321
Chaetognaths (Dec.)	0.558*	0.129	-0.427
Euphausiids (Jun.)	0.349	0.356	-0.550*
Euphausiids (Dec.)	0.768**	0.603*	-0.492
Amphipods (Apr.)	0.713**	0.504*	-0.395
Amphipods (Dec.)	0.712**	0.616*	-0.423

Sardine populations also seem to be influenced by the regime change that occurred at the late 1970s and late 1980s. Catches were very high between these two events. The fluctuation pattern of sardine catches in Korean waters generally coincided with that of Japanese sardine, although the Kuroshio Current off Japanese islands has no direct link to Korean populations. Zhang et al. (2000) hypothesized that the increased volume transport of Kuroshio in the East China Sea resulted in high catches of filefish and sardine due to the strong influx of the warm Tsushima Current into the Korean Strait and the East/Japan Sea (Fig. 8). Intensity of the Kuroshio Current has increased since 1976. As a result, the intensity and frequency of spring blooms decreased in the East/Japan Sea and the intensity of autumn blooms increased. Sardines are resident in the East/Japan Sea from June to November, therefore it was speculated that an autumn bloom was more advantageous to them than a spring bloom. This could account for the dramatic increase in the sardine stock after 1976 (Zhang et al., 2000). However, there is no clear understanding of historic fluctuations in the sardine population of the northwestern Pacific.

Anchovy occur primarily in the Korea Strait and the southern Yellow Sea during winter, and migrate shoreward to the southern coast of Korea to spawn in April-August. In many ecosystems around the world, abundance of anchovy and sardine alternate in time (Chavez et al., 2003). In Korean waters, however, this phenomenon did not occur. Figure 7 shows that the anchovy catch has increased continuously, even though sardine catches were high in 1980s. Elevated seawater temperatures in December showed a positive correlation with El Nino, and seemed to cause high growth for anchovy, and strong year-class for those populations in Korean waters (Kim and Kang, 2000). Statistically, the seasonal and long-term trends of embryonic mortality, egg production and spawning stock biomass of anchovy can be explained largely by spring warming, summer cooling and by less abundant zooplankton in the late 1980s (Kim and Lo, 2001) A matrix of simple correlation coefficients among planktonic organisms and fish indicated that the anchovy catch had a high correlation with chlorophyll-a concentrations in June and August, and with large zooplankton such as chaetognaths, euphausiids, and amphipods in fall and winter (Kim and Kang, 2000; Table 1).

Chub mackerel are widely distributed in the coastal areas of the northwestern Pacific. The Korean stock of this species stays in the northeastern East China Sea during December-February, and spawns between Jeju Island and Tsushima Islands from April to June (Hwang and Lee, 2005). The 1988 climate regime shift affected the habitat of chub mackerel by widening and moving it to the west of 128°E, and the distributional overlap of chub mackerel and Jack mackerel decreased after 1988 (Zhang et al., 2004). Mean length of chub mackerel in the fishery tended to decrease after the 1970s. Fork length was about 32 cm in the early 1970s, but decreased continuously, and was 29.2 cm in 2002. Recruitment was found to be correlated with salinity ($r=0.454$, $P<0.05$), zooplankton biomass ($r=0.692$, $P<0.01$), and copepod biomass ($r=0.815$, $P<0.01$) (Choi et al., 2000). Chub mackerel catches were also significantly correlated with SST in December, chlorophyll-a in June, and large zooplankton in the fall and early winter (Table 1).

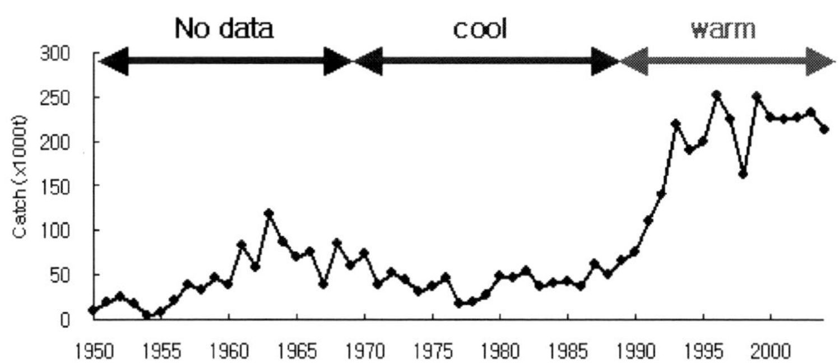

Figure 9. Total catch of common squid (*Todarodes pacificus*) in Korean waters. Alternation of warm and cool water temperatures during winter were indicated by arrows (Kim et al. 2007).

The life span of common squid in the northwestern Pacific has been assumed to be one year (Nakamura and Sakurai, 1993), and their spawning seasons in the East China Sea and the southern East/Japan Sea are spread throughout the year: i.e., summer (June-August), autumn (September-November) and winter (January-March) spawning groups

(Kim et al., 1997). Spawning grounds have not been identified precisely, but it is known that successful spawning locations are selected based on seawater temperature, bottom topography, and pycnocline depth (Sakurai et al., in press). In Korean waters, common squid catches increased abruptly during the early 1990s (Fig. 9), primarily due to fisheries in the East/Japan Sea. Seawater temperature might be important parameter controlling stock abundance. When the common squid catch increased in the early 1990s, the SST in Korean waters increased concurrently. Seawater temperature during winter seemed to be more important than other seasons. Winter temperature rose in the 1990s, and the winter population of common squid showed a rapid increase compared to the autumn population. A correlation analysis indicated that zooplankton biomass in October and December of the previous year (i.e., when common squid were small) showed an extremely significant correlation with common squid catches in September-December in present year. Euphausiids and amphipods are important food organisms for the survival of young common squid in the East/Japan Sea (Kang et al., 2002).

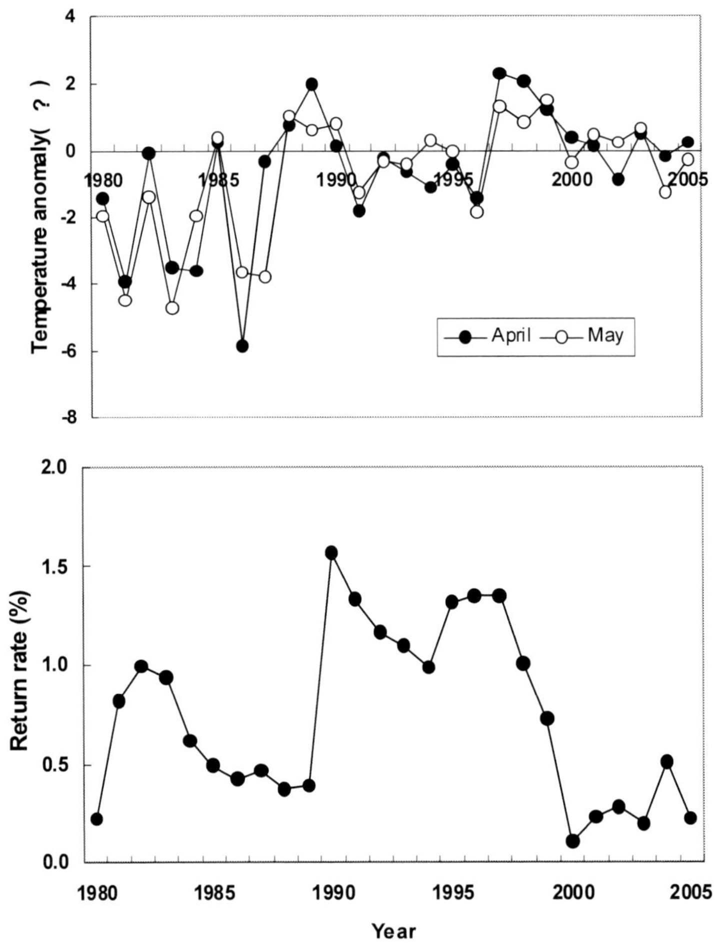

Figure 10. Seawater temperature anomalies near salmon hatcheries and the return rate of Korean chum salmon. Note that the return rate before 1990 does not include salmon catch in coastal areas of Korea (Kim et al. 2007).

Table 2. Means and standard deviations of seawater temperature in eastern coastal area of the Korean Peninsula during April and May, 1988~1990 and 1997~1999 periods (Kim et al., 2007)

	April	May
1988	11.2 ± 1.0 n = 28	14.7 ± 1.1 n = 30
1989	12.4 ± 0.7 n = 28	14.3 ± 0.9 n = 30
1990	10.6 ± 0.3 n = 27	14.5 ± 2.0 n = 30
1997	12.7 ± 1.0 n = 21	15.0 ± 0.4 n = 25
1998	12.5 ± 1.7 n = 24	14.5 ± 1.8 n = 23
1999	11.7 ± 1.5 n = 24	15.2 ± 1.1 n = 23

In the subarctic pelagic ecosystem of the East/Japan Sea, chum salmon are cold-water species spending in the surface layer of the ocean through most of their life. Because of the wide distribution of salmon over the course of their life, environmental conditions at specific areas influence salmon growth, migration route, and returning rate. Salmon hatcheries in East/Japan Sea coast are at the southern range of their distribution and would be expected to be negatively affected by a warming of the ocean surface. The homing success of chum salmon might depend on seawater temperature in the coastal areas because high temperatures have resulted in mass mortality of salmon fry when released. The returning rate of chum salmon released from Korean hatcheries is below 1.5 % which is very low compared to American and Japanese chum salmon. There was a negative correlation between return rates and SST during release time in the coastal areas. It has been known that 14 °C is the upper temperature limit for chum salmon fry in coastal areas (Mayama and Ishida. 2003), so that high temperatures in the spring of 1988-1990 and 1997-1999 appeared to be detrimental to the survival of released salmon fry, and resulted in the lower return rates of the spawners in 1991-1994 and 2000-2002 spawning periods, respectively (Fig. 10; Table 2). Actually, in the coastal areas of the East/Japan Sea, growth rates of fingerling chum salmon were higher in the 1990s than in the 1980s. The increased zooplankton abundance in the East/Japan Sea since the late 1980s might cause favorable growth conditions for young salmon in the 1990s (Seo et al., 2006).

Projection of Climate Changes and Possible Impacts to Fisheries

Based on the SRES A2 CO_2 emission scenario, future changes in air temperature and precipitation over the East Asia have been forecast (Fig. 11). The warming rate of annual mean temperature over the Korean Peninsula and the East Asia was projected as 0.61°C

/decade during the 2001-2100, so a warming of 5.5 °C is expected in the East Asia by the end of this century. Annual precipitation is expected to increase as much as 0.07 mm/day by 2100, and we could have about 2.6 % more precipitation annually in comparison with current annual precipitation (2.68 mm/day). However, the number of rainy days diminished by 19 days in 2100, illustrating that we may have more precipitation and fewer rainy days. Consequently, precipitation intensity (PI) will increase.

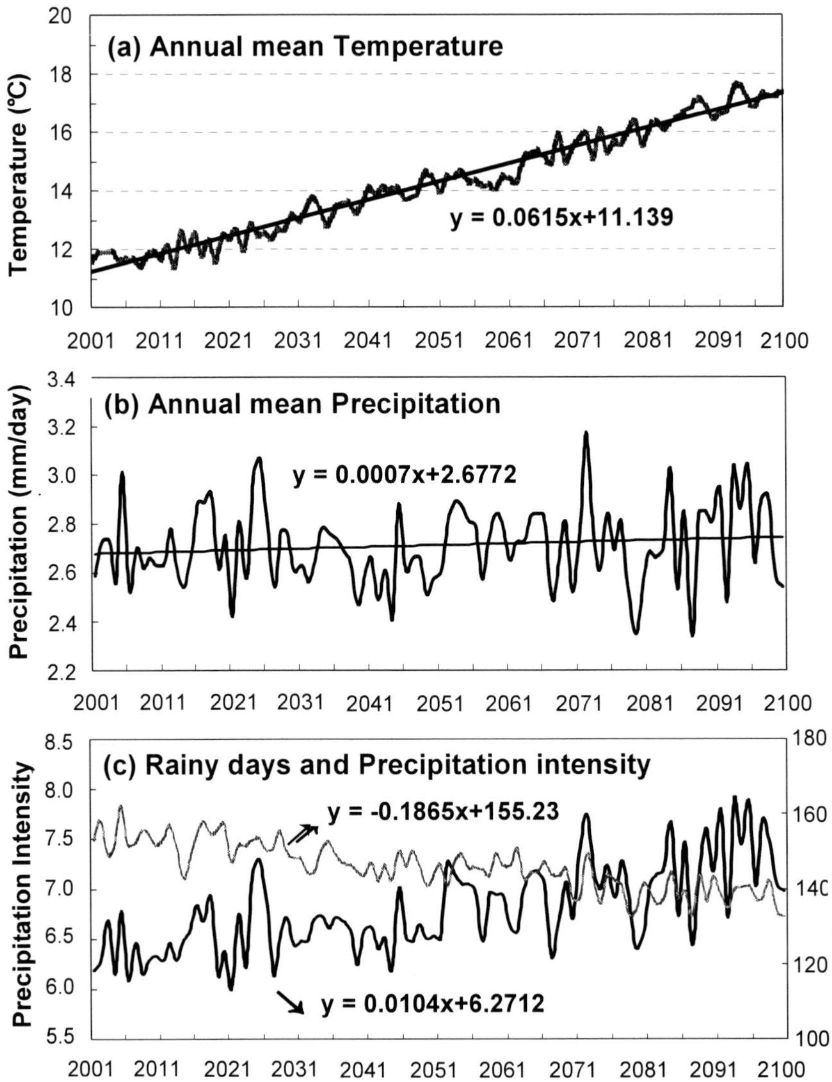

Figure 11. Changes in climatological parameters Korean Peninsula and the East Asia for the next 100 years (2001~2100) by the regional climate model based on the SRES A2 greenhouse emission scenario. (a) Annual mean temperature, (b) annual mean precipitation (mm/day), and (c) number of rainy days (light line, right scale) and precipitation intensity (thick line, left scale) over East Asia (Kim et al. 2007).

Based on the projection from Hahn's (1994) schematic temperature isotherms for the last 100 years, higher SST will prevail in the future due to the accumulation of CO_2 in atmosphere. Modeling predictions indicated a 2~3 °C increase in SST in the North Pacific Ocean within 50 years (Mikolajewicz et al., 1990). The climate chart based on SST observations in the Northwest Pacific during the 1960s~1980s period (JMA, 1991) showed that the effect of a 3 °C increase is equivalent to a 500 km northward shift of isotherms on the Pacific side of Japanese islands (Kim, 1995). Furthermore, thermal expansion of the sea surface due to warming might cause a different circulation pattern in the Northwest Pacific.

The marine ecosystem will probably be reorganized in accordance with climate/oceanographic changes. Because different fish species have different life cycles and habitat areas, the impact of large-scale climate change will vary with species. If the Aleutian Low intensifies under a global warming scenario, the oceanic circulation, consequently, the volume of the Kuroshio Current would be stronger, resulting in stronger plankton blooms in the autumn and favorable conditions for sardine and filefish productions. If the more frequent and intense Aleutian Lows observed in the 1980s occur in the future, periods of high sardine abundance may be more frequent.

Some other species such as saury and sandfish seem to be more related to ENSO than the Aleutian Low (Fig. 2). The period of more frequent and intense El Nino in the mid-1990s appeared to be favorable for saury production. Sporadic events such as high seawater temperature influence survival and recruitment of fish populations, although uncertainty in mechanistic processes between environment and biological phenomena still remains. Prediction for future fish stock conditions does not look like feasible given current scientific knowledge on the climate variability and cause-effect relationships among ecosystem components.

The retreat of cold water species and colonization of warm water species will be apparent in Korean waters in the 21st century. Actually, the abundance of small pelagic fish including common squid has been increasing in the Tsushima Warm Current system since the early 1990s when SST increased. Far southern species such as spotted mackerel (*Scomber australasicus*) comprised a higher proportion of the catch of mackerel-like pelagic species than before, and the winter catch of yellowtail (*Seriola quinqueradiata*) increased remarkably in the early 1990s when there were higher air temperatures, weak wind speeds, and warmer seawater at 50 m off the southern coast of the Korean Peninsula (Lee and Go, 2006). If wind intensity is reduced, with an SST increase, then common squid would be abundant, although the intensification of the Aleutian Low and El Nino cannot be predicted.

Coldwater species such as walleye pollock in the East/Japan Sea would tend to be reduced in abundance. Sakurai et al. (in press) and Kim et al. (1997) hypothesized an increase in common squid and small yellow croaker abundance, based on the observation that stable and warm ocean conditions during early life stages are beneficial for those populations. Although adults of walleye pollock show demersal behavior, young fishes (i.e., egg and larval stage and early juveniles) stay in the surface layer of the ocean where climate effects are maximal. One typical characteristic of the walleye pollock catches in Korea is the inclusion of immature juvenile walleye pollock (called 'small walleye pollock' of ages 1-2 in Korean statistics). Juvenile walleye pollock stay in upper layer near the coastal area. The proportion of immature pollock was higher than 90 % in the late 1970s, decreased continuously to 40 % in 1987, then increased to 63 % in 1990. A future warming trend of SST at spawning and nursery grounds could have a profound effect on the zooplankton community as well as early life stages of walleye pollock. Larvae and juveniles might benefit from the enhanced seawater temperature and prey organisms in coastal area, while high temperatures near spawning areas could have a negative effect on the population.

The salmon enhancement program plans to increase the number of hatchery fry released. Continuous warming in the East/Japan Sea at the time of fry release could reduce fry survival in coastal areas, so that a general warming trend in Korean waters would probably result in low chum salmon abundance and low returning rate, even though climate/environmental changes in the open ocean could improve growth (Seo et al., 2006). A re-location plan for hatcheries and replacement of chum salmon with species which are more resistant to higher temperature should be considered.

More than 90 % of world aquaculture products are from the northwestern Pacific area. In Korea, one third of marine products are currently obtained from marine aquaculture. The increase in seawater temperature accelerates physiological responses of organisms as well as degradation of environmental quality, which may result in the collapse of the aquaculture industry unless target species are changed. In particular, lower dissolved oxygen due to enhanced seawater temperature will be harmful to organisms in confined water mass such as lakes or cages.

Implications for Fisheries Management

Traditional Fisheries Management in Korea

Most fisheries resources in Korean waters are straddling stocks migrating among the waters of the three neighboring countries. Most fish stocks are currently known to be over-exploited, and some stocks are depleted due to the increase in fishing intensity and over-capitalization of fishing fleets. Other probable reasons for the depletion are land reclamations and coastal pollution, which destroy spawning and nursery grounds along the coastal regions of the three countries.

Although knowledge of the status of many species harvested in commercial fisheries is limited, many stocks are generally considered to be declining. Several commercially valuable species, such as small yellow croaker, white croaker, red seabream, sharp-toothed eel and some other demersal fish species are considered to be heavily exploited or even depleted (Zhang et al., 1992; Zhang et al., 1998; Zhang and Kim, 1999; Zhang et al., 1999; Lee, 1999).

For a long time Korea has relied on traditional control devices to manage its fisheries. These included mesh size restriction, minimum size of fish limits, closed areas/seasons, boat licenses, and gear limitations. There are more than 300 exploited species harvested by about eighty thousand fishing vessels using 37 (17 for coastal and 20 for offshore) gear types (Zhang and Marasco, 2003).

After extending its fisheries jurisdiction in 1994, the Korean government adopted a TAC-based fisheries management system. This system was implemented in 1999. Currently, TAC-based management applies to ten species (common mackerel, Jack mackerel, Pacific sardine, queen crab, snow crab, blue crab, common squid, hen cockle, spiny top shell and pen shell) for several related gear types (large purse seines, large offshore pots, divers, offshore gillnets, jigging, and community common fishery) operating under the system.

A number of difficulties are associated with traditional control devices and the TAC-based management system. For example, mesh and minimum landing size regulations were adopted to avoid the dangers of harvesting fish before they reach full maturity. Such measures alone, however, did not prevent fishing effort from increasing, since the number and power of vessels entering were not limited. In addition, regulations to increase mesh size further were not introduced because of industry resistance. Immediate economic pressures stemming from short-term losses resulted in the discounting of future benefits likely to be realized from the introduction of such measures. Closed areas were implemented to allow stocks to reproduce and grow by reducing fishing mortality. However, adequate funds for monitoring were not available. The TAC-based management system also faces significant problems in the estimation of acceptable biological catch, monitoring and enforcement. The system relies on assessments of fisheries resources based on traditional single-species approaches (Fig.12), rather than ecosystem-based approaches (Fig. 13).

Ecosystem-Based Fisheries Assessment and Management

As the twenty-first century unfolds, the world now faces unpredented challenges and problems. The ongoing struggle over fisheries resources is merely a microcosm of a much larger struggle in which diverse interests compete for the use of water for fishing, industrial needs, waste desposal, recreation, and transport. The three basic and interrelated problems of fisheries are: the threat of over-exploiting the resource, overcapitalization or over-expansion of fishing fleets, and the negative consequences of fisheries on the survival of marine life and associated habitats. The diverse, dynamic, and complex nature of fishery management problems necessitates the development of a process that increases interdisciplinary and interagency interactions, as well as the definition of biological indicators to be applied. Resource managers, researchers, user groups and other interested parties must be involved in the process to minimize conflict and maximize commitment to sustainable management (Zhang and Marasco, 2003).

The world's oceans are at or near maximum sustainable fishery yields. The number of overexploited stocks increased by 2.5 times between 1980 and 1990 (Alverson and Larkin, 1994). Much of the global sustainable yield is maintained by increased fishing for species at progressively lower trophic levels (Pauly et al., 1998) and the prospect of increasing total sustainable yield is unlikely (Pauly and Christensen, 1995).

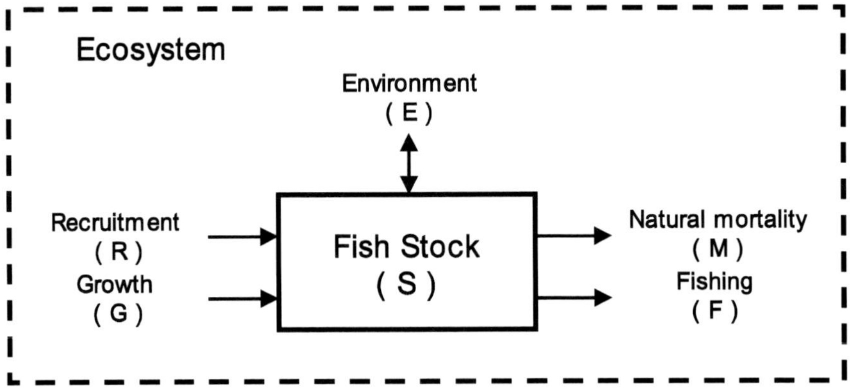

Figure 12. Factors related to abundance of fish stock (Zhang, 2002).

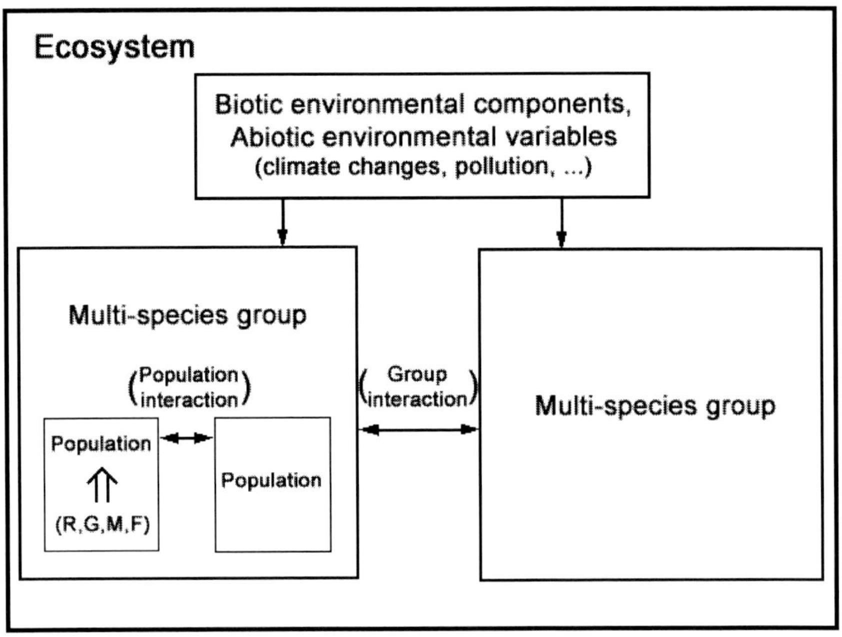

Figure 13. Interactions of organisms with biotic and abiotic environments in a marine system (Zhang, 2002).

In this situation concern is growing over the effects of fishing on the ecosystem. Fisheries are managed within a setting that lacks full information on, for example, fish population dynamics, interactions among species, effects of climate environmental factors and human activitiy on fish and their ecosystem. Recognition of this uncertainty and its potential consequences has led to the adoption of a precautionary approach in UNCED Rio Declaration (Principle 15), the 1995 UN Agreement on Straddling and Highly Migratory Fish Stocks, the 1995 FAO Code of Conduct for Responsible Fisheries and the 2002 WSSD's Implementation Plan. The precautionary approach is focused on reducing the likelihood of fisheries having adverse impacts on marine resources and host ecosystems (FAO, 1995).

The consequences of overfishing are expressed in social, economic, cultural, and ecological changes. However, the ecological consequences of overfishing are often undocumented and may be poorly known or overlooked. In many cases, the ecological correlates of changing fish populations could have served as evidence of intensified exploitation effects. Frequently, we recognize ecological consequences after the fact, because we do not consider them in our management decision and do not monitor ecosystem changes properly.

Fisheries scientists and managers of marine living resources are recognizing the need for a paradigm shift in fisheries management from traditional single-species approaches to ecosystem-based approaches in recent years. Traditional single-species approaches only consider four population factors, namely recruitment, growth, natural mortality and fishing (Fig. 12) and assume a constant environment. Ecosystem-based approaches consider all ecological interactions among related animals at various levels (Fig. 13). The stocks exploited by fisheries are parts of marine ecosystems where many species feed on one another. Therefore, sustainable fisheries must consider a holistic view as shown in Fig. 13. An ecosystem view involves the use of other tools of management in addition to fisheries regulation, such as stock and productivity enhancement, provision of physical structure, or marine protected areas.

To understand the mechanisms behind climate change and variations in biological productivity it is necessary to carry out proper cooperative interdisciplinary research on climate, as well as physical and biological aspects of ocean ecosystems. In the early 1990s when it had become obvious that interdecadal environmental changes profoundly affected marine ecosystems in the North Pacific. Of particular interest was the impact of climate on recruitment, abundance, and distribution of commercial fishes, which suggested that early recognition of such events could be useful for fishery management (Wooster and Zhang, 2004). However, identifying a regime shift is much easier than understanding the process determining it. Unraveling the processes behind multi-decadal variability and how they affect ocean ecosystems and biogeochemical cycling will require an integrated observational and modeling effort. Such efforts should be expanded to global scales. Measurement networks will be required for ocean physics, ecology, and biogeochemistry (Chavez et al., 2003).

There is some evidence that appropriate ecosystem data, which integrate the higher frequency variation, when combined with physical data might have advantages for early detection of major environmental change. Fishery data have limited value for this purpose because of inadequate sampling, length of fish life cycles, and the complicating effects of fishing. Selected zooplankton species might be useful early indicators of environmental change if sampling coverage could be increased and speedy means of identification developed (Wooster and Zhang, 2004).

Table 3. Indicators and reference point for detecting climate changes in the ecosystem-based resource assessment system (Zhang et al., in prep)

Attribute	Indicator	Reference point		
		Target (0)	Limit (1)	Beyond Limit (2)
Recruitment	Deviation from S-R curve (S-R)	$\delta(S-R) \leq 0.05S-R_{Normal}$	$0.05S-R_{Normal} < \delta(S-R) \leq 0.10S-R_{Normal}$	$\delta(S-R) > 0.10S-R_{Normal}$
	Position of S and R	$\delta(S \text{ or } R) \leq 0.05(\text{mean } S \text{ or mean } R)$	$0.05(\text{mean } S \text{ or mean } R) < \delta(S \text{ or } R) \leq 0.10(\text{mean } S \text{ or mean } R)$	$\delta(S \text{ or } R) > 0.10(\text{mean } S \text{ or mean } R)$
	Anomaly of pre-recruits	$\delta(R) \leq -0.05\left\lvert\frac{R_i-\overline{R}}{\overline{R}}\right\rvert$	$-0.05\left\lvert\frac{R_i-\overline{R}}{\overline{R}}\right\rvert < \delta(R) \leq -0.10\left\lvert\frac{R_i-\overline{R}}{\overline{R}}\right\rvert$	$\delta(R) > -0.10\left\lvert\frac{R_i-\overline{R}}{\overline{R}}\right\rvert$
Integrity of functional group	Subtropical/Temperate species of catch (I/T)	$\delta(I/T) \leq 0.05\overline{I/T}$	$0.05\overline{I/T} < \delta(I/T) \leq 0.10\overline{I/T}$	$\delta(I/T) > 0.10\overline{I/T}$
Fishing ground	Position of fishing ground (FG)	$\delta(FG) \leq 0.05\overline{FG}$	$0.05\overline{FG} < \delta(FG) \leq 0.10\overline{FG}$	$\delta(FG) > 0.10\overline{FG}$

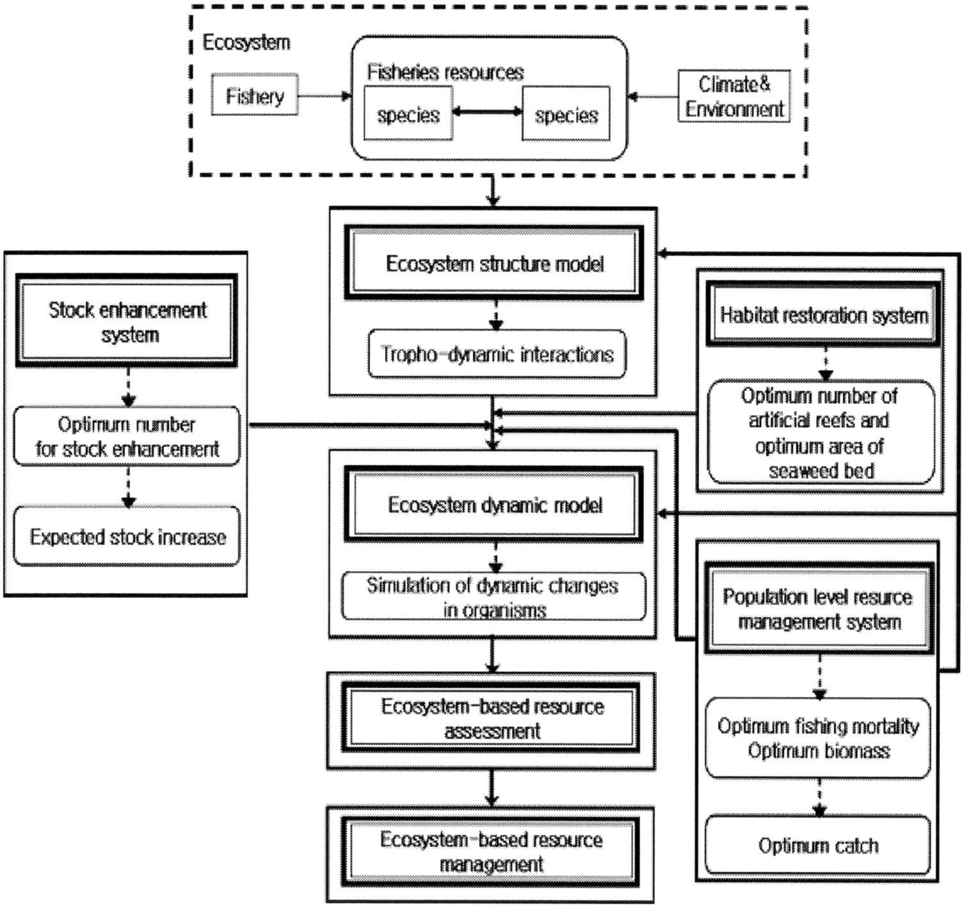

Figure 14. Flow chart showing the ecosystem-based fishery resource assessment and management system (Zhang et al., in prep).

As an alternative, Zhang et al. (in prep) have developed an ecosystem-based resource assessment system for Korean fisheries (Fig. 14) with indicators which detect climate changes indirectly (Table 3). These indicators include deviations from a normal spawner-recruit relationships or pre-recruit anomalies (Fig. 15), the position of fishing grounds and the ratio of temperate and subtropical species in the catch (Table 3).

Joint studies of fisheries resources are needed among the neighboring countries in the NW Pacific to maintain and to sustainably utilize fisheries resources in the region. These studies should include exchanges of fishery data and information on the distribution, biomass and ecological characteristics of fish stocks. In addition, stock assessments should include relationships between stock parameters and ocean climate environmental factors (Zhang and Lee, 2001).

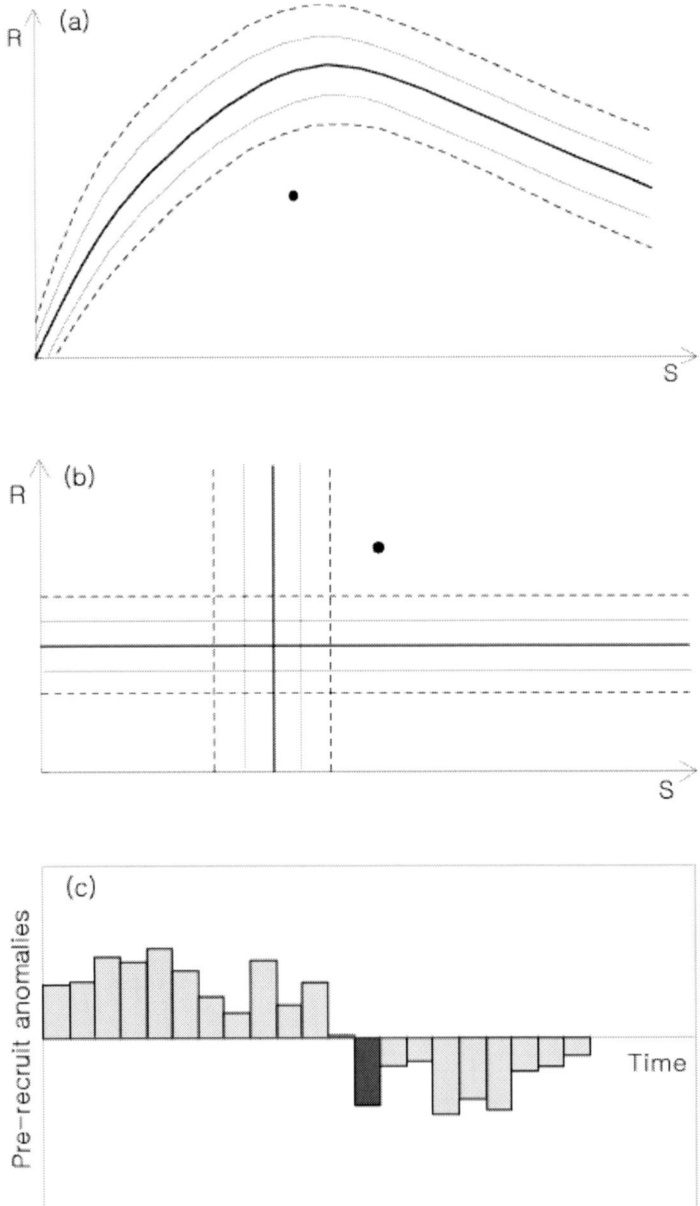

Figure 15. Schematic diagram showing deviations from spawner-recruits(S-R) curve (a), deviations from mean S and mean R (b), and pre-recruit anomalies (c) detecting climate changes (Zhang et al., in prep).

Conclusion

Environmental conditions change frequently, while each fish population has its own migration route and life history schedule which has been developed on evolutionary time scale. Fish spawning is influenced by abiotoc environments such as temperature and salinity, and their growth and survival are also controlled by the abundance of prey and predators. Water properties are influenced by climate variability, especially in the coastal zone where spawning and nursery grounds of marine fish species are usually found. The degradation of coastal habitats and climate changes could threaten the survival of some species (Kim et al., 2007). Marine ecosystems in the Northwest Pacific are influenced by factors such as the Kuroshio Current, seasonal monsoon and wind, river runoff, and interactions between species. Interdisciplinary region-wide research will be requested to understand the relationship between climate change and fisheries productivity.

Although the current state of scientific knowledge is not sufficient to fully implement the principles of ecosystem management, Fluharty and Cyr (2001) have suggested three research themes required to implement this approach:: (1) increase understanding of the ecosystem effects of fishing, (2) monitor trends and dynamics of marine ecosystems, and (3) identify alternative governance systems and approaches. More recently, Jamieson and Zhang (2005) outlined an immediate need to: 1) improve predictive capability with regard to climate and human impacts on ecosystems, 2) develop a more explicit definition of ecosystem-based management objectives, 3) develop objective criteria and sensitive indicators to measure the success in achieving desired ecosystem state or condition, and 4) develop a more formalized decision-making framework.

The research strategies for these themes should be developed and implemented as international (i.e., GLOBEC or PICES Working Group on Ecosystem-based Management Science and its Application to the North Pacific) or national interagency programs involving academic as well as government scientists. The existing international organization or programs could coordinate to the above tasks and resolve problems. Because ecosystem principles apply globally, each country should participate in international programs that further fisheries management objectives. A significant enhancement in funding, research man power, and fishery research vessels will be required if these research themes are to be fulfilled.

A comprehensive ecosystem-based fisheries management approach would require managers to consider all interactions that a target fish stock has with predators, competitors, and prey species; the effects of weather and climate on fisheries biology and ecology; the complex interactions between fishes and their habitat; and the effects of fishing on fish stocks and their habitat. However, the approach need not be endlessly complicated. An initial step may require only that managers consider how the harvesting of one species might impact other species in the ecosystem. Fishery management decisions made at this level of understanding can prevent significant and potentially irreversible changes in marine ecosystems caused by fishing and climate changes.

An ecosystem-based fisheries management approach is an important complement to existing fisheries management approaches. This approach should be implemented immediately despite current uncertainties about ecosystems and their responses to human actions, since the potential benefits of implementation are as large as or greater than the potential risks of inaction (Pikitch et al., 2004). When fishery managers understand the complex ecological and socioeconomic environments in which fish and fisheries exist, they may be able to anticipate the effects that fishery management will have on the ecosystem and the effects that ecosystem change will have on fisheries. There is an urgent need for strong governmental action to stop overfishing, protect habitat, and support expanded research and monitoring programs in order to implement an ecosystem-based fisheries management.

References

Alverson, D.L. and P.A. Larkin. 1994. Fisheries: Fisheries science and management. Pages 150-167 in C.D. Voigtlander (ed.) *The state of the world's fishery resources: Proceedings of the World Fisheries Congress*, Plenary Session, Oxford and IBM Publishing, New Delhi.

Chang, K.-I., S. Kim, Y.-G. Park, D. Jeon, J.H. Lee, and C.-H. Kim. In press. Climate variability in neighboring seas of Korea. Korea Meteorology Agency.

Chiba, S., Hirota, Y., Hasegawa, S., Saino, T., 2005. North-south contrast in decadal scale variations in lower trophic ecosystems in the Japan Sea. *Fisheries Oceanography*, **14**: 401-412.

Chavez, F.P., J. Ryan, S.E. Lluch-Cota, and M. Niquen. 2003. From anchovies to sardines and back: Multidecadal change in the Pacific Ocean. *Science*, **299**, 217-221.

Choi, Y.M., J.B. Lee, C.I.Zhang, C.I.Baik, and J.H. Park. 2000. Assessment and management of common mackerel (*Scomber japonicus*) in Korean waters, based on the relationship between recruitment and the ocean environmental factors. p. 52. Abstracts of Korea-Japan Joint GLOBEC Symposium: Long-term Variations in the Northwestern Pacific Ecosystems.

Cole, J. and J. McGlade. 1998. Clupeoid population variability, the environment and satellite imagery in coastal upwelling systems. *Reviews in Fish Biology and Fisheries*. **8**: 445-471.

FAO. 1995. Report of the expert consultation on guideline for responsible fisheries management. *FAO Fisheries Report* No. **519**. FAO, Rome, Italy.

FAO. 2007. Universal software for fishery statistics time series. FAO website.

Fluharty, D. and N. Cyr. 2001. Implementing ecosystem-based management of fisheries in the context of U.S. regional fisheries management: Recommendations of the NMFS Ecosystem Principles Advisory Panel. *CalCOFI Report* **42**: 66-73.

Hahn, S.B., 1994. SST warming of Korean coastal waters during 1881-1990. *KODC Newsletter*, **24**: 29-37.

Hwang, S.-D. and T.-W. Lee. 2005. Spawning dates and early growth of chub mackerel *Scomber japonicus* as indicated by otolith microstructure of juveniles in the inshore nursery ground. *Fisheries Science*, **71**: 1185-1187.

Jamieson, G. and C.I. Zhang (Eds.). 2005. Report of the Study Group on Ecosystem-based Management Science and its Application to the North Pacific. *PICES Scientific Report* No. **29**, 77p.

JMA. 1991. *Climate chart of sea surface temperatures of the western North Pacific and the global ocean.* Japan Meteorological Agency.

Kang, S., S. Kim, and S.-W. Bae. 2000. Changes in ecosystem components induced by climate variability off the eastern coast of the Korean Peninsula during 1960-1990. *Progress in Oceanography*, **47**: 205-223.

Kang, Y.S., J.Y. Kim, H.G. Kim, and J.H. Park. 2002. Long-term changes in zooplankton and its relationship with squid, *Todarodes pacificus*, catch in Japan/East Sea. *Fisheries Oceanography*, **11**: 337-346.

Kim, J.Y., Kim, S. Choi, Y.M., Lee, J.B., 2006. Evidence of density-dependent effects on population variation of Japanese sardine (*Sardinops melanosticta*) off Korea. *Fisheries Oceanography*, **15**: 345-349.

Kim, J. Y. and N.C. H. Lo. 2001. Temporal variation of seasonality of egg production and the spawning biomass of Pacific anchovy, *Engraulis japonicus*, in the southern waters of Korea. *Fisheries Oceanography*, **10**: 297-310.

Kim, S. 1995. Climate change and the fluctuation of fishery resources in the North Pacific. *Ocean Policy Research*, **10**: 107-142. (in Korean with English abstract)

Kim, S. 2003. Changes in fisheries resources in relation to variability of oceanic environments. *J. Korean Society of Fisheries Resources*, **6**: 11-20. (in Korean with English abstract)

Kim, S., S.G. Jung, C.I. Zhang, 1997. The effect of seasonal anomalies of seawater temperature and salinity on the fluctuation in yields of small yellow croaker, *Pseudosciaena polyactis*, in the Yellow Sea. *Fisheries Oceanography*, **6**: 1-9.

Kim, S. and S. Kang. 1998. The status and research direction for fishery resources in the East Sea/Sea of Japan. *J. of Korean Society of Fisheries Resources*, **1**: 44-58. (in Korean with English abstract)

Kim, S. and S. Kang. 2000. Ecological variations and El Nino effects off the southern coast of the Korean Peninsula during the last three decades. *Fisheries Oceanography*, **9**: 239-247.

Kim, S. and S. H. Khang. 2000.Yellow Sea. Pages 487-497 in C. Sheppard (ed.) *Seas at the millennium: an Environmental Evaluation.* Elsvier.

Kim, S., C.I. Zhang, J.Y. Kim, J.H. Oh, S. Kang, and J.B. Lee. 2007. Climate variability and its effects on major fisheries in Korea. *Korean Journal of Oceanography.*

Lee, J.-U. 1999. The exploitation and management of fisheries resources in adjacent economic zone. With an emphasis on shared stocks around Korean waters. Pages 151-161 in *Ocean science and technology: vision* **21**. Ministry of Maritime Affairs and Fisheries, Seoul, Korea.

Lee, S.-J. and Y.-B. Go. 2006. Winter warming and long-term variation in catch of yellowtail (*Seriola quinqueradiata*) in the South Sea, Korea. *Korean J. Ichthyology*, **18**: 319-328.

MacCall, A.D. 1979. Population estimates for the waning years of the Pacific sardine fishery. *CalCOFI, Rep.* **20**: 72-82.

Mantua, N.J. and S.R. Hare. 2002. The Pacific decadal oscillation. *Journal of Oceanography*, **58**: 35-44.

Mantua, N.J., Hare, S.R., Zhang, Y., Wallace, J.M., Francis, R.C., 1997. A Pacific interdecadal climate oscillation with impacts on salmon production. *Bull. of American Meteorological Society*, **78**: 1069-1079.

Mayama, H., and Ishida. Y. 2003. Japanese studies on the early ocean life of juvenile salmon. *N. Pac. Anadr. Fish Comm. Bull.*, **3**:41-67.

Mikolajewicz, U., B.D. Santer, and E. Maier-Reimer. 1990. Ocean response to greenhouse warming. *Nature*, **345**: 589-593.

Nakamura, Y. and Y. Sakurai. 1993. Age determination from daily growth increments in statoliths of some groups of the Japanese common squid *Todarodes pacificus*. p. 337-342. *Recent Advances in Fisheries Biology*, Tokai University Press, Tokyo.

Otterson, G., S. Kim, G. Huse, J.J. Polovina, and N.C. Stenseth. in press. Major routes by which climate signals force marine populations. *Journal of Marine Systems*.

Overland, J.E. and M. Wang. 2007. Future climate of the North Pacific Ocean. *Transactions of American Geophysical Union*, **88**(16): 178-179.

Pauly, D., V. Christensen, J. Dalsaar, R. Frose, and F. Torres, Jr. 1998. Fishing down marine food webs. *Science*, **279**: 860-863.

Pauly, D. and V. Christensen. 1995. Primary production required to sustain global fisheries. *Nature*, **374**: 255-257.

Park, W.-S. and I.S. Oh. 2000. Interannual and interdecadal variations of sea surface temperature in the East Asian Marginal Seas. *Progress in Oceanography*, **47**: 191-204.

PICES. 2004. Marine Ecosystems of the North Pacific. *PICES Special Publication*, **1**, 280 p.

Pikitch, E.K. and 16 co-authors. 2004. Ecosystem-based fishery management. *Science*, **305**: 346-347.

Sakurai, Y., S. Miyanaga, J. Yamamoto, K. and Mori. in press. How environmental factors affect stock size of ommastrephid squid, Todarodes pacificus – A possible scenario. PICES/GLOBEC Symposium on Climate variability and ecosystem impacts on the North Pacific: a basin-scale synthesis. April 19-21, 2006, Honolulu, USA. p. 62.

Seo, H., Kim, S., Seong, K., Kang, S., 2006. Variability in scale growth rates of chum salmon (*Oncorhynchus keta*) in relation to climate changes in the late 1980s. *Progress in Oceanography*, **68**: 205-216.

Tian, Y., Kidokoro, H., Watanabe, T., Iguchi, N., in press. The late-1980s regime shift in the ecosystem of Tsushima Warm Current in the Japan/East Sea: Evidence of historical data and possible mechanisms. PICES/GLOBEC CCCC Symposium, April 19-21, 2006, Honolulu, USA.

Yatsu, A., Watanabe, T., Ishida, M., Sugisaki, H., Jacobson, L.D., 2005. Environmental effects on recruitment and productivity of Japanese sardine *Sardinops melanosticta* and chub mackerel *Scomber japonicus* with recommendations for management. *Fisheries Oceanography*, **14**: 263-278.

Wooster, W.S. and C.I. Zhang. 2004. Regime shifts in the North Pacific: early indications of the 1976-1977 event. *Progress in Oceanography*, **60**: 183-200.

Zhang, C.I. 2002. Prospect of ecosystem-based fisheries management. *J. Korean Society of Fisheries Resources*, 5: 73-90. (in Korean with English abstract)

Zhang, C.I. and Y. Gong. 2005. Effect of ocean climate changes on the Korean stock of Pacific saury, *Cololabis saira* (Brevoort). *J. of Oceanography*, **61**: 313-325.

Zhang, C.I. and S. Kim. 1999. Living marine resources of the Yellow Sea ecosystem in Korean waters: status and perspectives. In: Sherman, K. and Q. Tang (eds.), Large

Marine Ecosystems of the Pacific rim: assessment, sustainability, and management. Blackwell Science. 465 p.

Zhang, C. I. and J. B. Lee, 2001. Stock assessment and management implications of horse mackerel (*Trachurus japonicus*) in Korean waters, based on the relationships between recruitment and the ocean environment. *Progress in Oceanography*, **49**: 513-537.

Zhang, C. I. and S. K. Lee, 2004. Trophic levels and fishing intensities in Korean marine ecosystems. *Journal of Korean Society of Fisheries Resources*, **6**: 194-206.

Zhang, C.I. and R.J. Marasco. 2003. New approaches in fishery assessment and management under the exclusive economic zone regime in Korea. *American Fisheries Society Symposium*, **38**: 685-693.

Zhang, C. I., C. I. Baik, and S. I. Lee. 1999. A population ecological study of the white croaker in Korean waters. *Journal of Korean Society of Fisheries Resources*, **2**(1): 77-83.

Zhang, C. I., S. Kim, and S. B. Yoon. 1992. Stock assessment and management implications of small yellow croaker in Korean waters. *Journal of Korean Fisheries Society*, **25**(4): 282-290.

Zhang, C. I., Y.J. Kang, C.S. Park, and M.H. Sohn. 1998. A study on the stock management of the sharp-toothed eel in Korean waters. V. Stock assessment and management implications. *Journal of Korean Society of Fisheries Resources*, **1**(1): 36-43.

Zhang, C. I., S. Kim, J.B. Lee, D. Gunderson, H.W. Park, and J.H. Lee. In prep. Biomass-based resource assessment and management in Korea. Unpublished manuscript, 26p.

Zhang, C. I., J. B. Lee, S. Kim, and J.H. Oh. 2000. Climatic regime shifts and their impacts on marine ecosystem and fisheries resources in Korean waters. *Progress in Oceanography*, **47**: 171-190.

Zhang, C. I., J. B. Lee, Y. I. Seo, S.C. Yoon, and S. Kim. 2004. Variations in the abundance of fisheries resources and ecosystem structure in the Japan/East Sea. *Progress in Oceanography*. **61**: 245-265.

Zhang, C.I., V. Radchenko, T. Sugimoto, and S.M. Hyun. 2006. Interdisciplinary physical and biological processes of the Sea of Okhotsk and the Japan/East Sea. Pages 549-636 in A Robinson and K. Brink (eds.), *The Sea*. Harvard University Press.

Zhang, C. I., S. C. Yoon, and J. B. Lee. in press. Effects of the 1988/89 climate regime shift on the structure and function of the southwestern Japan/East Sea ecosystem. *Journal of Marine Systems*, doi:10.1016/j.jmarsys.2006.05.015

In: Climate Change Research Progress
Editor: Lawrence N. Peretz, pp. 217-241
ISBN 978-1-60021-998-6

Chapter 6

DECADAL–TO–CENTENNIAL SCALE CLIMATE–CARBON CYCLE INTERACTIONS FROM GLOBAL CLIMATE MODELS SIMULATIONS FORCED BY ANTHROPOGENIC EMISSIONS

Igor I. Mokhov, Alexey V. Eliseev and Andrey A. Karpenko
Obukhov Institute of Atmospheric Physics,
Russian Academy of Sciences, Pyzhevski, Moscow, Russia

Abstract

Simulations of the climate–carbon cycle interaction are discussed in comparison with observationally–based estimates for the global carbon cycle characteristics. Since the beginning of the industrial era, the storage of the carbon dioxide in the atmosphere is smaller than the corresponding anthropogenic emissions. This is due to uptake of the atmospheric carbon dioxide to the terrestrial biota and ocean. Moreover, during the the 20th century, the sink of the carbon dioxide from the atmosphere to the terrestrial ecosystems became larger, due to CO_2 fertilisation effect. However, in the 21st century, the global climate models with the carbon cycle project that interactions between climate and carbon cycle basically lead to the stronger growth of the carbon dioxide burden in the atmosphere. These interactions were in the focus of the Coupled Climate–Carbon Cycle Model Intercomparison Project (C^4MIP). One of the basic outcome of this and related activities is positive climate–carbon cycle feedback leading to enhanced buildup of CO_2 in the atmosphere due to response of climate and corresponding changes in the terrestrial and oceanic uptakes of carbon. In particular, the soil respiration growth overcompensates the increase of the net primary production and leads to the diminishing sink of the carbon dioxide from the atmosphere top the living biota and soil. In some models and emission scenarios the terrestrial biota even eventually becomes the source of the carbon dioxide. The parameter of the feedback between climate and carbon cycle changes non-monotonically in the 20th and 21st centuries, depicting characteristic periods of the emission growth and the respective climate response. On a century timescale, climate–carbon cycle feedback may saturate. The 20th century observations of the carbon cycle are insufficient to constrain future evolution of the coupled carbon–cycle system. Nevertheless, they are able to narrow the respective uncertainty range. This can be illustrated, in particular, by employing the Bayesian statistical treatment of the respective ensemble. However, for the most probable values of the governing parameters of the coupled system, climate–carbon cycle interactions enhance global warming in the 21st century by about 10% or even more under the SRES marker emission scenarios.

1. Introduction

During the industrial era, anthropogenic emissions of carbon dioxide (basically related to to the fossil fuel, cement manufacture, and land use) led to the increase of the atmospheric concentration of carbon dioxide $pCO_{2,a}$ from $276 - 284$ ppmv [55] in (these range is characteristic for the whole part of the last millenium before an industrialisation has started around 1750) to 379 ppmv in 2005 (as measured at the Mauna Loa Observatory) [57]. It is the carbon dioxide buildup in the atmosphere which is assumed to be basically responsible for the global warming of about $0.5 - 0.7$ K as observed in the 20th century.

Until recently, the most historical simulations and future projections of the climate changes with coupled climate models were perfromed with specified concentrations of greenhouse gases (CO_2, in particular) in the atmosphere [24, 46]. However, the corresponding anthropogenic emissions only partly stored in the atmosphere with the rest taken up by the oceans and by the terrestrial ecosystems. For example, according to [54], total (fossil fuel and land use) emissions for 1800–1994 amount $320 - 460$ PgC. Only between one third and half of these emissions are stored in the atmosphere (165 ± 4 PgC). Oceanic uptake for this period has amounted 118 ± 19 PgC, and the terrstrial biosphere sink is estimated to be in the range $61 - 141$ PgC [54]. The atmospheric buildup of the carbon dioxide for a given time interval is

$$c_0 \Delta pCO_{2,a} = E_c - U_{oc} - U_l, \quad (1)$$

with $c_0 = 2.123$ GtC/yr, E_c is cumulative anthropogenic CO_2 emission for this time interval, and U_{oc} and U_l are corresponding oceanic and terrestrial uptakes of carbon. These fluxes, in turn, depend on the climate state and may be quite different for the future, anthropogenically forced, trajectory of the coupled climate–carbon cycle system.

As a result, starting from the papers [10] and [16], a lot of climate modellers' efforts are devoted to the simulations of the global carbon cycle, interactively coupled with the three-dimensional climate models [5, 6, 18, 11, 30, 41, 19, 40, 48, 47, 12, 14, 13]. It was shown in those papers, that this interactive coupling changes the buildup of the carbon dioxide in the atmosphere in comparison to the hypothetical case, when the carbon cycle does not feel the climate changes. As a result, the so called climate–carbon cycle feedback (C^3F) term has been introduced. This feedback can be measured, for instance, as a difference of the atmospheric CO_2 concentration $pCO_{2,a}$ in some prechosen year under given carbon dioxide emission scenario between two sets of the numerical experiments. In the first set, the coupled (cpl) simulation is performed with the climate–carbon cycle model forced by the CO_2 emissions. In the second set (uncoupled simulation, ucpl), the climate model is forced by the output of the carbon cycle model; in turn, the carbon cycle model is forced by the CO_2 emissions, but not taking into account an influence of climate changes on the carbon cycle dynamics (or simply treating carbon dioxide as a radiatively non–active gas). The climate–carbon cycle feedback can be measured either in terms of the feedback parameter

$$f = \Delta p\mathrm{CO}_{2,\mathrm{a}}^{\mathrm{cpl}} / \Delta p\mathrm{CO}_{2,\mathrm{a}}^{\mathrm{ucpl}}. \tag{2}$$

or in terms of the feedback gain $g = (f - 1)/f$ [16, 18].

Up to date, most coupled models implement only carbon cycle processes relevant to the decadal–to–centennial scale feedbacks between climate and carbon cycle. Among those processes are atmosphere–ocean exchange of carbon, terrestrail plant photosynthesis, their respiration and mortality, and soil respiration. The processes operating at much longer scales (e.g., silicate weathering) are not included in these models. This is consistent with typical length of simulations with such models ranging from decades up to a few millenia forced either by historical anthropogenic emissions or by anthropogenic emissions projected into the future.

For the oceanic part, it is broadly accepted that the ocean exchanges the inorganic carbon with the atmosphere based simply on the solubility of the carbon in the sea water, while the oceanic chemistry can be treated as a background process [56].

A convenient and natural assumption employed in the most of the mentioned above simulations is the preindustrial climate–carbon cycle state was equilibrated, e.g., [58].

In any simulation, the system inertia delays the climate response to early anthropogenic emissions of carbon dioxide. As a result, at the early part of simulations, coupled and uncouped runs should be almost indistinguishable between each other. In turn, f should be unity at initial time and close to unity slightly afterwards. If $\mathrm{C}^3\mathrm{F}$ is positive, f has to increase generally in the course of integration. However, as it will be shown below, its non–monotonic temporal behaviour of feedback parameter is possible due to temporal inhomogeneities of anthropogenic emissions and because of the near–saturated behaviour of the carbon dioxide absorption lines in the Earth's atmosphere.

The above–mentioned positive feedback between climate and carbon cycle is basically associated with an enhanced soil respiration in warmer climate (and to a smaller amount — to an enhanced plant respiration). This effect overwhelms the direct fertilisation effect of the carbon dioxide. However, the response of gross plant photosynthesis to climate changes is ambiguous. Starting from the present day climate, warmer temperature and increased precipitation generally lead to the enhanced plant photosynthesis [1, 27]. However, further climate changes may decrease this photosynthesis due to, e.g., the dieback of living biota [1, 27]. Furthermore, this effect can be regionally dependent; in particular, the strong dieback of the Amazonian forests is simulated [9, 28]. As will be shown below, response of the oceanic carbon cycle to climate changes is ambiguous as well.

The goal of the present paper is to review recent efforts in modelling the interactions between climate and carbon cycle (Secs. 2. and 3.). In addition, based on simulations with climate model of intermediate complexity, variations in characteristics of $\mathrm{C}^3\mathrm{F}$ will be studied (Sec. 4.). In particular, it will be shown that climate–carbon cycle feedback eventually saturates. Finally (Sec. 5.), a structural uncertainty of the coupled system will be assessed by employing large ensemble with the same model of intermediate complexity.

2. Quantification of the Climate–Carbon Cycle feedback

In [18], a quantification of C^3F is suggested. Namely, for the uncoupled runs, oceanic and terrestrial uptakes are linearly related to the corresponding atmospheric buildups of CO_2

$$U_{\mathrm{x}}^{\mathrm{ucpl}} = \beta_{\mathrm{x}} \Delta p\mathrm{CO}_{2,\mathrm{a}}^{\mathrm{ucpl}}, \tag{3}$$

with $\mathrm{x} = \mathrm{oc}, \mathrm{l}$. The corresponding uptakes in the coupled runs are linearly related to the $\Delta p\mathrm{CO}_{2,\mathrm{a}}^{\mathrm{cpl}}$ and changes in the global annual surface air temperature $\Delta T_{\mathrm{g}}^{\mathrm{cpl}}$

$$U_{\mathrm{x}}^{\mathrm{cpl}} = \beta_{\mathrm{x}} \Delta p\mathrm{CO}_{2,\mathrm{a}}^{\mathrm{cpl}} + \gamma_{\mathrm{x}} \Delta T_{\mathrm{g}}^{\mathrm{cpl}}, \tag{4}$$

with β's substituted from (3). From (1), (3), and (4), the resulting feedback gain reads [18, 17]

$$g = -\alpha \left(\gamma_{\mathrm{l}} + \gamma_{\mathrm{oc}} \right) / \left(1 + \beta_{\mathrm{l}} + \beta_{\mathrm{oc}} \right) \tag{5}$$

with transient temperature sensitivity $\alpha = \Delta T_{\mathrm{g}} / \Delta p\mathrm{CO}_{2,\mathrm{a}}$ [18, 17]. One notes, that this definition of temperature sensitivity is different from the commonly applied climate sensitivity employing with logarithm of $p\mathrm{CO}_{2,\mathrm{a}}$ rather than with $p\mathrm{CO}_{2,\mathrm{a}}$ itself.

In this, β's figure the direct of $p\mathrm{CO}_{2,\mathrm{a}}$ on the terrestrial plants (e.i., fertilisation) and on the atmosphere–ocean CO_2 exchange. In contrast,γ's are relevant for the climate–carbon cycle feedback [13].

3. Coupled Climate–Carbon Cycle Intercomparison Project (C^4MIP)

Up to date, the most comprehensive intercomparison of the coupled climate– carbon cycle models is the Coupled Climate–Carbon Cycle Intercomparison Project (C^4MIP) [17]. Eleven models were participating in this Project, among them seven general circulation models and four models of intermediate complexity. All models were forced by the historical emissions of carbon dioxide for 1860–2000 (according to [39] for the fossil fuel emissions and to [25] for the land use emissions) and by the SRES emission scenario A2 [24] for the 21st century. Changes in physical and biogeochemical properties of the vegetation were neglected.

In the coupled C^4MIP simulations, $p\mathrm{CO}_{2,\mathrm{a}}^{\mathrm{cpl}}$ in 2100 reaches $720 - 1020$ ppmv depending on the model. All models exhibit positive C^3F. However, qualitative characterteristics and underlying mechanisms differ drastically among the participating models. In particular, additional (due to carbon–climate cycle interactions) atmospheric buildup of carbon dioxide for the year 2100 varies between 20 ppmv and 200 ppmv (with six of eleven models falling into the range $50 - 100$ ppmv). The simulated respective feedback gains are 0.04–0.31 (model average is 0.15, eight of eleven models simulate gains in the interval 0.10–0.20).

Even larger spread is exhibited the particular components of the carbon budget are considered. To year 2100, terrestrial CO_2 flux ranges among the models from the sink of 11 PgC/yr to the source of 6 PgC/yr. While the models agree between each other that oceans remain CO_2 sink till the end of the 21st century, the simulations figure rather broad interval $4 - 10$ PgC/yr for F_{oc} in this time. In the intercomparison, a mutual compensation between oceanic and terrestrial fluxes of carbon is noted: the models with large oceanic sink of CO_2 simulate the smaller corresponding terrestrial sink (or the higher respective terrestrail source). As a result, eight of eleven models relate C^3F to changes in terrestrial carbon flux, while three attribute it to the ocean.

In the coupled simulations, the airborne fraction of the anthropogenically emitted carbon dioxide r_a to the end of the 21st century is in the range 0.42–0.71 (eight of eleven models fall into the range 0.47–0.60). The respective landborne fraction r_l is between 0.01 and 0.45 (eight of eleven models simulate the range 0.14–0.27). In the coupled C^4MIP simulations, oceanborne fraction r_{oc} of the anthropogenic CO_2 emissions is between 0.15 and 0.36 (with most models being in the range 0.20–0.26).

Coefficients β's of direct effect of CO_2 change on carbon uptakes are positive for both oceanic and terrestrial fluxes indicating enhancement of both uptakes under growing $pCO_{2,a}$. The simulated intervals of these coefficients $\beta_l = 0.2 - 2.8$ PgC/ppmv (mean value is 1.35 PgC/ppmv, nine of eleven models fill the range $0.8 - 1.6$ PgC/ppmv) and $\beta_{oc} = 0.9 - 1.5$ PgC/ppmv (mean value is 1.13 PgC/ppmv), respectively.

In contrast, coefficients γ's of climate–carbon cycle feedback are negative for both terrestrial and oceanic fluxes pointing to supression of these fluxes under CO_2–induced warming. The simulated ranges are $\gamma_l = -(20 - 177)$ PgC/K and $\gamma_{oc} = -(14 - 67)$ PgC/K, correspondingly, with ensemble averaged values $\gamma_l = -79$ PgC/K and $\gamma_{oc} = -30$ PgC/K, respectively.

As a summary, current generation of the coupled climate–carbon cycle models converges in positive C^3F, but attributes it to different mechanisms, either oceanic or terrestrial, and with a large diversity in quantitative characteristics of this feedback.

4. IAP RAS Climate Model of Intermediate Complexity

4.1. Description of the Model and Its Validation for the 20th Century Observations

The climate module of the Earth system model of intermediate complexity developed at the A.M. Obukhov Institute of Atmopsheric Physics RAS (IAP RAS CM) is described in [51, 21, 45, 46]. It includes modules for redistribution of the shortwave and longwave radiation, convection, cloud and precipitation formation. Large–scale atmospheric and oceanic dynamics (with the scales larger than those corresponding to the synoptic processes) are resolved explicitly. The synoptic–scale processes are treated as Gaussian ensembles. Sea ice in the IAP RAS CM is diagnosed from current surface air and sea surface temperatures. Soil hydrology in the model is prescribed and tuned to reproduce the observed climate. The IAP RAS CM horizontal resolution is 4.5 degrees latitude and 6.0 degrees longitude with 8 vertical layers in the atmosphere (up to 80 km) and 4 layers in the ocean. The IAP RAS CM simulates the present–day climate quite realistically [45, 46, 50]. The sensitivity of this model version to the doubling the carbon dioxide in the atmosphere is 2.2^o C [46, 50]. It lies in the lower part of such estimates for the current generation of the climate models [24, 50].

Carbon uptake by the terrestrial vegetation–soil system is simulated with the use of zero–dimensional equations [58, 34, 36]

$$\frac{dC_{\rm v}}{dt} = NPP - L - D, \qquad (6)$$
$$\frac{dC_{\rm s}}{dt} = L - R_{\rm s},$$

where $C_{\rm v}$ and $C_{\rm s}$ are carbon terrestrial vegetation and soil stocks, NPP is the terrestrial vegetation net primary productivity, L is litterfall, $R_{\rm s}$ is heterotrophic (soil) respiration. The terms on the right hand sides of Eqs. (6) depend on the corresponding carbon stocks and on the anomaly $\Delta T_{\rm g}$ of globally averaged annual mean surface air temperature (GSAT) from the reference value (the latter will be specified later) [36]

$$\begin{aligned} NPP &= P - R_{\rm p}, \\ P &= A_{\rm p} g_{\rm f}(p{\rm CO}_{2,{\rm a}}) C_{\rm v(s)} Q_{10,{\rm p}}^{\Delta T_{\rm g}/\Delta T_0} \\ R_{\rm p} &= A_{\rm r} C_{\rm v} Q_{10,{\rm r}}^{\Delta T_{\rm g}/\Delta T_0}, \\ L &= A_{\rm l} C_{\rm v}, \\ R_{\rm s} &= A_{\rm s} C_{\rm s} Q_{10,{\rm s}}^{\Delta T_{\rm g}/\Delta T_0}. \end{aligned} \qquad (7)$$

Here P is carbon production rate of photosynthesis, $R_{\rm p}$ is autotrophic (biota) respiration rate, $C_{v(s)}$ is defined below, and $A_{\rm p}$, $A_{\rm r}$, $A_{\rm l}$, $A_{\rm s}$ are constants. The temperature dependences of the carbon fluxes are proportional to $Q_{10,{\rm x}}^{\Delta T_{\rm g}/\Delta T_0}$, where $\Delta T_0 = 10$ K, $Q_{10,{\rm x}}$ are constants, ${\rm x} = {\rm p, r, s}$ [37]. While this approach performs unsatisfactorily for the model employing regional and/or seasonal resolution [37], it is considered to describe the climate dependences of the globally averaged annual mean carbon fluxes [53, 36] quite realistically. The reason for this is due to the relative smallness of $\Delta T_{\rm g}$ for the climate change expected at decadal–to–centennial scales in comparison to the regional and seasonal variations of the surface air temperature. In Eqs. (7), $g_{\rm f}$ is the fertilisation factor which is formulated according to the Michaelis–Menten law (see, e.g., [36])

$$g_{\rm f}(p{\rm CO}_{2,{\rm a}}) = \begin{cases} 0 & \text{for } p{\rm CO}_{2,{\rm a}} < k_{\rm c}, \\ \frac{p{\rm CO}_{2,{\rm a}} - {\rm k_c}}{k_{\rm M} + p{\rm CO}_{2,{\rm a}} - k_{\rm c}} & \text{otherwise}, \end{cases}$$

where $k_{\rm M}$ is the half–saturation point, $k_{\rm c}$ is the compensation point (which is the threshold concentration of the carbon dioxide in the atmosphere, needed to start the photosynthesis). $C_{\rm v(s)}$ is the steady state living biomass, corrected for the agriculture harvesting. This variable is modelled using an additional prognostic equation [36]:

$$\frac{dC_{\rm v(s)}}{dt} = -k_{\rm D} D, \qquad (8)$$

$k_{\rm D}$ is constant. In the absense of land use emissions, total mass of carbon, stored in the living biomass would stay unchanged in this particular model. In the presense of the positive land use emissions, this mass decreases. To avoid it becomes negative, $C_{\rm v,s}$ is constrained not to change if it crosses zero level. The carbon flux to the ocean due to runoff is neglected, because its contribution is small, $0.5 - 1.0$ GtC/yr [23]. Finally, the carbon uptake by the terrestrial vegetation–soil system is a difference

$$F_l = dU_l/dt = NPP - R_s. \tag{9}$$

There is considerable uncertainty in the choice of the mentioned above Q_{10}'s. In [36], based on a literature review, the following "best guesses" were suggested: $Q_{10,p} = 1.50$, $Q_{10,r} = 2.15$, $Q_{10,s} = 2.40$. In the current IAP RAS CM version, these values are adopted for the standard implementation of the zero–dimensional carbon cycle model.

For k_M, any of the values 150 ppmv [36] or 450 ppmv [15, 7] are adopted. Hereafter, these two versions are denoted as Fl150 and Fl450, respectively. The constant k_c is set to 29 ppmv. For the constant k_D, the value 0.27 is assigned [36].

The values of other model parameters are tuned to simulate the preindustrial carbon cycle state with $pCO_{2,a,0} = 280$ ppmv, $C_{s,0} = 1500$ PgC, $C_{v,0} = 550$ PgC, $C_{v(s),0} = 550$ PgC, $R_{p,0} = 50$ PgC/yr, $L_0 = 50$ PgC/yr [23]. The adopted values correspond to interannual–to–decadal time scales $A_l^{-1} = 11$ yr, $A_r^{-1} = 11$ yr, $A_s^{-1} = (A_l C_{lb,0}/C_{s,0})^{-1} = 30$ yr, $A_p^{-1} = (A_l + A_r/g_f(pCO_{2,a,0})) = 3.4$ yr. Correspondingly, ΔT_g in the model is defined as an anomaly from the equilibrated preindustrial GSAT.

In the simplest setting, the carbon uptake by the ocean is related linearly to the globally averaged annual mean SST $T_{oc,g}$ and to the atmospheric concentration of carbon dioxide $pCO_{2,a}$ [59, 18]:

$$F_{oc} = \frac{dU_{oc}}{dt} = u_c \frac{dpCO_{2,a}}{dt} - u_T \frac{dT_{oc,g}}{dt}. \tag{10}$$

The values u_c and u_T are tuned to reproduce the observational estimations for the values of the oceanic carbon uptakes during 1980's and 1990's [24, 52, 26, 35] given the observed trends of GSAT [4] and the atmospheric concentration of the crabon dioxide [33] for the 20th century. As a result, the chosen values of the coefficients in Eq. (10) are $u_c = 1.31$ PgC/ppmv, $u_T = 0.3$ PgC/K. Hereafter, this version is denoted as FocL.

Formulation of F_{oc} in form (10) posesses a physically necessary condition to be zero in an equilibrium state. However, it could behave in a wrong way if emissions stop abruptly. After such emission stop, in reality, $pCO_{2,a}$ is expected to decrease due to continuing ocean uptake. In the model used here, in contrast, it would start to increase, because ocean would start to outgas CO_2. As a result, Eq. (10) is expected to be realistic only under continuing emissions of CO_2 and can not be used in other circumstances.

Motivated by this reasoning, another formulation of the oceanic uptake of CO_2 has been implemented in the IAP RAS CM based on the nonlinear model [3] but with chemical constants computed as functions of temperature in accordance to [44]. In this, oceanic salinity is prescribed and, similar to [3], the dependence of evasion factor of partial pressure of CO_2 between atmosphere and ocean is fitted by

$$E(pCO_{2,a}, T_{oc,g}) = E_0 \left(E_1(T_{oc,g}) p_n^2 + E_2(T_{oc,g}) p_n + E_3(T_{oc,g}) \right) \tag{11}$$

with $E_0 = 10$, $p_n = pCO_{2,a}/290$ ppmv, $E_1 = -2.19T_n^2 + 3.73T_n - 1.57$, $E_2 = 18.9T_n^2 - 32.4T_n + 14.2$, $E_3 = -19.8T_n^2 + 35.8T_n - 15.8$, $T_n = T_{oc,g}/300$ K. Hereafter, this version is denoted as FocNL.

Table 1. Basic characteristics of the simulated carbon cycle for the two IAP RAS CM versions, FocL–Fl150 and FocNL–Fl450, in comparion with recent observational estimates.

	FocL–Fl150	FocNL–Fl450	obs.
F_{oc}, 1980–1989, PgC/yr	2.3	2.1	1.9 ± 0.6 [24] 1.8 ± 0.8 [26] 1.8 ± 0.8 [57]
F_{oc}, 1990–1999, PgC/yr	2.7	2.3	2.1 ± 0.7 [26] 1.9 ± 0.5 [49] 1.9 ± 0.6 [38] 1.8 ± 0.8 [26] 2.2 ± 0.4 [57]
U_{oc}, 1860-1994, PgC	107	110	118 ± 19 [54]
U_{oc}, 1980-1999, PgC	49	44	37 ± 8 [54]
F_l, 1980–1989, PgC/yr	1.3	2.2	$(-0.3) - (+3.8)$ [24] $0.3 - 4.0$ [26] $(-0.2) - (+3.4)$ [57]
F_l, 1990–1999, PgC/yr	1.5	2.6	$1.6 - 4.8$ [26] 1.2 ± 0.7 [49] 1.2 ± 0.8 [38] $(-0.9) - (+4.3)$ [57]
U_l, 1860-1994, PgC	85	110	$61 - 141$ [54]
U_l, 1980-1999, PgC	31	48	39 ± 18 [54]

Finally, the atmosperic part of the carbon cycle is formulated based on the differential form of Eq. (1)

$$c_0 d\Delta p\mathrm{CO}_{2,\mathrm{a}}/dt = E - F_{\mathrm{oc}} - F_{\mathrm{l}}, \quad (12)$$

with E standing for annual anthropogenic emissions of CO_2.

For the standard settings of parameters, the IAP RAS CM forced by the historical fossil fuel and land use emissions [39, 25] generally reproduces basic observed characteristics of the carbon cycle, see Tab. 1 and Figs. 1–3. An improvement is noted in the oceanic part for the version FocNL–Fl450 in comparison to the older version FocL–Fl150. Both versions reproduce the observed variations of $p\mathrm{CO}_{2,\mathrm{a}}$ accurately enough, but underestimate (overestimate) $p\mathrm{CO}_{2,\mathrm{a}}$ slightly at the beginning (end) of this period. The absolute error does not exceeds 8 ppmv (Fig. 1). Such deviations are typical for the current generation of the coupled climate–carbon cycle models [10, 5, 11, 18, 30, 41, 40]. The globally averaged temperature change in the 20th century is reproduced with a reasonable accuracy as well (Figs. 4, 5).

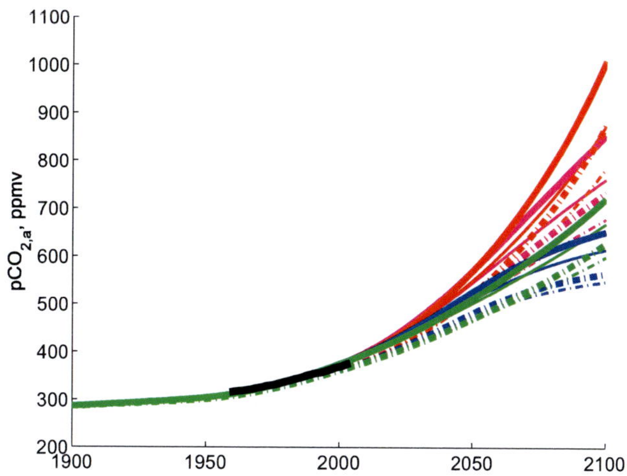

Figure 1. Atmospheric concentration of CO_2 simulated by the IAP RAS CM versions FocL–Fl150 and FocNL–Fl450 (thin and thick lines, respectively) in the coupled and uncoupled runs (solid and dash–dotted lines, respectively), for the emission scenarios A2, A1B, B2, and B1 (red, magenta, green, and blue curves, correspondingly) compared with the values measured at the Mauna Loa Observatory [33] (black line).

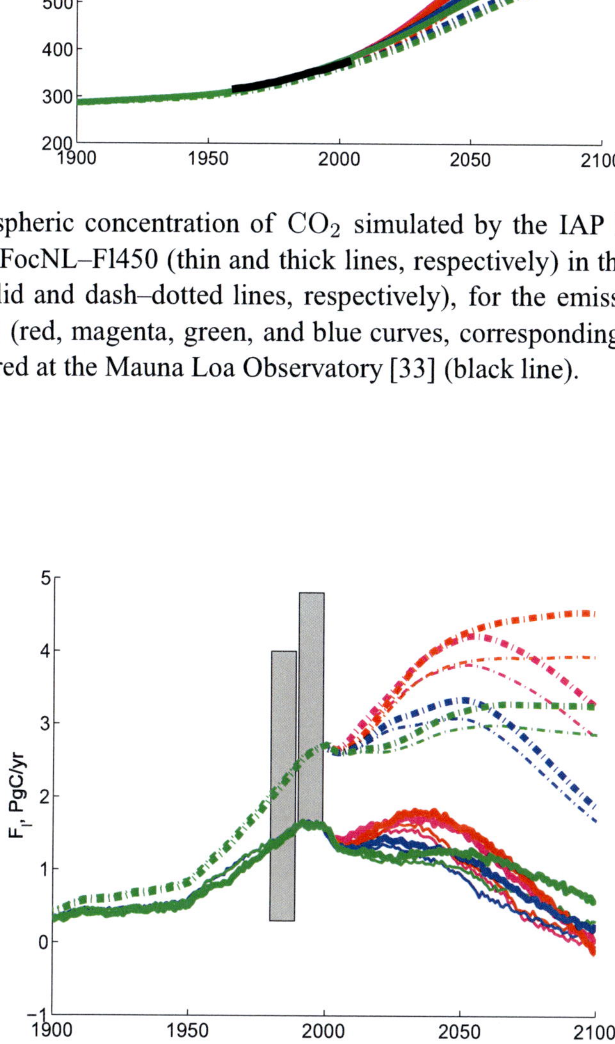

Figure 2. Terrestrial uptakes of CO_2 simulated by the IAP RAS CM versions FocL–Fl150 and FocNL–Fl450 (thin and thick lines, respectively) in the coupled and uncoupled runs (solid and dash–dotted lines, respectively), for the emission scenarios A2, A1B, B2, and B1 (red, magenta, green, and blue curves, correspondingly) compared with uncertainty ranges figured in [26] for 1980’s and 1990’s (gray rectangles).

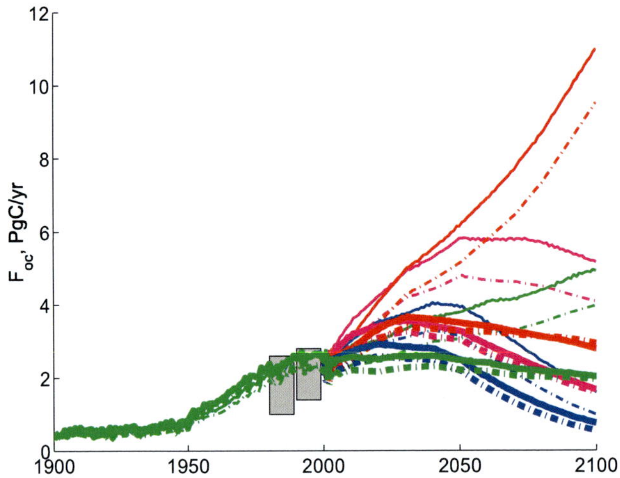

Figure 3. Oceanic uptakes of CO_2 simulated by the IAP RAS CM versions FocL–Fl150 and FocNL–Fl450 (thin and thick lines, respectively) in the coupled and uncoupled runs (solid and dash–dotted lines, respectively), for the emission scenarios A2, A1B, B2, and B1 (red, magenta, green, and blue curves, correspondingly) compared with uncertainty ranges figured in [26] for 1980's and 1990's (gray rectangles).

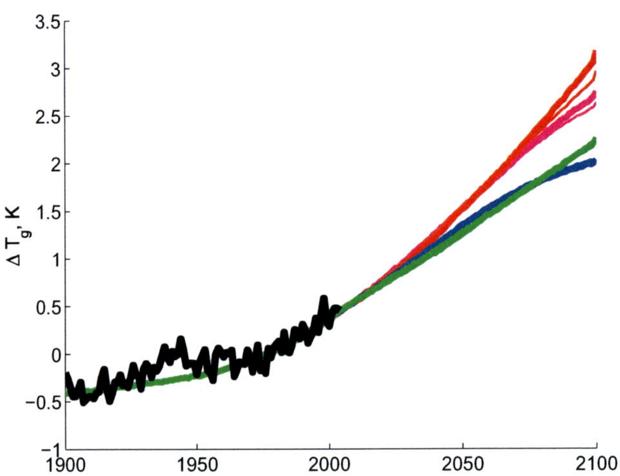

Figure 4. Anomaly (with respect to the preindustrial state) of the global annual surface air temperature simulated by the IAP RAS CM versions FocL–Fl150 and FocNL–Fl450 (thin and thick lines, respectively) for the emission scenarios A2, A1B, B2, and B1 (red, magenta, green, and blue curves, correspondingly) compared with the data from the University of East Anglia (Climate Research Unit) [4] (black line).

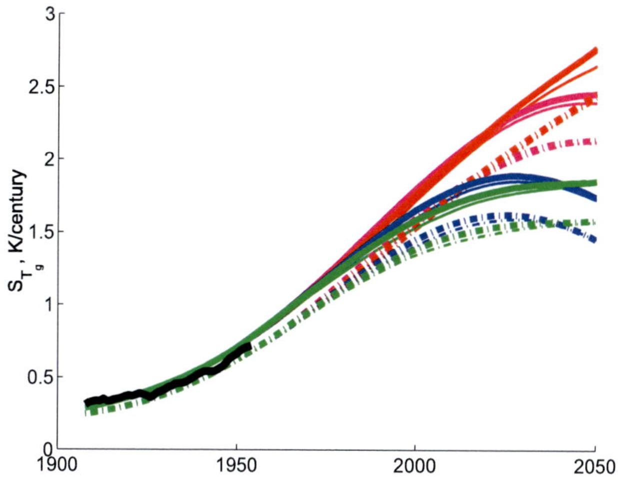

Figure 5. Linear trends of the globally averaged annual mean surface air temperature for the running 100–yr window centered at the year indicated at the x–axis (e.g., the value for 2050 represents 2001–2100). Shown are values simulated by the IAP RAS CM versions FocL–Fl150 and FocNL–Fl450 (thin and thick lines, respectively) for the emission scenarios A2, A1B, B2, and B1 (red, magenta, green, and blue curves, correspondingly) and data from the University of East Anglia (Climate Research Unit) [4] (black line).

4.2. Climate–Carbon Cycle Interactions in 20th and 21st Centuries in IAP RAS CM

For the 21st century, experiments with the IAP RAS CM were performed under the anthropogenic CO_2 emissions prescribed according to the SRES emission scenarios A2, A1B, B2, and B1 [24] (ordered here according to decrease of the cumulative emissions). The results are presented in Tab. 2 and plotted in Figs. 1–10. The atmospheric concentration of carbon dioxide in the coupled simulations grows to the late 21st century up to $615 - 875$ ppmv in the version FocL–Fl150 and up to $652 - 1010$ ppmv in the version FocNL–Fl450. The corresponding temperature increases relative to the preindustrial state are $2.4 - 3.4$ K and $2.5 - 3.7$ K. The climate–carbon cycle interactions increase the buildup of the atmnospheric carbon dioxide on $67 - 90$ ppmv and $87 - 134$ ppmv in the late 21st century for the versions FocL–Fl150 and FocNL–Fl450, respectively. This increase is stronger for the more agressive emissions scenarios.

As a result, global warming is stronger in the coupled runs than in their uncoupled counterparts by $0.31 - 0.35$ K for the version FocL–Fl150 and by $0.39 - 0.41$ K for the version FocNL–Fl450. This additional warming may be further studied making use the linear trends of T_g computed for the running 100–yr window. These trends are systematically higher for the coupled than for the uncoupled runs. Generally, there is little difference between versions FocL–Fl150 and FocNL–Fl450. For scenario A2, trends of T_g increase monotonically during the whole course of integration. For scenarios A1B and B2, the trends become flatter to the end of simulations, and for the scenario B1 they even diminish during that period. This reflects the course of emissions during these scenarios leading to the respective changes in $pCO_{2,a}$ under different scenarios.

Table 2. Key values simulated by the two versions of the IAP RAS CM (FocL–Fl150 and FocNL–Fl450) under marker SRES emission scenarios A2, A1B, B2, and B1: (i) atmospheric concentration of carbon dioxide $pCO_{2,a}$ in year 2100 for the coupled simulations together with difference between coupled and uncoupled simulations (in brackets); (ii) global warming ΔT_g in year 2100 relative to the equilibrated preindustrial state together with difference between coupled and uncoupled simulations (in brackets); (iii) value of climate–carbon cycle feedback parameter f in year 2100; (iv) airborne r_a, landborne r_l, and oceanborne r_{oc} fractions in year 2100 of anthropogenically emitted carbon dioxide for the coupled (uncoupled) simulations.

		A2	A1B	B2	B1
$pCO_{2,a}$, ppmv	FocL–Fl150	875 (90)	762 (83)	669 (69)	615 (67)
	FocNL–Fl450	1010 (134)	854 (90)	721 (93)	652 (87)
ΔT_g, K	FocL–Fl150	3.38 (0.31)	3.05 (0.34)	2.65 (0.34)	2.43 (0.35)
	FocNL–Fl450	3.65 (0.41)	3.22 (0.41)	2.73 (0.37)	2.50 (0.39)
f	FocL–Fl150	1.18	1.21	1.22	1.25
	FocNL–Fl450	1.22	1.26	1.27	1.31
r_a	FocL–Fl150	0.57 (0.48)	0.56 (0.46)	0.55 (0.45)	0.54 (0.43)
	FocNL–Fl450	0.70 (0.57)	0.66 (0.52)	0.62 (0.46)	0.59 (0.49)
r_l	FocL–Fl150	0.09 (0.25)	0.10 (0.26)	0.12 (0.28)	0.13 (0.31)
	FocNL–Fl450	0.09 (0.24)	0.11 (0.27)	0.13 (0.29)	0.14 (0.32)
r_{oc}	FocL–Fl150	0.34 (0.29)	0.34 (0.28)	0.33 (0.27)	0.33 (0.26)
	FocNL–Fl450	0.21 (0.19)	0.23 (0.21)	0.25 (0.22)	0.26 (0.23)
β_l, PgC/ppmv	FocL–Fl150	0.87	1.06	1.21	1.39
	FocNL–Fl450	0.79	0.98	1.15	1.34
γ_l, PgC/K	FocL–Fl150	-124	-137	-142	-157
	FocNL–Fl450	-130	-148	-154	-170
β_{oc}, PgC/ppmv	FocL–Fl150	1.3 (by construction)			
	FocNL–Fl450	0.61	0.70	0.81	0.87
γ_{oc}, PgC/K	FocL–Fl150	-0.3 (by construction)			
	FocNL–Fl450	-18	-20	-20	-21

The larger difference in the temperature and CO_2 concentration responses between the coupled and uncoupled runs for the model version FocNL–Fl450 in comparison to the version FocL–Fl150 is due to nonlinearity of the oceanic sink of carbon from the atmosphere. Temperature rise suppresses F_{oc} for the version FocNL stronger than for the version FocL. Despite of the different values for the half–saturation points, terrestrial carbon uptakes behave quite similarly between two model versions (see Fig. 2) with eventual decrease in the 21st century. In contrast, oceanic uptakes monotonically increase in the 21st century for the version FocL–Fl150 and eventually decrease for the version FocNL–Fl450 (Fig. 2). As a result, C^3F is attributed basically to the terrestrial pool for the former version, and both to terrestrial and oceanic pools for the latter version.

For both feedbacks, however, typical time intervals may be determined by using on these uptakes. Till the mid 20th century, for both model versions, terrestrial uptakes of

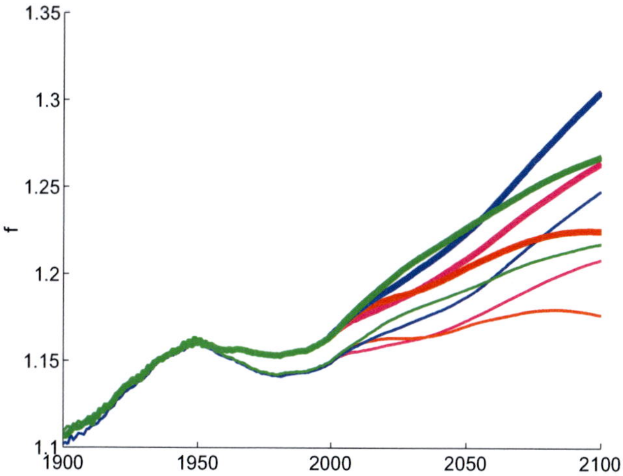

Figure 6. Parameter of the climate–carbon cycle feedback simulated by IAP RAS CM versions FocL–Fl150 and FocNL–Fl450 (thin and thick lines, respectively) for the emission scenarios A2, A1B, B2, and B1 (red, magenta, green, and blue curves, correspondingly). Growth of $pCO_{2,a}$ is computed with respect to year 1860.

carbon differ only slightly between the coupled and uncoupled runs. For oceanic uptakes, this statement is valid even till the late 20th century. From this, one inferres that, in the 20th century, climate–carbon cycle feedback is not too strong. In this period, changes in F_l and F_{oc} are basically due to CO_2–induced fertilisation of terrestrial productivity and due to increase in difference between partial pressures of carbon dioxide in surface air and surface oceanic water. In the IAP RAS CM coupled simulations, F_l peaks around year 2000 and F_{oc} reach its maximum in the first half of the 21st century (the latter maximum is found only for the version FocNL–Fl450). In any model version, terrestrial uptakes in the coupled runs deviate strongly from their counterparts in the uncoupled runs for the most part of the 21st century. To a smaller extent, this statement is valid also for the oceanic uptakes in the second part of the 21st century. As a result, one concludes that, in the 21st century, climate–carbon cycle strongly affects the results of simulations.

Temporal behaviour of the climate–carbon cycle interaction can be further quantified via temporal dependence of the respective parameter f (Eq. 2), see Fig. 6 [47, 14]. In the early industrial period, it is close to unity and grows monotonically till reaching the maximum in the middle of the 20th century. Afterwards, f slightly diminishes and attains shallow minimum in the early 1980's. After that, the C^3F generally increases.

The maximum of f in the mid 20th century is due to drastic increase of anthropogenic emissions in this period. It results in the enhanced buildup of CO_2 in the atmosphere. This buildup is accompanied by diminished influence of the climate–carbon cycle interactions because of the climate system inertia and associated delay in climatic response. This diminished influence is reflected in the C^3F parameter decrease. However, the delay in the climate response becomes smaller in a few decades, and f starts to grow again. It is notable that evolution of the coupled system in terms of the C^3F parameter is almost indistinguishable between the versions FocL–Fl150 and FocNL–Fl450 till the maximum in the middle of the 20th century. Afterwards, f differs substantially among the model versions. In particular, shallow manimum in the late 20th century is even shallower in the latter model version. When system moves from maximum in the middle of the 20th century to shallow minimum in the early 1980's, $pCO_{2,a}$ increases from about 310 ppmv to about 340 ppmv. Non–monotonic behaviour of f (as a function of $pCO_{2,a}$ which is, at a decadal time scale, a monotonic function of time) within this range of the atmospheric carbon dioxide concentration may be also seen in [19] (Figs. 1a,b there).

While both model versions exhibit a general increase of f in the 21st century for most emission scenarios, they show slight decrease of this parameter in the late 21st century under the most agressive emission scenario A2. This decrease is more characteristic for the version FocL–Fl150 than for FocNL–Fl450 and may be termed a C^3F saturation. Physically, it is related to the nearly saturated absorption lines of CO_2 in the atmosphere leading to the well–known logarithmic–type dependence of respective radiative forcing on $pCO_{2,a}$ [61]. As a consequence, when atmospheric carbon dioxide concentration is high, any additional atmospheric buildup of CO_2 results in a relatively small additional raditive forcing. In particular, for the aggressive emission scenarios $pCO_{2,a}^{cpl} - pCO_{2,a}^{ucpl}$ does not lead to substantional additional radiative forcing in the presence of high values of $pCO_{2,a}$ and climate–carbon cycle feedback saturates [14]. In turn, given the length of integration, high concentrations of carbon dioxide in the atmosphere may be achieved only if the respective emission scenario is agressive enough. This is the reason why the saturation is exhibited only for the scenario A2 which is most agressive among those used here. However, if relatively moderate emissions scenarios (A1B, B2, and B1) are extended beyond year 2100, this saturation is expected to be visible for these scenarios as well.

As a final note on the eventual saturation of the climate–carbon cycle feedback, one may distinguish its peculiarities in terms of feedback parameter, on one hand, and in terms of feedback gain, on the other one. In particular, an eventual C^3F saturation is visible more clearly in terms f than in terms of g. This is because of $dg/dt = f^{-2}df/dt$ and f attains higher values at the late part of the 21st century in comparison to its earlier part. In turn, it is a plausible reason why an eventual C^3F saturation is not clearly exhibited in the C^4MIP integrations where only temporal changes in g, not in f, were examined. Even in this, one C^4MIP model exhibits decrease of g in the late 21st century, and a few others show a slower decrease in the second part of this century in comparison to the first part (see Fig. 2b in [17]).

Decadal–to–centennial variations in the climate–carbon cycle interaction may be understood via temporal behaviour of β's and γ's at these timescales. These coefficients were computed for the running 100–yr windows.

In this computtaion, coefficient β_l monotonically decreases with time for both model versions FocL–Fl150 and FocNL–Fl450. For the earliest available period 1860–1959, it amounts $1.5 - 1.6$ PgC/ppmv. Then it decreases to $0.9 - 1.4$ PgC/ppmv for the version FocL–Fl150 and to $0.8 - 1.3$ PgC/ppmv for the version FocNL–Fl450 (with smaller values for the more agressive scenarios), see Tab. 2. Basically, temporal variations of this coefficient reflect interplay between the Michaelis–Menten law (as a function of $pCO_{2,a}$) and $dpCO_{2,a}/dt$.

Similarly, in the version FocNL–Fl450, coefficient β_{oc} decreases with time from about 1.5 PgC/ppmv in 1860–1959 to $0.6 - 0.9$ PgC/ppmv in 2001–2100 (Tab. 2), again due to

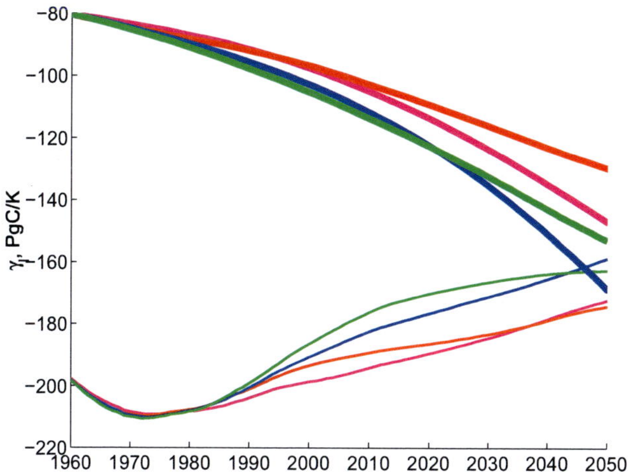

Figure 7. Coefficients γ_l computed for the running 100–yr window centered at the year indicated at the x–axis (e.g., the value for 2050 represents 2001–2100). Shown are values simulated by the IAP RAS CM versions FocL–Fl150 and FocNL–Fl450 (thin and thick lines, respectively) for the emission scenarios A2, A1B, B2, and B1 (red, magenta, green, and blue curves, correspondingly).

interplay between $dpCO_{2,a}/dt$ and the dependence of evasion factor on $pCO_{2,a}$. In the version FocL–Fl150, the value of $\beta_{oc} \equiv 1.3$ PgC/ppmv by construction.

In contrast, temporal behaviour of γ_l differs markedly between the two model versions (Fig. 7). For the version FocL–Fl150, coefficient γ_l attains high values even during the earliest part of integration and then slightly diminishes. This decrease is plausibly not significant due to smallness of T_g in this period. For the version FocNL–Fl450, the value of γ_l is about -80 PgC/K in 1860–1959 and then increases by magnitude amounting $-(130 - 170)$ PgC/K in 2001–2100 with smaller values for more aggressive scenarios in comparison to the less agressive ones (Tab. 2). The latter dependence on scenario is exhibited by the version FocL–Fl150 as well.

Even stronger difference between two model versions on the decadal–to–centennial scale is exhibited by γ_{oc}, see Fig. 8. This coefficient is constant by construction for the version FocL–Fl150. For the version FocNL–Fl450, coefficient γ_{oc} steadily increases in magnitude. Its value for this model version is -6 PgC/K in 1860–1959 and $-(18 - 21)$ PgC/K in 2001–2100 (see also Tab. 2).

Difference in the climate–carbon cycle interactions is also reflected in temporal behaviour of landborne and oceanborne fractions (r_l and r_{oc}, respectively) of the anthropogenically emitted carbon dioxide, see Figs. 9 and 10. The former variable strongly decreases in the 20th–21st centuries. This decrease is stronger in the coupled runs in comparison to their uncoupled counterparts with a little difference between the versions FocL–Fl150 and FocNL–Fl450. The decrease of r_l with time in uncoupled runs is due to enhanced fertilisation in the CO_2–enriched atmosphere. This results in higher NPP, larger litterfall, larger storage of carbon in soil, and, in turn, in higher soil respiration which overcompensates an increase of NPP.

In contrast, r_{oc} differs drastically between the two model versions. For FocL–Fl150,

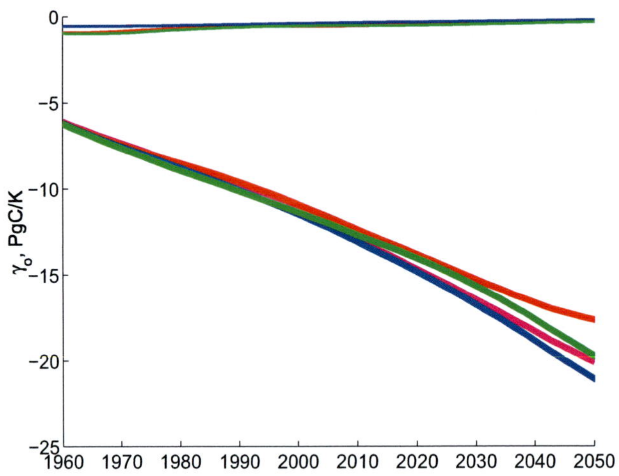

Figure 8. Coefficients γ_{oc} computed for the running 100–yr window centered at the year indicated at the x–axis (e.g., the value for 2050 represents 2001–2100). Shown are values simulated by the IAP RAS CM versions FocL–Fl150 and FocNL–Fl450 (thin and thick lines, respectively) for the emission scenarios A2, A1B, B2, and B1 (red, magenta, green, and blue curves, correspondingly).

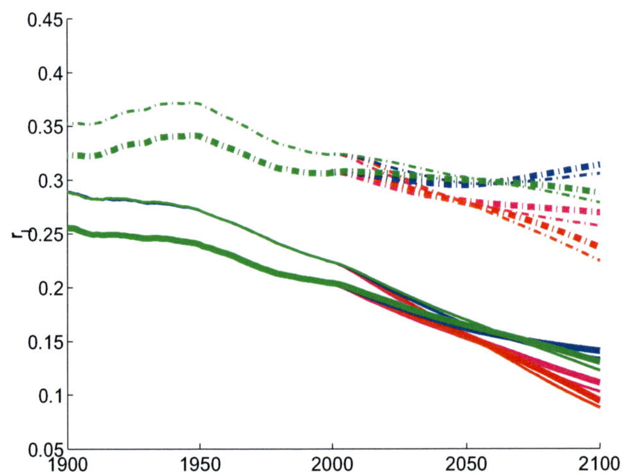

Figure 9. Landborne fraction r_l of anthropogenically emitted CO_2 as simulated by IAP RAS CM versions FocL–Fl150 and FocNL–Fl450 (thin and thick lines, respectively) for the emission scenarios A2, A1B, B2, and B1 (red, magenta, green, and blue curves, correspondingly).

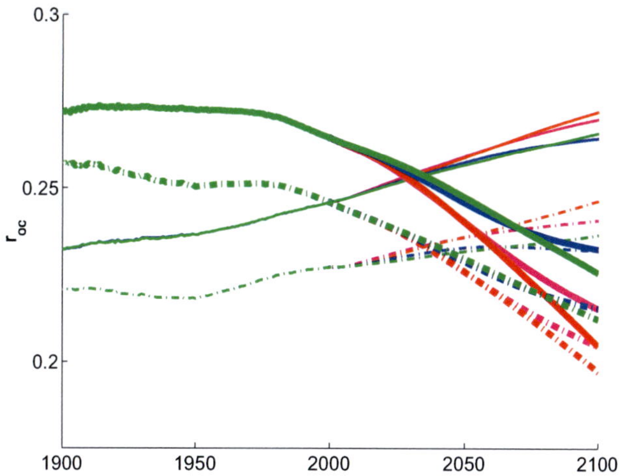

Figure 10. Oceanborne fraction r_{oc} of anthropogenically emitted CO_2 as simulated by IAP RAS CM versions FocL–Fl150 and FocNL–Fl450 (thin and thick lines, respectively) for the emission scenarios A2, A1B, B2, and B1 (red, magenta, green, and blue curves, correspondingly).

this variable generally increases during the course of integration, with higher increase exhibited in the coupled runs. For the version FocNL-Fl450, oceanborne fraction r_{oc} basically decreases with time. This difference reflects smallness of γ_l for the former version in comparison to the latter one.

5. Structural Uncertainty of Climate–Carbon Cycle Interaction

As it was mentioned in Introduction, there are uncertainties in the response of plant photosynthesis to warming temperatures. In addition, in [42], based on the results of the field measurements, it was noticed, that the nitrogen limitation of the organic matter decomposition in the soils may lead to the final decrease of the soil respiration under growing temperatures. The Q_{10}'s for different processes are quite uncertain. In particular, in [36], based on the literature review, the following limits for these quantities were suggested: $Q_{10,\mathrm{p}} = 1.1 - 2.66$, $Q_{10,\mathrm{r}} = 1.4 - 3.0$, $Q_{10,\mathrm{s}} = 1.3 - 3.8$. In [29], the range for $Q_{10,\mathrm{s}}$ was estimated as $1.4 - 2.8$ based on the climate variations due to El Niño events and volcano activity. While the generally adopted range of half–saturation point is $400 - 460$ ppmv [15, 7], in some studies different values are adopted (e.g., $k_{\mathrm{M}} = 150$ ppmv in [36, 13, 14, 60]). The latter is caused basically either by simplifications employed in the terrestrial carbon cycle model (see the respective discussion in [13]) or by neglection of other anthropogenically induced effects (e.g., due to sulphate aerosols) in tuning the model to the 20th century observations. Moreover, intensity of climate–carbon cycle interations depend on the overall model's sensitivity to greenhouse forcing [20]. Different fomulations of the coupled models lead to the diversity of the simulations with a striking example of such diversity exhibited in the C^4MIP integrations.

An objective approach to treat this diversity is to perform ensemble integrations with a climate model by perturbing different parameters of the carbon cycle in a systematic way. In this section, structural uncertainty due to terrestrial carbon cycle is explored for such ensemble integrations with Ithe AP RAS CM forced by the historical emissions for the 19th and 20th centuries and by the SRES A2 emissions for the 21st century.

In this integrations, a subset of parameters $Q_{10,\mathrm{p}}$, $Q_{10,\mathrm{r}}$, $Q_{10,\mathrm{s}}$, k_{M} and k_{D} is perturbed. Namely, $Q_{10,\mathrm{p}}$ was assigned one of four values (0.8, 1.0, 1.5, and 2.0). $Q_{10,\mathrm{r}}$ was also varied between three values, 1.4, 2.15, and 3.0. The corresponding constant for soil respiration, $Q_{10,\mathrm{s}}$, was assigned one of the four values (1.0, 2.0, 2.4, and 3.0). The half saturation point k_{M} was pertubed among the values 100 ppmv, 150 ppmv, 200 ppmv, and 450 ppmv. In these simulations, constants A's in (7) are also changed in order to keep the realistic preindustrial state in the model. To test the model sensitivity to the choice of the constant k_{D} regulating the living biomass with respect to the agricultural activity, this constant was also assigned one of the values 0 and 0.27.

To test the uncertainty due to sensitivity of the model to the growth of the carbon dioxide in the atmosphere, simulated temperature anomalies entering the carbon cycle routine were multiplied by a constant factor in the course of integrations. This factor was assigned one of the values 0 (for the carbon cycle of the IAP RAS CM, it is analogous to the uncoupled simulations), 1.0 (the standard model setting), 1.7, and 2.0. One notes, that the latter two multiplication factors were choosed in order to denser populate the phase space of the model rather than to represent any particular climate model or ensemble of such models.

All the model simulations were started from the preindustrial equilibrated model state. In every simulation, the first model year with the non–zero carbon dioxide emissions corresponds to the julian year 1859. Simulations end in the year corresponding to the julian year 2100. A subset of these simulations was considered earlier in [12, 13].

The sources of uncertainties different from the mentioned above, e.g., due to uncertainties in CO_2 emissions, governing parameters of the model's oceanic carbon cycle, influence of other external forcings (methane, ozone, sulphates, carbon aerosols, volcanic aerosols, etc) are not considered here. As characteristics of these forcings also associated with some uncertainty ranges, taking them into account result likely in a wider uncertainty ranges than those presented below (e.g., [2]). Further, in the present paper, the ensemble simulation is performed only for the version FocL–Fl150 without accounting for structural uncertainty between FocL–Fl150 and FocNL–Fl450. Even in this simplified case, the undertaken study presumes large computational burden. The cumulative length of the simulations performed in the present paper is 371,712 model years. Such long simulations are precluded currently for the state–of–the–art general circulation model due to technical reasons [8, 50]. This advocates the usage of the climate model of intermediate complexity for the purposes of the present study.

If the performed simulations are subjected to the constraint of the maximum allowed deviation $\epsilon_{p\mathrm{CO}_{2,\mathrm{a}}}$ of the simulated $p\mathrm{CO}_{2,\mathrm{a}}$ from the observed values (measured at the Mauna Loa observatory during the second half of the 20th century) and/or to the constraint that simulated terrestrial and oceanic uptakes in 1980's and 1990's must be in the range published in [26], it is possible to rule out some of the simulations, performed in the present paper, and narrow the corresponding uncertainty range.

The results of such analysis for values of $p\mathrm{CO}_{2,a}$ in year 2100 are presented in Tab. 3.

Table 3. Ranges of atmospheric CO_2 concentration (ppmv) for different values of the allowed deviations $\epsilon_{pCO_{2,a}}$ of the simulated carbon dioxide concentration in the atmosphere from the Mauna Loa observations for 1959–2000, and for different constraints on the simulated uptakes.

	constrained uptakes			
$\epsilon_{pCO_{2,a}}$, ppmv	none	F_l	F_{oc}	F_l and F_{oc}
none	496 ... 1379	585 ... 991	496 ... 1379	585 ... 991
10	496 ... 1241	705 ... 988	496 ... 1241	705 ... 988
5	496 ... 1197	789 ... 896	496 ... 1197	789 ... 896

As seen from this Table, constraining the simulations based on the 20th century observations narrows the respective uncertainty range substantially. If no constraint is imposed, the uncertainty range for $pCO_{2,a}(2100)$ is $496 - 1379$ ppmv. If all above–mentioned characteristics are constrained (e.g., both F_l and F_{oc} are in the range depicted in [26], and the tolerable deviation of the simulated atmospheric concentration of carbon dioxide from the Mauna Loa observations is $\epsilon_{pCO_{2,a}} = 2$ ppmv), the corresponding range shrinks to $789 - 896$ ppmv. Thus terrestrial uptakes of carbon constrain simulations more efficiently than the oceanic ones. This is possibly due to neglection of the structural uncertainty between versions FocL–Fl150 and FocNL–Fl450 because in the former version F_{oc} changes only due to different evolutionary trajectories of $pCO_{2,a}$.

The sampled probability density functions (PDFs) $P(pCO_{2,a})$ are non–symmetrical with respect to the modal values with a heavier right tail. Even heavier right tail is exhibited by the sampled probability density function $P(\ln pCO_{2,a})$ closely related to the PDF of the radiative forcing of CO_2. The latter also implies skewed PDF of the temperature response to the growth of $pCO_{2,a}$ with a larger probability for the response stronger than the modal one.

An analysis of this ensemble simulation may be further performed by usage of the Bayesian averaging [32, 22]. The latter approach allows one to quantify the structural uncertainty of the simulation by weighting the members of such an ensemble with a smaller (larger) weights attached to less (more) realistic simulations. For this analysis, prior probability distribution functions for terrestrial and oceanic uptakes are chosen to be non–informative (i.e., uniform) [32] with a total range of variation corresponding to [26]. The respective prior function for $pCO_{2,a}$ is chosen to be Gaussian with a standard deviation $\Sigma_{pCO_{2,a}} = 2 - 15$ ppmv. The latter values are chosen subjectively in order to represent "tolerable" deviation of simulated atmospheric concentration of CO_2 from the Mauna Loa observations. The value of $\Sigma_{pCO_{2,a}}$ does not affect the Bayesian ensemble mean too much, see Fig. 11. Moreover, in the presence of other constraints, the ensemble standard deviation of simulated $pCO_{2,a}$, $\sigma_{pCO_{2,a}}$, also shows only weak dependence on $\Sigma_{pCO_{2,a}}$. For instance, for year 2100, $\sigma_{pCO_{2,a}}$ increases from 46 ppmv at $\Sigma_{pCO_{2,a}} = 2$ ppmv to 61 ppmv at $\Sigma_{pCO_{2,a}} = 15$ ppmv. These values are much smaller than the value $\sigma_{pCO_{2,a}} = 182$ ppmv which attained by simple, equally weighted, averaging of the members within this ensemble (see also Fig. 11). Interesting, the equally weighted ensemble mean is substantially higher than the respective Bayesian means.

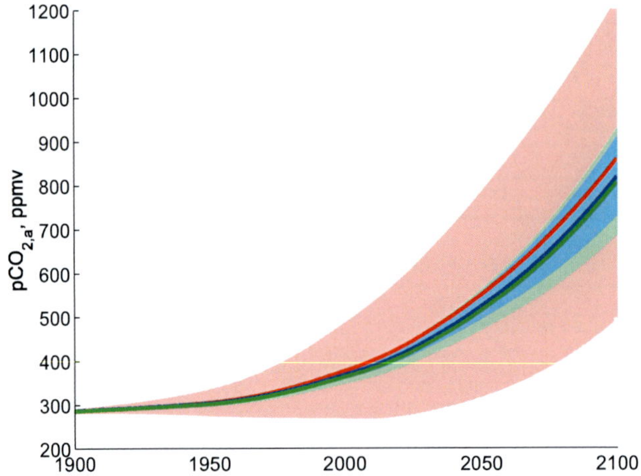

Figure 11. Uncertainty propagation for $p\mathrm{CO}_{2,\mathrm{a}}$ in the ensemble simulations with IAP RAS CM forced by the historical+SRES A2 emissions. Shown are ensemble means (lines) and uncertainty ranges (shaded areas) defined here as a doubled standard deviations for the equally weighted averaging (red) and for the Bayesian averaging with $\Sigma_{p\mathrm{CO}_{2,a}} = 8$ ppmv and $\Sigma_{p\mathrm{CO}_{2,a}} = 15$ ppmv (blue and green, correspondingly). See text for further details.

Irrespective of the approach to narrow the uncertainty within the ensemble (either selecting only realisitc simulations or employing Bayesian averaging) the resulting uncertainty is still substantial. It reflects relative smallness of the 20th century changes in comparison to those expected in the 21st century [13, 31]. In particular, in [31] it was stated that "the observational record proves to be insufficient to tightly constrain carbon cycle processes or future feedback strength with implications for climate–carbon cycle model evaluation". Moreover, in [43] the 20th century course of the carbon cycle characteristics was reproduced accurately with a model accounting only direct CO_2 influence on F_l and F_oc without considering respective climate feedbacks. The latter is consistent with [13] where it was shown that the changes in carbon cycle observed during the 20th century may be reproduced even by the model with a negative climate–carbon cycle feedback.

This point is further illustrated in Fig. 12. In this figure, uncertainty range includes both positive and negative $\mathrm{C}^3\mathrm{F}$. This range becomes even wider during the course of integration. Nonetheless, positive feedback is more probable than the negative one, and the respective ensemble means of $\Delta p\mathrm{CO}_{2,\mathrm{a}}^{\mathrm{cpl}} - \Delta p\mathrm{CO}_{2,\mathrm{a}}^{\mathrm{ucpl}}$ are positive. These ensemble means in year 2100 amount 136 ppmv for the equally–weighted averaging, 64 ppmv for Bayesian averaging with $\Sigma_{p\mathrm{CO}_{2,a}} = 15$ ppmv, and 57 ppmv for Bayesian averaging with $\Sigma_{p\mathrm{CO}_{2,a}} = 8$ ppmv. The last two means are statistically indistiguishable between each other taking into account ensemble standard deviations of $\Delta p\mathrm{CO}_{2,\mathrm{a}}^{\mathrm{cpl}} - \Delta p\mathrm{CO}_{2,\mathrm{a}}^{\mathrm{ucpl}}$ amounting 107 ppmv and 98 ppmv, correspondingly. In addition, these two ensemble averages are within the most probable range derived from the C^4MIP simulations, $50 - 100$ ppmv (see 3.).

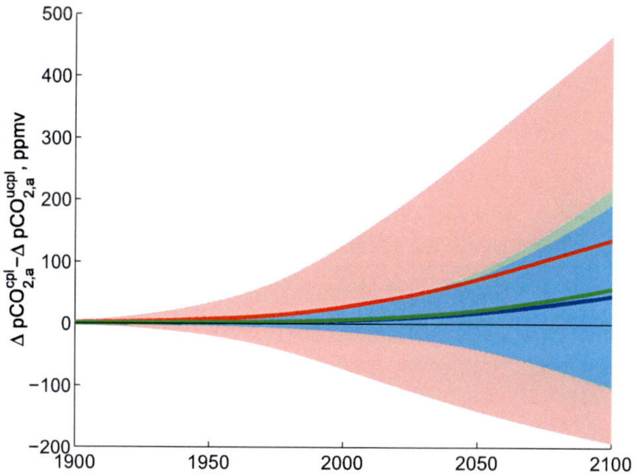

Figure 12. Uncertainty propagation for the C^3F intensity $\Delta pCO_{2,a}^{cpl} - \Delta pCO_{2,a}^{ucpl}$ in the ensemble simulations with the IAP RAS CM forced by the historical+SRES A2 emissions. Shown are ensemble means (lines) and uncertainty ranges (shaded areas) defined here as a doubled standard deviations for the equally weighted averaging (red) and for the Bayesian averaging with $\Sigma_{pCO_{2,a}} = 8$ ppmv and $\Sigma_{pCO_{2,a}} = 15$ ppmv (blue and green, correspondingly). Black line depicts zero intensity of C^3F. See text for further details.

6. Conclusion

At the decadal–to–sentennial scale, the climate–carbon cycle interaction is set basically by an interplay in responses of net primary productivity and soil respiration to climate changes, and by analogous response in the oceanic carbon uptake. The current generation of the coupled models converges at the positive sign of the resulting feedback (C^3F) with a typical enhancement of the global warming by about 10% in the 21st century. However, there is a large diversity among these models, with some of them attributing the climate–carbon cycle feedback to the terrestrial pools while the others trace C^3F to the oceanic reservoir. Even larger diversity is found at regional level.

Unfortunately, future evolution of the coupled system can not be effectively constrained by the 20th century observations due to relative smallness of the changes in the state of the coupled system in comparison to those expected in future. In particular, uncertainty ranges for the 20th century observations do not preclude zero–intensity or even negative climate–carbon cycle feedback. It is illustrated employing large ensemble of simulations with a climate model of intermediate comlexity IAP RAS CM. The members of this ensemble were stratified in accordance with a realism of their simulation of the carbon cycle characteristics observed in the 20th century. Bayesian ensemble averaging and/or selection of the ensemble members "tolerably" deviating from the observations narrow the above–mentioned uncertainty range but still unable to rule out negative C^3F. However, the corresponding ensemble means lead to the positive climate–carbon cycle feedback.

In the IAP RAS CM runs, characteristics of the climate–carbon cycle interaction undergo substantial changes during the 20th and, especially, during the 21st century. In particular, coefficients γ_l and γ_{oc} (computed for the running 100–yr window) basically increase in magnitude indicating a strengthening of C^3F at a century timescale. This contrasts with diminishing direct influence of $pCO_{2,a}$ growth on terrestrial and oceanic uptakes of carbon as manifested in β_l and β_{oc} which both decrease with time. General strengthening of climate–carbon cycle interaction at a century timescale is also supported by general increase of the feedback parameter f during the course of integration.

At a decadal timescale, however, deviations from this tendency are observed. During the 20th century, there are periods when climate–carbon cycle feedback either weakens or strengthens. This is related to the abrupt changes in the intensity of anthropogenic emissions and to the delay in response of the coupled system to these emissions.

In addition, at a century timescale, an eventual saturation of C^3F is observed. Physically, it is related to the weak (logarithmic) dependence of the radiative forcing on $pCO_{2,a}$. As a result, additional buildup of carbon dioxide in the atmosphere does not enlarge this forcing too much. As a result, additional (with respect to the simulation neglecting C^3F) warming is small, and climate–carbon cycle feedback saturates.

However, this saturation may not be expected during next few decades when C^3F is expected to markedly modify the behaviour of the coupled system.

References

[1] Adams B. ; White A. ; Lenton T.M. , *Ecol. Mod.*, 2004, 177, 353–391.

[2] Allen M.R. ; Stott P.A. ; Mitchell J.F.B. ; Schnur R. ; Delworth T.L. , *Nature*, 2000, 407, 617–620.

[3] Bacastow R. . Numerical evaluation of the evasion factor. In Bolin B. , Ed, *Carbon Cycle Modelling, SCOPE–16*, pages 95–101. J. Wiley and Sons, N.Y., 1981.

[4] Brohan P. ; Kennedy J.J. ; Harris I. ; Tett S.F.B. ; Jones P.D. , *J. Geophys. Res.*, 2006, 111, D12106.

[5] Brovkin V. ; Bendtsen J. ; Claussen M. ; Ganopolski A. ; Kubatzki C. ; Petoukhov V. ; Andreev A. , *Glob. Biogeochem. Cycles*, 2002, 16, 1139.

[6] Brovkin V. ; Sitch S. ; Bloh W. von; Claussen M. ; Bauer E. ; Cramer W. , *Glob. Change Biol.*, 2004, 10, 1253–1266.

[7] Budyko M.I. ; Izrael Y.A. , Eds. *Anthropogenic climate change*. Arizona Univ. Press, Tucson, 1991.

[8] Claussen M. ; Mysak L. ; Weaver A. ; Crucifix M. ; Fichefet T. ; Loutre M.-F. ; Weber S. ; Alcamo J. ; Alexeev V. ; Berger A. ; Calov R. ; Ganopolski A. ; Goosse H. ; Lohmann G. ; Lunkeit F. ; Mokhov I. ; Petoukhov V. ; Stone P. ; Wang Z. , *Clim. Dyn.*, 2002, 18, 579–586.

[9] Cox P.M. ; Betts R.A. ; Collins M. ; Harris P.P. ; Huntingford C. ; Jones C.D. , *Theor. Appl. Climatol.*, 2004, 78, 137–156.

[10] Cox P.M. ; Betts R.A. ; Jones C.D ; Spall S.A. ; Totterdell I.J. , *Nature*, 2000, 408, 184–187.

[11] Dufresne J.-L. ; Friedlingstein P. ; Berthelot M. ; Bopp L. ; Ciais P. ; Fairhead L. ; Le Treut H. ; Monfray P. , *Geophys. Res. Lett.*, 2002, 29, 1405.

[12] Eliseev A.V. ; Mokhov I.I , *Comp. Tech.*, 2006, 11, 14–19.

[13] Eliseev A.V. ; Mokhov I.I. , *Theor. Appl. Climatol.*, 2007, 89, 9–24.

[14] Eliseev A.V. ; Mokhov I.I. ; Karpenko A.A. , *Izvestiya, Atmos. Ocean. Phys.*, 2007, 42, 1–14.

[15] Farquhar G.D. ; Caemmerer S. von; Berry J.A. , *Planta*, 1980, 149, 78–90.

[16] Friedlingstein P. ; Bopp L. ; Ciais P. ; Dufresne J.-L. ; Fairhead L. ; Le Treut H. ; Monfray P. ; Orr J. , *Geophys. Res. Lett.*, 2001, 28, 1543–1546.

[17] Friedlingstein P. ; Cox P. ; Betts R. ; Bopp L. ; Bloh W. von; Brovkin V. ; Doney S. ; Eby M. ; Fung I. ; Govindasamy B. ; John J. ; Jones C. ; Joos F. ; Kato T. ; Kawamiya M. ; Knorr W. ; Lindsay K. ; Matthews H.D. ; Raddatz T. ; Rayner P. ; Reick C. ; Roeckner E. ; Schnitzler K.-G. ; Schnur R. ; Strassmann K. ; Weaver A.J. ; Yoshikawa C. ; Zeng N. , *J. Climate*, 2006, 19, 3337–3353.

[18] Friedlingstein P. ; Dufresne J.-L. ; Cox P.M. ; Rayner P. , *Tellus*, 2003, 55B, 692–700.

[19] Fung I.Y. ; Doney S.C. ; Lindsay K. ; John J. , *Proc. Nat. Acad. Sci.*, 2005, 102, 11201–11206.

[20] Govindasamy B. ; Thompson S. ; Mirin A. ; Wickett M. ; Caldeira K. ; Delire C. , *Tellus*, 2005, 57B, 153–163.

[21] Handorf D. ; Petoukhov V.K. ; Dethloff K. ; Eliseev A.V. ; Weisheimer A. ; Mokhov I.I. , *J. Geophys. Res.*, 1999, 104, 27253–27275.

[22] Hoeting J.A. ; Madigan D. ; Raftery A.E. ; Volinsky C.T. , *Stat.Sci.*, 1999, 14, 382–401.

[23] Houghton J.T. ; Callander B.A. ; Varney S.K. , Eds. *Climate Change: The Supplementary Report to the IPCC Scientific Assessment, Intergovernmental Panel on Climate Change.* Cambridge University Press, Cambridge, 1992.

[24] Houghton J.T. ; Ding Y. ; Griggs D.J. ; Noguer M. ; Linden P.J. van der; Dai X. ; Maskell K. ; Johnson C.A. , Eds. *Climate Change 2001: The Scientific Basis. Contribution of Working Group I to the Third Assessment Report of the Intergovernmental Panel on Climate Change.* Cambridge University Press, Cambridge/New York, 2001.

[25] Houghton R.A. , *Tellus*, 2003, 55B, 378–390.

[26] House J.I. ; Prentice I.C. ; Ramankutty N. ; Houghton R.A. ; Heimann M. , *Tellus*, 2003, 55B, 345–363.

[27] Huntingford C. ; Cox P.M. ; Lenton T.M. , *Ecol. Mod.*, 2000, 177, 41–58.

[28] Huntingford C. ; Harris P.P. ; Gedney N. ; Cox P.M. ; Betts R.A. ; Marengo J.A. ; Gash J.H.C. , *Theor. Appl. Climatol.*, 2004, 78, 177–185.

[29] Jones C.D. ; Cox P.M. , *Atmos. Sci. Lett.*, 2001.

[30] Jones C.D. ; Cox P.M. ; Essery R.L.H. ; Roberts D.L. ; Woodage M.J. , *Geophys. Res. Lett.*, 2003, 30, 1479.

[31] Jones C.D. ; Cox P.M. ; Huntingford C. , *Tellus*, 2006, 58B, 603–613.

[32] Kass R.E. ; Raftery A.E. , *J. Amer. Stat. Assoc.*, 1995, 90, 773–795.

[33] Keeling C.D. ; Chine J.F.S. ; Whorf T.P. , *Nature*, 1996, 382, 146–149.

[34] Kwon O. ; Schnoor J.L. , *Glob. Biogeochem. Cycles*, 1994, 8, 295–305.

[35] Le Quéré C. ; Aumant O. ; Bopp L. ; Bousquet P. ; Ciais P. ; Francey R. ; Heimann M. ; Keeling R.F. ; Kheshgi H. ; Peylin P. ; Piper S.C. ; Prentice I.C. ; Rayner P.J. , *Tellus*, 2003, 55B, 649–656.

[36] Lenton T.M. , *Tellus*, 2000, 52B, 1159–1188.

[37] Lloyd J. ; Taylor J.A. , *Func. Ecol.*, 1994, 8, 315–323.

[38] Manning A.C. ; Keeling R.F. , *Tellus*, 2006, 58B, 95–116.

[39] Marland G. ; Boden T.A. ; Andres R.J. . Global, regional, and national CO2 emissions. In *Trends: A Compendium of Data on Global Change*. Carbon Dioxide Information Analysis Center, Oak Ridge National Laboratory, U.S. Department of Energy, Oak Ridge, Tenn., 2005.

[40] Matthews H.D. ; Weaver A.J. ; Meissner K.J. , *J. Climate*, 2005, 18, 1609–1628.

[41] Matthews H.D. ; Weaver A.J. ; Meissner K.J. ; Gillett N.P. ; Eby M. , *Clim. Dyn.*, 2004, 22, 461–479.

[42] Melillo J.M. ; Steudler P.A. ; Aber J.D. ; Newkirk K. ; Lux H. ; Bowles F.P. ; Catricala C. ; Magill A. ; Ahrens T. ; Morrisseau S. , *Science*, 2002, 298, 2173–2176.

[43] Melnikov N.B. ; O'Neill B.C. , *Geophys. Res. Lett.*, 2006, 33, L02705.

[44] Millero F.J. , *Geophys. Cosmophys. Acta*, 1995, 59, 661–677.

[45] Mokhov I.I. ; Demchenko P.F. ; Eliseev A.V. ; Khon V.Ch. ; Khvorostyanov D.V. , *Izvestiya, Atmos. Ocean. Phys.*, 2002, 38, 555–568.

[46] Mokhov I.I. ; Eliseev A.V. ; Demchenko P.F. ; Khon V.Ch. ; Akperov M.G. ; Arzhanov M.M. ; Karpenko A.A. ; Tikhonov V.A. ; Chernokulsky A.V. ; Sigaeva E.V. , *Doklady Earth Sci.*, 2005, 402, 591–595.

[47] Mokhov I.I. ; Eliseev A.V. ; Karpenko A.A. , *Comp. Tech.*, 2006, 11, 156–165. [in Russian with an English abstract].

[48] Mokhov I.I. ; Eliseev A.V. ; Karpenko A.A. , *Doklady Earth Sci.*, 2006, 407, 424–428.

[49] Patra P.K. ; Maksyutov S. ; Ishizawa M. ; Nakazawa T. ; Takahashi T. ; Ukita J. , *Glob. Biogeochem. Cycles*, 2005, 19, GB4013.

[50] Petoukhov V. ; Claussen M. ; Berger A. ; Crucifix M. ; Eby M. ; Eliseev A.V. ; Fichefet T. ; Ganopolski A. ; Goosse H. ; Kamenkovich I. ; Mokhov I.I. ; Montoya M. ; Mysak L.A. ; Sokolov A. ; Stone P. ; Wang Z. ; Weaver A. , *Clim. Dyn.*, 2005, 25, 363–385.

[51] Petoukhov V.K. ; Mokhov I.I. ; Eliseev A.V. ; Semenov V.A. . *The IAP RAS Global Climate Model*. Dialogue–MSU, Moscow, 1998.

[52] Plattner G.-K. ; Joos F. ; Stocker T.F. , *Glob. Biogeochem. Cycles*, 2002, 16, 1096.

[53] Raich J.W. ; Schlesinger W.H. , *Tellus*, 1992, 44B, 81–99.

[54] Sabine C.L. ; Feely R.A. ; Gruber N. ; Key R.M. ; Lee K. ; Bullister J.L. ; Wanninkhof R. ; Wong C.S. ; Wallace D.W.R ; Tilbrook B. ; Millero F.J. ; Peng T.-H. ; Kozyr A. ; Ono T. ; Rios A.F. , *Science*, 2004, 305, 367–371.

[55] Siegenthaler U. ; Monnin E. ; Kawamura K. ; Spahni R. ; Schwander J. ; Stauffer B. ; Stocker T.F. ; Barnola J.-M. ; Fischer H. , *Tellus*, 2005, 57B, 51–57.

[56] Siegenthaler U. ; Sarmiento J.L. , *Nature*, 1993, 365, 119–125.

[57] Solomon S. ; Qin D. ; Manning M. ; Marquis M. ; Averyt K. ; Tignor M.M.B. ; LeRoy Miller H. ; Chen Z. , Eds. *Climate Change 2007: The Physical Science Basis*. Cambridge University Press, Cambridge/New York, 2007. [in press].

[58] Svirezhev Y.U. ; Krapivin V.F. ; Tarko A.M. . Modeling of the main biosphere cycles. In Malone T.F. ; Roederer J.S. , Eds, *Global Change*, pages 298–313. Cambridge University Press, Cambridge, 1985.

[59] Thomas H. ; England M.H. ; Ittekkot V. , *Geophys. Res. Lett.*, 2001, 28, 547–550.

[60] Williamson M.S. ; Lenton T.M. ; Shepherd J.G. ; Edwards N.R. , *Ecol. Mod.*, 2006, 198, 362–374.

[61] Zuev V.E. ; Titov G.A. . *Atmosphere Optics and Climate*. Spektr, Tomsk, 1996. [in Russian].

In: Climate Change Research Progress
Editor: Lawrence N. Peretz, pp. 243-268
ISBN: 978-1-60021-998-6

Chapter 7

CLIMATE CHANGE AND POPULATION HEALTH

Ying Zhang, Peng Bi and Janet E. Hiller
Discipline of Public Health, School of Population Health and Clinical Practice, Faculty of Health Sciences, University of Adelaide. Address: Level 9, 10 Pulteney Street, SA 5000. Australia

Abstract

Climate change has already affected many aspects of the earth ecological system. Population health should be one of the most concerned. However, the relationship between climate and population health, including both chronic diseases and infectious diseases, is far from clear. A critical review of previous studies on the relationship between climate change and population health has been conducted with focuses on research methodology, including study population, study design and statistical method, and their results. Further research trends on climate change and population health will be addressed based on the critical review.

Introduction

There is substantial evidence showing that global climate has been changing and will continue to change. There is a clear increase in the world surface temperature during the last century [1]. It is projected that global average temperature will increase by up to 6ºC by 2100 compared to 1990 [1]. Climate change has already affected a diverse set of physical and biological systems worldwide, with the costs associated with climate change estimated to be about US$ 60 billion in 2003 [2].

Among the potential impacts of climate change, public health may be one of the most important. In addition to the excess deaths, injuries and illness directly attributable to extreme weather events and climate-induced natural disasters, meteorological variables (such as ambient temperature, rainfall, relative humidity and air pressure), and El Nino Southern Oscillation (ENSO) may indirectly cause increased risk of infectious diseases by altering the distribution of vector species, influencing the reproduction of the pathogens (vector-borne, food-borne and water-borne), affecting the frequency and level of contamination of drinking water and affecting the behaviours of human beings to increase chances of contact [3].

Understanding the relationship between climate change and diseases poses an unusual challenge to the scientific community for research. Firstly, it is difficult to collect high quality health surveillance, climatic and demographic data on a large population scale over a long period of time, particularly in developing countries for historic data analysis, which is essential for determining the association and for prediction. Secondly, advanced mathematic and statistical models with few uncertainties have yet to be developed to quantify this association. Thirdly, it is highly possible that the effects of climate change on nature of diseases may vary from context to context, given various climatic conditions, different vector or pathogen species, and the socio-economic status of populations in different regions. Moreover, so far, there have been few published studies examining the burden attributable to climate change at a local (sub-national) level. Furthermore, there have been very few studies to explore possible adaptation and mitigation strategies in the response to climate change in population health area.

This chapter has systematically reviewed current knowledge in studying the relationship between climate change and population health to suggest future study directions in this research field to achieve a better understanding of the relationship between climate change and population health and to provide scientific evidence for local policy makers and communities to develop public health strategies to prevent or reduce future risks as climate change continues.

I. Current Knowledge about the Relationship between Climate Change and Population Health

1. Introduction

The relationship between climate change and population health has been of considerable concern recently. In the early 1990s, there was little public awareness of the health risks posed by global warming and climate change [1], which was reflected in the first report of the United Nation's Intergovernmental Panel on Climate Change (IPCC) [4]. Although there was little empirical data in the first assessment, three global assessments of health impact (in 1991, 1996 and 2001) and the latest report from the IPCC in 2007 have been conducted as a result of a growing awareness of climate change [5].

According to the latest report from the IPCC in 2007, during the 20th century the world average surface temperature has increased approximately 0.74°C [6]. Scientific evidence has shown that climate change in the last century could be, at least partly, attributable to human activities, and that further changes in climate are inevitable at both global and regional levels. It is projected that global average temperature will increase in the range of 1.4 to 5.8 °C by 2100, compared to 1990 [7]. At a regional level, in Australia for instance, annual average temperatures will be 0.4~2.0°C higher by 2030, than those of 2002 [8]. China also has shown a similar trend in warming and the average temperature there will have increased 2.2 °C by 2050 [9]. Additionally, average annual temperatures over Europe increase at a rate of between 0.1 and 0.4°C per decade and mean sea level rises by up to 68 cm by the 2050s. The rates of mean annual temperature changes in the Latin American region for the next century are projected to be 0.2–2°C. [1]

The effects of climate change - direct and indirect - have been reported in many publications. There is now good evidence that regional change in climate, particularly increases in temperature, have already affected a diverse set of physical and biological systems in many parts of the world [8]. The United Nations Environment Programme (UNEP) has reported that, in 2003, the worldwide cost of climate change was about US$ 60 billion, 10% higher than in 2002 [2]. Extreme temperature, heavy precipitation, and climate-induced natural disasters may directly cause excess deaths, injuries and illness. For example, McMichael et al [10] assessed that, in 2002, extreme temperatures contributed to the death of about 1,100 people over 65 years old in 10 Australian and two New Zealand cities. Moreover, many other effects may result from climate change, such as the degradation of the local environment, reduction in yields of agricultural production and the loss of biodiversity [11].

The indirect impact of climate change on population health, including the effect of climate variation on diseases, is not clear. Although evidence exists suggesting the association between some diseases, such as malaria, and meteorological variables, there are limitations and uncertainties in previous studies in terms of data availability and reliability, study design, statistical methods and confounder control. Furthermore, the effects of climate change may vary in different regions with varying levels of economic development. The association between enteric infection and meteorological variables has received relatively little attention.

2. Changing Climate

There are many factors including physical, chemical and biological processes that interact to affect the climate on the earth's surface, including factors intrinsic to the climate system itself and external factors derived from human activities. There is strong evidence to show that most of the warming observed over the last 50 years is the result of human activities [12].

It has been demonstrated that climate change is happening at a global level. According to the IPCC 2007 report [6], the global average surface temperature has increased approximately 0.74°C during the last century. In addition, the warming trend has nearly double over the last 50 years. Observed changes in precipitation suggest that annual precipitation has continued to increase in the middle and high latitudes of the Northern Hemisphere (likely to be 0.5 to 1% per decade), except eastern Asia. The rate of the global mean sea level rise during the 20th century is in the range of 1.0-2.0 mm per year. The activity of the El Nino-Southern Oscillation (ENSO) has been unusual since the mid-1970s compared with the previous 100 years, with warm phase ENSO episodes being relatively more frequent, more persistent and intense than the opposite cool phase.

A projected range of global average surface warming of 1.4 to 5.8 °C by 2100, compared to 1990, suggests a rate of warming almost two to ten times higher than that observed during the last century. Global average precipitation and evaporation are projected to increase by about 1% to 9% by 2100, depending on which scenario and climate change model is used. Regionally, projected rainfall changes vary, making some regions wetter and others drier [1].

3. Impact of Climate Change on Population Health

The climate change described above may have an impact on many aspects of human and natural systems, including population health, agriculture and biological systems [13]. Some of the effects may be positive, such as cold-related deaths in colder regions [14]. Globally, the impacts are adverse to most people in the world. Both direct and indirect impact of climate change is discussed in this section. Table 1 summarizes the potential impact of climate change on population health (Table 1).

3.1. Direct Impact of Climate Change on Population Health

The most direct thermal effect is heat stress, which can lead to heat exhaustion and heatstroke. Many studies have shown that there is a U-shape, V-shape or J –shape relationship between extreme temperature and mortality [15]. For example, during 1979-1999, 8,015 extra deaths in the United States were heat-related [16]. The summer heatwave of 2003 in Europe was exceptional for the extensive loss of life, with 14,800 deaths in France during 9 days of extreme temperatures [17]. In Australia, it is stated that heatwaves kill more people than any other natural hazard [18]. South Australia had the highest death rate (1.14/100,000) caused by excessive heat between 1807 and 1994 [19]. Interaction with humidity can increase the hazards of heat stress [20].

Table 1. Potential effects of climate change on population health

Exposure Factors	Health outcome
Direct effects	
Thermal extremes	Heat or cold-related illness and deaths
Other extreme events	Deaths, injuries, psychological disorders, damage to public health infrastructure
Indirect effects	
Effect on range and activity of vectors and infective parasites	Changes in geographical ranges and incidence of vector-borne disease
Changed local ecology of water-borne and enteric infective agents	Changed incidence of diarrhoeal and other infectious diseases
Changed food productivity through changes in climate and associated pests and diseases	Malnutrition and hunger, and consequent impairment of child growth and development
Sea level rise with population displacement and damage to infrastructure	Increased risk of infectious disease, psychological disorders
Biological impact of air pollution changes	Asthma and allergies; other acute and chronic respiratory disorders and deaths
Social, economic and demographic dislocation through effects on economy, infrastructure and resource supply	Wide range of public health consequences: mental health and nutritional impairment, infectious disease, civil strife

(Modified from McMichael and Haines, 1996 [8])

Other extreme weather events including floods, storms and drought may have a direct impact on population health. Immediate effects are mostly deaths and injuries. For example, river floods in central Europe in 1997 killed over 100 people. A tropical cyclone in Bangladesh in 1991 affected 10 million people, with 138,000 dead and 460,000 injured [21]. Drought affects population health mainly via its impact on food production. ENSO-related extreme weather events, which are associated with extremely dry conditions in many areas and extensive flooding in others, may also have a significant impact on population health. Nearly 2000 people died in the Hurricane Katrina and the subsequent floods in the USA in 2005. This storm cost more than 80 billion US dollars in damage [22].

Different people may have varying sensitivity to extreme weather. It has been observed that the elderly and children, people with low socio-economic status, and people under intense physical stress or with pre-existing disease are the most vulnerable populations [1]. For example, in the Chicago heatwaves of 1994 and 1995, people over 65 years old accounted for 63% and 72% of the deaths, respectively [23].

Populations in developing countries may be especially vulnerable because of a lack of the resources and infrastructure to adapt to heat waves. In China, for example, an increase in mortality from heatstroke was observed when maximum daily temperatures in Nanjing exceeded 36°C for 17 consecutive days in July 1988 [24]. It is estimated that in China the annual incremental costs for adaptation to climate change would be from 0.8 billion US dollars in 2000 to 3.48 billion US dollars in 2050 [25]. Despite these significant implications, there has been relatively little research in developing countries on the health impact of climate change.

In summary, most research on the direct impact of climate variation on population health has been conducted in North American and European countries, and few studies have been performed in Asia, Australasia and Africa. Most previous studies focus on mortality rather morbidity, while the later is more sensitive in the assessment of health loss. Furthermore, potential confounders, such as air pollution, health facilities and socio-economic status have not been considered in the above studies. In addition, most of the research described above is cross-sectional and conducted within one country alone.

3.2. Indirect Impact of Climate Change on Population Health

Many studies have reported the impact of climate variation on infectious diseases, such as vector-borne diseases and enteric infections, including malaria, dengue fever, Ross River virus infection (RRV), Japanese encephalitis, haemorrhagic fever with renal syndrome (HFRS), salmonellosis and *Campylobacter* infection as well as other diarrhoeal diseases [25-31].

3.2.1. Vector/Rodent-Borne Diseases

Climate conditions primarily affect the transmission of vector-borne diseases in three ways [32,33]. Firstly, climate variation may alter the distribution of vector species depending on whether conditions are favourable or unfavourable for their breeding places and reproductive cycles. An increase in temperature will accelerate the propagation of vector species. In addition, rainfall and relative humidity in various ways all have effects on the life cycle of vectors. Thus, the transmission patterns of these diseases may be affected. Secondly, temperature can also influence the reproduction of pathogens within the vector organism (external incubation period, EIP). Thirdly, changes in weather can affect the behaviour of human beings, eg increasing exposure opportunities to vectors and thereby cases of infection.

Malaria is considered the most important vector-borne disease worldwide, with cases occurring in more than 92 countries [34]. Human malaria is caused by four identified species of Plasmodia, namely, *Plasmodium vivax, P. falciparum, P. ovale* and *P. malariae*. About 70 species are vectors of malaria under natural conditions found most abundantly in tropical and subtropical regions, with female *Anopheles* mosquitoes as the major vectors [35]. It is re-emerging as the top infectious killer and the top tropical disease according to WHO priorities [34]. Climate change may increase the risk of the reintroduction of malaria to non-epidemic countries, such as Australia, because of the frequency of international travel to and from endemic countries, eg Papua New Guinea, unless programs of vector control are maintained or enhanced [21].

The relationship between malaria and meteorological variables has been assessed in many regions, including Africa, Europe, Asia, South America and Australia [36-50]. Positive association between temperature and rainfall and malaria cases has been reported by most studies. Projective studies suggest that the exposure to malaria may have a 16%-28% increase in Africa by 2100 [39]. However, some studies show negative or no association between meteorological variables and malaria incidence [42, 45]. The inconsistent results in previous studies could be due to different study designs, target populations, study areas and statistical analyses.

Potential risk of malaria in current non-epidemic regions has been reported. For example, by 2050, the increase in temperature may lead to an 8%-14% increase in malaria transmission in England [37]. In Australia, the length and intensity of wet seasons has a significant effect on the distribution of the main vector in Australia, *Anopheles farauti* [49]. Modelling shows that global warming will enlarge the potential range of this vector; by the year 2030 it could extend to Gladstone along the Queensland coast, 800 km south of its present limit [47].

Dengue Fever is the most common and widespread arthropod-borne viral infection in the world. There are four distinct virus serotypes, ranging from a sub-clinical infection to the severe disease known as dengue haemorrhagic fever, even dengue shock syndrome and death within 24 hours, without proper hospital care [51]. Currently, vector control is the major approach for disease prevention and control. *Aedes aegypti,* the principal vector, has adapted well to urban environment conditions of poor housing, overcrowding and inadequate sanitation [52]. The number of dengue cases has increased 30-fold globally during the last 50 years. A pandemic in 1998, in which 1.2 million cases of dengue fever and dengue haemorrhagic fever were reported from 56 countries, was unprecedented [51].

Climate change would substantially increase the magnitude and spatial extent of the risk of dengue fever occurring in New Zealand under present climatic conditions [53]. In Australia, by 2100, if no effective policy and public health measures have been developed, the zone of potential transmission of dengue fever may expand 1,800 kilometres south, as far as Sydney [54]. In contrast, by markedly constraining greenhouse gas emissions immediately, this southward extension could be limited to 600 km [54].

Only a few studies have linked meteorological data with this disease. It has been observed that *Ae. aegypti* mosquitoes, the only dengue vector currently present in mainland

Australia, increases directly with temperature [55]. Climate change may increase the area of land with a wet and warm climate suitable for *Ae. Aegypti*. Additionally, the southern Australian mainland, with a slight increase in temperature, may begin experiencing epidemics of dengue fever [56]. In south China and Malaysia, the association between meteorological variables and dengue fever has also been documented [57,58].

In Australia, Bi et al [59] have investigated the association between an outbreak of dengue in north Queensland and meteorological variables, but a one-year study period for a time series analysis is too short. Pontes et al [60] found that the vector population developed independently of rainfall in a Brazilian city, which could be due to the living habits of the mosquito. However, only a graphic assessment was conducted between precipitation and density of the vector in this study and no statistical analysis was applied.

Ross River virus (RRV) is the most notified mosquito-borne pathogen in Australia, with approximately 5,000 notified human cases annually [28]. During the last decade, there have been more than 41,000 cases notified, with the most serious epidemic occurring in 1996 [61]. RRV usually incubates for 7 to 9 days. RRV infection causes a characteristic syndrome, including constitutional effects, rash and rheumatic manifestations. There are multiple and diverse vectors of RRV infection with over 40 species from the genera *Aedes* recorded. The most important vectors are *A. vigilax*, *A. camptorhynchus* and *C. annulirostris* [62]. The former two species are present along the Australian coastline with *A. vigilax* in South Australia, Western Australia and Tasmania and *A. camptorhynchus* along the Queensland coast. *C. annulirostris,* which breeds in freshwater habitats, is widely distributed in inland Australia, particularly in the Murray River regions [62]. Different vectors with different sensitivity to climate may have varied transmission when climate changes. Therefore, the impact of climate variability on RRV infection should be investigated in various climatic regions.

The influence of temperature, rainfall and tides on the transmission of RRV varies within and between regions because of different vector species. Tong et al [63] conducted a series of studies of RRV and climate change in Queensland. They claimed that rainfall, temperature and tides have been associated with the monthly incidence in Queensland. Climate variability may also contribute to the spatial change in the distribution of disease in Queensland. A different response of RRV to climate variability between coastline and inland cities in Queensland has also been identified, indicating a greater impact of maximum temperature on RRV infection in coastal rather than inland cities, while minimum temperature and relative humidity inland seem to affect RRV infection more than at the coastline [64]. Additionally, the lag times and determinants of RRV infection were different between the Cairns and Townsville regions [65,66]. In Western Australia, different climatic scenarios for meteorological data projected that RRV would increase in activity and distribution [67,68]. During 1988-1989, epidemics of RRV infection in Western Australia occurred after unusually heavy rainfall and high tide heights [67].

Japanese encephalitis is an arboviral disease, with its incidence and geographic distribution mainly in rural areas of tropical and temperate Asia and the Pacific region. Japanese encephalitis is transmitted by mosquitos with pigs being the reservoir host and the source of infection [69]. The illness can progress to encephalitis with a fatality of 30%. As a mosquito-borne disease with a seasonal distribution, external environmental factors including weather variables may play a significant role in its transmission. Studies in different rural areas of Asia have suggested that weather variables such as temperature, rainfall and relative humidity may influence the incidence of Japanese encephalitis [70-73]. In addition to rural areas, significant correlations between weather variables and monthly number of Japanese encephalitis cases, with certain threshold of temperatures, were reported recently in a Chinese city with temperate climate [74].

Haemorrhagic fever with renal syndrome (HFRS) is a zoonosis caused by the Hantaan or Hantaan-related virus, with characteristics of fever, haemorrhage and hypotension [29]. Rodents, mostly mice, act as a reservoir and the source of infection. Humans are infected when they come into contact with excreta from infected rodents. The epidemic situation is serious in China, where there were over 1 million cases from 1953-1996. In 2003, the nationwide incidence rate was 1.7/100,000, a 6% increase compared with the previous year [75].

The transmission of HFRS is influenced by environmental, occupational and reservoir factors. However, the potential effects of meteorological factors on rodent-borne diseases are more uncertain and have received less attention than vector-borne diseases. Except for the work of Bi et al [29,76], few quantitative estimates have been made of the possible impact of long-term climate change on rates of rodent-borne infections. Examination of the relationship between rainfall and the incidence of HFRS in a low-lying region of China showed a significant inverse association between the amount of precipitation and the incidence of HFRS when the density of rodents and the opportunities for human contact were considered. However, studies were only conducted in a low-lying region, and further research needs to be done to confirm the situation in other regions.

3.2.2. Enteric Infections

Enteric infection is a public health problem in both developing and developed countries. In the United States, about 9 million cases of water-borne disease occur every year [77]. Food-related diseases account for about 10% of morbidity and mortality in the UK with costs of about UK£ 6 billion annually [78]. The cost of enteric infection was estimated at over 2.6 billion Australian dollars each year in Australian communities [79]. According to the WHO, in developing countries infectious diseases transmitted by contaminated food and water are a constant and frequently fatal threat to health [52]. Diarrhoeal diseases are estimated to cause an annual 2.7 billion cases and 1.9 million deaths, mostly in infants and young children [52].

Many cases of enteric infection peak in case numbers during summer. Temperature and relative humidity directly influence the rate of replication of the pathogens and the survival of enteroviruses in the environment. Rainfall, especially heavy rainfall events, may affect the frequency and level of contamination of drinking water. Moreover, climate change may influence water resources and sanitation so that water supply is effectively reduced. Such water scarcity may necessitate using sources of fresh water of poorer quality, such as rivers, which are often contaminated. All these factors may result in an increased incidence of such diseases [77].

There is inadequate research in food- and water-borne contaminants and diseases. As a result, it is very difficult to identify where and how a change in climatic conditions might alter the hazards posed by these diseases [80]. The relationship between enteric infection and climate variation has been documented by limited studies conducted in Europe, Australia, the

USA and Asia [30,31,77,81-86]. A dose-response relationship suggested that each 1°C increase in temperature is related to a 5% increase in the risk of severe diarrhoea [81]. The Australian study, based in Queensland, also showed a positive association between salmonellosis and temperature [82]. Recent studies have reported the association between *Campylobacter* infection and temperature [30,84,87]. However, published results are not consistent. Some studies report positive correlation while others claim a negative or no association between temperature and *Campylobacter* infection. Limitations in interpreting the results exist as a consequence of inappropriate statistical analysis or considering only one climatic variable (Table 2). Only one study has included the contamination of consumable animals and birds in the analysis of climate and enteric infection [84]. There have been few quantitative analysis of the association between hepatitis A and climate change [83].

So far, few studies have been conducted which examine the impact of climate on enteric infection. There is a lack of consistency in published results. Moreover, there are difficulties in studying the impact of climate variation on such diseases. Firstly, the under-reporting of infectious diseases is a common but inevitable problem in disease surveillance, especially for enteric infection. Secondly, the criteria for diagnosis may not be identical across different regions. These factors may lead to the lack of consistency in the research results.

4. Attributable Health Burden to Climate Change

The Global Burden of Disease (GBD) concept, first published in 1996 [96], aimed to draw a worldwide picture of disease burden to answer the question: 'How big is this health problem?' by using international uniform frameworks or calculation methods. The burden of disease (BoD) is a quantitative index to measure health status and relevant attributed risk factors in a given population. BoD studies have been carried out at global, national and regional levels, using summary measures such as Disability Adjusted Life Years (DALYs) to combine prevalence of disease, quality of life and life expectancy in a single health indicator.

The global Environmental Burden of Disease (EBD) study estimated the burden of disease from 26 major environmental risk factors, including climate change [97]. Any global environmental changes that arise in the future will tend to appear first in local settings where decision making usually takes place. Therefore studies at this level are necessary. Within the EBD study, the assessment of climate change attributable burden is proposed to estimate DALYs lost in the future rather than today [98].

The GBD has made efforts to answer this question by incorporating more and more potential risk factors. Climate change is one of the selected risk factors in the category of 'environmental risks' in the GBD study 2000. It was estimated that 2% of DALYs for diarrhoeal diseases and 2% of DALYs for malaria could be attributed to climate change in the world in 2000 [98]. In total, the mortality attributed to climate change was 154,000 deaths and the attributable burden was 5,500,000 DALYs. In addition, the attributable DALYs in the high mortality, developing countries (defined by WHO) was 2,535,000 for males and 2,667,000 for females, but was relatively low in developed countries [98].

Table 2. Review of original studies on climate change and enteric infections

Reference	Study population	Study period	Indicators		Statistical methods	Major findings
			Meteorological	Disease		
Tam, 2006 [30]	England	1989-1999	Weekly humidity, sunlight hours, mean central England temperature	Case counts of *Campylobacter* infection	Time series adjusted negative binomial regression	$1^{\circ}C$ rise in mean temperature accounting for 5% increase in number of cases with a threshold for $14^{\circ}C$.
Kovats, 2004 [31]	10 European countries	varying for each country, 3-18 yrs	A single national temperature series	Weekly counts of salmonellosis (outbreak- linked cases)	Poisson regression adapted for time series analysis	Above a threshold for 6C there was a linear association between temperature and the number of cases of salmonellosis, strongest in the 15-64 age group.
Checkley, 2000 [81]	Lima, Peru	1993-98	El Nino episode, mean ambient temperature	Number of diarrhoeal cases	Time series linear regression, Poisson regression, square-root transformation of the daily number of cases	The increase in admission for diarrhoea in Lima was associated with an increase in ambient temperature and a decrease in relative humidity.
D'Sonza, 2004 [82]	Five Australian cities	1991-2001	Mean monthly temperature and rainfall	Cases of salmonellosis 'Outbreak month'	Log-linear models, Poisson regression	Positive association between salmonellosis and temperature.
Hu, 2004 [83]	Australia	1991-2000	SOI	Monthly cases of Hep A	Correlation analysis	Correlation between incidence of Hep A and SOI was observed.

Table 2. Continued

Reference	Study population	Study period	Indicators		Statistical methods	Major findings
			Meteorological	Disease		
Patrick, 2004 [84]	Denmark	1998-2001	Average temperature, maximum temperature, rainfall, humidity and amount of sunlight	Weekly incidence of *Campylo-bacter* infection	Linear regression	Temperature and sunlight or humidity predicted the incidence well.
Zhang, 2007 [85]	Chinese	1991-2003	SOI	Case counts of bacillary dysentery	General linear models (GLM) adjusted for autocorrelation and seasonality	A one-standard-deviation decrease in the SOI might cause up to 207 more dysentery cases per month in the study region.
Zhang, 2007 [86]	Chinese cities	1987-2000 for temperate city, 1996-2003 for subtropical city	Minimum and maximum temperatures, rainfall, relative humidity, air pressure	Case counts /incidence of bacillary dysentery	SARIMA models	Maximum temperature, minimum temperature, rainfall, relative humidity and air pressure were significantly correlated with the incidence of dysentery. However, different relationship was detected in different climatic areas.
Kovat, 2005 [87]	Countries in Europe and Oceania	1989-2002, varying for each country	Weekly representative temperature series in each country (author-defined)	Incidence of *Campylo-bacter* infection	Poisson generalised linear models	No effect of temperature or rainfall on campylobacter was detected.

Table 2. Continued

Reference	Study population	Study period	Indicators		Statistical methods	Major findings
			Meteorological	Disease		
Curriero, 2001 [88]	USA	1948-94	Rainfall	'Outbreak' of diarrhoea	Fisher test	Extreme rainfall was associated with the outbreaks.
Singh, 2001[89]	18 Pacific Island countries	1986-94 1978-88 (Fiji)	Temperature, water availability	Average annual rates of diarrhoea in adults	Cross-sectional study, Poisson regression analysis	Positive association between temperature, extreme rainfall and diarrhoea reports, negative association between water availability and diarrhoea rates.
Bentham, 1995 [90]	England and Wales	1982-91	Monthly temperature	Weekly number and monthly incidence of food poisoning	Regression analysis, log-transformation	Association between temperature and monthly incidence was observed and projected to 2010, 2030 and 2050.
Nath, 1992 [91]	Varanasi, Indian (<5 age)	1988-89	Monthly relative humidity, rainfall, maximum and minimum temperature	Children under five diagnosed with rotavirus diarrhoea	Case control study, chi-square test and Fisher test	Diarrhoea due to rotavirus was not correlated with rainfall and relative humidity.
Pinfold, 1991 [92]	Northeast Thailand	1982-87	Rainfall, temperature	Reported incidence of acute diarrhoea	Descriptive study	Incidence of diarrhoea appeared to be inversely related to a sharp decrease in temperature.
Konno, 1983 [93]	Yamagata, Japan	1974-81	Monthly relative humidity, temperature	Acute diarrhoea admissions	Descriptive study	The infection frequency was related to temperature but not to humidity. The association was stronger in cooler months.

Table 2. Continued

Reference	Study population	Study period	Indicators		Statistical methods	Major findings
			Meteorological	Disease		
Louis, 2005 [94]	England and Wales	1990-1999	Minimum and maximum temperatures, rainfall, hrs of sunshine	Weekly incidence of *Campylo-bacter* infection	Multiple regression adjusted for auto-regression	Campylobacter rates were correlated with temperature, particularly for children under the age of 5.
Fleury, 2006. [95]	Canadian cities .	1992-2000	ambient temperature	weekly reports of confirmed cases of three pathogens in Canada: Salmonella, pathogenic Escherichia coli and Campylobacter	generalized linear models (GLMs) and generalized additive models (GAMs)	a strong non-linear association between ambient temperature and the occurrence of all three enteric pathogens.

An increasing trend in the health burden of vector-borne diseases and enteric infections related to climate change has been projected, if other factors remain constant [99]. For example, in China, considering both future climatic and population scenarios, if other factors remain constant the Years Lost due to Disability (YLDs) for bacillary dysentery might double by 2020 and triple by 2050 in some Chinese cities, compared with the YLDs observed in 2000 [99]. Indeed, the extra burden of disease caused by future climate change would be much higher than the estimates given in this study if all other climate-related diseases are included.

Population growth and ageing will have an adverse impact on the health burden. Although the interaction between social and environmental factors cannot be identified, it is clearly indicated that the combined impact of global warming and demographic change on the burden of disease would be much higher than the impact of global warming alone. Young children and the elderly would be the most vulnerable subpopulations to rising temperature and demographic change.

The scarcity of health environmental data and the significant number of knowledge gaps concerning the relationship between climate and population health has resulted in many uncertainties in this study. These gaps need to be addressed in order to facilitate more precise national and international assessments on public health vulnerability to climate change. Current scientific methods cannot identify the threshold health effects of climate change, so that the burden of disease attributable to climate change is not "tolerable" [100]. The significant uncertainty about the future effects of climate change on human population health is a reason to reduce greenhouse gas emissions, and not a reason for inaction.

5. Summary

Due to climate change at global and regional levels, human beings will suffer from more hot days, more extreme rainfall, and more floods and droughts, which may have effects, direct and indirect, on population health. In addition, a large number of papers relating to climate change and diseases reflect increasing research interest in this field. As a result, much scientific evidence has shown the impact of climate change, particularly rise in temperature and changing rainfall patterns. However, many potentially adverse effects remain uncertain. Thus, further research needs to be conducted to better understand the relationship between climate change and population health.

II. Challenges in Examining the Relationship between Climate Change and Population Health

Empirical studies are the most commonly used in examining the relationship between climate and population health, in which historical data of climate observations and disease surveillance have been analysed in order to examine their association. However, inappropriate methods, inconsistent definitions of exposure and statistical analysis without the control of confounders, have limited the research.

1. Quality and Availability of Data

Generally, at least three types of data, including meteorological data, demographic information and health outcomes are necessary for studying climate/disease relationships. Sometimes, information on potential confounding factors, such as occupational factors and density of vectors, is included. The time span of data analysed is commonly about 10 years, but may be up to 50 years [88], depending on data availability. Data on a long-term scale are also needed to distinguish inter-annual variation from long-term changes. Due to different disease surveillance systems, the availability and quality of data are usually not guaranteed.

In general, developed countries have more reliable data kept for a longer time than those in developing countries. For example, a high quality weather record is available in Australia with over 100 years continuous recording of climate observations [6]. For health surveillance data, under-reporting is inevitable for infectious diseases in both developed and developing countries. However, data quality is better in developed countries due to legally registration systems and better quality of public health service. One of the reasons for relatively few such studies in developing countries may be the lack of reliable data over a long-term period. Although data quality is less reliable in less developed countries, more studies are encouraged to be conducted in these countries by the WHO, depending on what research questions will be asked, because much is unknown about the impact of climate change in such areas [13].

It should be acknowledged that reliable data are essential for ecological studies. In addition to health surveillance data, other datasets, including density of vector populations, socio-economic status and host behaviors, are also important in estimating the effects of climate variability on population health. Collecting data of reliable quality is more challenging in developing countries than in developed countries.

2. Difficult to Quantify Exposure-Effect Relationship

Exposure of global climate change is difficult to define at a local level. The definitions of climatic exposure and the use of indices of exposure and outcome vary in previous studies. For example, a Dutch study claimed that a cold spell could cause greater mortality than a heatwave, which was most likely due to the much longer time period in 'cold spells' than in heatwaves [15]. Additionally, the use of different meteorological variables, such as maximum temperature, minimum temperature or mean temperature, has not been justified in most studies. Due to the high correlation among meteorological variables, it may be helpful to adopt some combined meteorological indices such as the Southern Oscillation Index (SOI) [85], although there are limitations in these indicators. The assessment of climatic exposure should be consistent to make results comparable across various regions.

While the indicators of health outcomes depend on the aims of each study, the number of cases of a disease is frequently used as an outcome. Incidence rates, sometimes with logarithmic transformations, may be more reasonable over a long period of time in the presence of fluctuations in population size. Mortality is another commonly used index in examining the impact of extreme weather events. It should be noted that summary measures of the burden of disease, such as DALY, associated with climate change have only been assessed by the WHO [97]. Burden of disease attributable to climate change should be estimated at more local regions to provide valuable evidence for health policy makers.

3. Appropriate Statistical Analysis

3.1. Time-Series Regression Models

The regression methods to quantify the relationship between climate and diseases vary in published studies. Some used time-series analysis [31,40,41,63-66,81], while others applied logistic regression [57,101,102]. Although time series analysis has its advantages, many issues need to be taken into account, including seasonality, inter-annual fluctuation, autocorrelation, lag times and multicollinearity. Such issues should be carefully considered in examining the association between meteorological variables and diseases otherwise it may not be clearly detected within a mixture disturbing by the other factors.

Several statistical analysis methods have been developed to deal with time-series data. Poisson regression models have been commonly adopted in public health research because most health problems are assumed having Poisson distribution. In terms of dealing with time-series data in this research field, the autoregressive integrated moving average (ARIMA) or seasonal autoregressive integrated moving average (SARIMA) model [103] have been recently applied to quantify the relationship between meteorological variables and population health and demonstrate better statistical power [104]. These models have the advantages of controlling autocorrelation of time-series data, potential long-term trends and seasonal distribution. Although logistic regression has advantages in predicting epidemics, some information may be lost when translating continuous data into categorical data [101]. No perfect model has been developed for all studies so that the choice of an appropriate model is of importance and depends on the nature of studied health conditions and the characteristics of time-series datasets.

Scenario-based modelling is an essential tool for climate projection, using existing statistical or theoretical knowledge to forecast possible health impact due to climate change. However, most projections assume that all other factors, including socio-economic status and the population distribution, remain constant. In many studies, the validity and sensitivity of models have been assessed by comparing real and simulated data [39].

3.2. Threshold Detection

Evidence shows that the relationship between meteorological variables and diseases is not always linear, at least within a range. For example, the study of malaria in England demonstrates a 14.5°C temperature threshold above which transmission of vivax malaria can occur, indicating a climate above which temperatures are generally favourable for transmission [105]. A study examining the relationship between temperature and *Campylobacter* infection indicates a 1°C rise related to a 5% increase in the number of cases up to a threshold of 14°C [30]. In addition, climate thresholds could vary in different climatic areas, depending on local climatic conditions, as illustrated in recent studies in China about the association between temperature and bacillary dysentery [86].

Climate thresholds exist to facilitate the growth of pathogens and vectors in favourable environment and to affect human beings and their behaviours increasing the opportunities to have diseases. More importantly, threshold detection has great significant implications for public health in improving the effectiveness of disease prevention and control by providing a climate threshold in public health early warning system.

4. Interaction between Meteorological Variables and Other Environmental Risk Factors

The causal relationship between climate and health is very complex, with many factors such as socio-economic status, environmental hygiene, vaccines, drug-resistance, human behaviors and vectors playing a part. It is not fully understood what role these factors may play in the causal relationship, and the nature of possible synergistic counteractive effects. An assumption in most studies is that changes in the number of cases would occur with changes in meteorological variables only if other factors remain constant over a certain study period.

Prevailing meteorological conditions often influence the concentration of air pollutants in an airshed [106]. Temperature has been considered a possible confounder of air pollution effects [107,108] suggesting that a complex mode of interaction may exist between air pollution, weather variables and health outcomes. Some studies have found the health effects of air pollution to be compounded during the warm season [109-112] with cardiovascular or pulmonary health outcomes having significant associations only in summer or winter respectively [112]. As heat stress can place strain on the heart and exacerbate the cardiotoxicity of chemicals, interactive health effects between temperature and air pollution are biologically plausible. However some studies have found no evidence of interaction [113,114]. The temperature-air pollution relationship is evidenced by the prevalence of photochemical smogs in the warmer months [111] often associated with temperature inversions. In general, the possible interaction or effect modification of air pollution/climate and health association has not been clarified.

5. Summary

There are several levels of uncertainty inherent in the assessment of climate-related health risks. Statistical models vary to the extent to which they allow for non-climatic confounders, such as socio-economic development and public health interventions. The estimation of the exposure to climate change may not be accurate due to the lack of complete knowledge about how the climate system will respond to continuing change in the composition of gases in the atmosphere [115]. Assumptions have to be made for the projection of future burden related to climate change because it cannot be known in advance how technology, society, and human behavior will change over the coming decades.

Considering so many uncertainties, no model or statistical method is perfect to detect the potential and actual impact of climate change on population health. However, approaches based on a combination of scientific methods or tools, careful design and current understanding can be presented. The level of uncertainty can be reduced in several ways by maximizing statistical power (large sample sizes, data gathered over a large region and over a long time period) and using study designs to control for major confounding factors (time series study) [6].

Given the above challenges in this research field, the interpretation of findings and results from such studies should be made with care. Based on an incomplete understanding of the relationship between climate and population health, the results can only be presented in terms of probabilities, or given confident intervals at a certain statistical level. More importantly, researchers should be cautious in generalizing the results, as there may not be similar climatic and socio-economic conditions in other locations.

III. Future Research Trends in Providing Scientific Evidence to Prevent and Reduce Health Burden Associated with Climate Change

Climate variation and climate change may pose new risks to population health, including both chronic and infectious diseases. These risks include additional cases of diseases, spatial change in diseases, re-emergence of infectious diseases and the introduction of new diseases. By addressing the challenges in this field of study, the recommendations for a better adaptation and mitigation to climate change are as follows.

1. More Climate-Related Conditions

In addition to diseases that have demonstrated certain relationship with meteorological variables, more climate-related diseases should be investigated in future studies, such as asthma and *Campylobacter* infection, to have an overall picture of the association between climate variation and population health. Furthermore, investigation of the association between climatic variables and serotypes of infectious diseases, such as the subtypes of salmonellosis, *Salmonella typhimurium* and *Salmonella enteritidis,* will clarify the relationship between climate variation and infectious diseases, given the different transmission pathways for various serotypes.

2. Various Climatic Regions

Future studies are encouraged to be conducted in various climatic regions, including regions in tropical, subtropical and temperate climate zones, because the effects of climate on population health vary depending on different local climatic situations. Given the difference in local health services, living conditions and environmental hygiene, large attention should be paid in developing countries with large vulnerable populations exposed to climate change.

3. Subgroup Populations

Subgroup population, such as children, elderly and indigenous populations, are vulnerable population and may suffer more from future climate change. Studies focusing on the impact of climate change on these subgroups are rather few. Future studies targeting subgroup populations would certainly improve public health measures.

4. Advanced Projective Modelling

Advanced statistical models, particularly time-series models with consideration of the control of seasonality, secular trend, autocorrelation and other potential confounders, are necessary to detect the association between climate change and population health. Projective studies based on future climate scenarios depend on advanced models to generate precise data in the future for better adaptation and mitigation. In Addition, models which include additional influencing factors, such as vector populations and socio-economic status, would certainly enhance the power of the statistical analysis, thus rendering the results more reliable.

5. Local Health Burdens Attributable to Climate Change

Projections of the mortality and morbidity burden of disease attributable to climate change would be of great value in current policy making. More projective studies should be conducted, based on scenarios of future climate, including changes in temperature, rainfall, humidity and sea level, as well as changes in demography and socio-economic status, as long as necessary data are available.

6. Climate Change Adaptation and Mitigation

In order to reduce the vulnerability of human beings, adaptation facilitated by public agencies should play a very important role in response to climate change. Mitigation measures, in a short-term or long-term, to climate change have yet to be recognized and evaluated at global and local levels. Cost-effective analysis should be applied broadly to forcing strategies and policies for climate change adaptation and mitigation. To win the battle against climate change is not the task only for health sector. Policies, measures and instruments to mitigate climate change needs close collaboration between governmental sectors, nationally and internationally. Climate change adaptation and mitigation should be addressed in future studies for a sustainable development.

Conclusion

The relationship between climate change and population health is of great concern recently as climate change is unquestionable, reflecting in the rising number of published paper in this research field. Evidence suggests climate change could be one of the causes of extra health burden. However, given the challenges in examining the relationship between climate and population health, further studies are necessary to achieve a better understanding of the relationship between climate and population health to prevent and reduce the health risk from future climate change.

References

[1] McCarthy JJ, Canziani OF, Leary NA, Dokken DJ and White KS, Editors. Climate change 2001: impacts, adaptation, and vulnerability. Contribution of working group II to the third assessment report of the Intergovernmental Panel on Climate Change. 2001. Cambridge: Cambridge University Press.

[2] United Nations Environment Program. UNEP annual report 2003. [Cited 5 May 2004]; Available from: http://www.unep.org/AnnualReport/2003/index.asp

[3] Martens W, Niessen L, Rotmans J, Jetten T and McMichael A. Potential impact of global climate change on malaria risk. *Environmental Health Perspectives*. 1995. 103: 458-64.

[4] Intergovernmental Panel on Climate Change. *The regional impacts of climate change: an assessment of vulnerability*. 1991. Cambridge: Cambridge University Press.

[5] Intergovernmental Panel on Climate Change homepage. IPCC reports 2007. [Cited 20 Feb 2007]; Available from: http://www.ipcc.ch/

[6] Intergovernmental Panel on Climate Change. Climate change 2007: the physical science basis. *Summary for policymakers*. Contribution of Working Group I to the Fourth Assessment Report of the IPCC.[Cited 20 March 2007]; Available from: http://www.ipcc.ch/

[7] Pittock B, Editor. *Climate change: an Australian guide to the science and potential impacts*. 2003. Canberra: Commonwealth of Australia.

[8] McMichael AJ, Haines A, Slooff R and Kovats S, Editors. *Climate change and human health*. 1996. Geneva: World Health Organization.

[9] China Meteorological Bureau. Extreme weather issues in China, 2003. [Cited 6 June 2004]; Available from: http://www.wmo.gov.cn

[10] McMichael AJ, Woodruff R, Whetton P, et al., Editors. *Human health and climate change in Oceania: a risk assessment* 2002. 2003. Canberra: Commonwealth of Australia.

[11] Patz J. Global warming. *BMJ*. 2004. 328: 1323-4.

[12] Watson R, Patz J, Gubler D, Parson E and Vincent J. Environmental health implications of global climate change. *Journal of Environmental Monitoring*. 2005. 7: 834-43.

[13] Ahmad Q and Warrick RA. Methods and Tools. in Climate change 2001: Impacts, adaptation and vulnerability. *Intergovernmental Panel on Climate Change*. 2001. Cambridge: Cambridge University Press.

[14] Moore TG. Global warming could improve health. *Consumers' Research Magazine*. 1998. 81: 15-20.

[15] Huynen M, Martens P, Schram D, Weijenberg MP and Kunst AE. The impact of heat waves and cold spells on mortality rates in the Dutch population. *Environmental Health Perspectives*. 2001. 109: 463-70.

[16] Donaghue ER, Nelson M, Rudis G, et al. Heat-treated deaths - Chicago, Illinois, 1996-2001, and United States, 1979-1999. *Morbidity and Mortality Weekly Report*. 2003. 52: 610-4.

[17] Bouchama A. The 2003 European heat wave. [Cited 3 May 2004]; Available from: http://www.springerlink.com/media/MsTJVKOXRNWHLR6LWVTP/ Contributions

[18] Coates L. *An overview of fatalities from some natural hazards in Australia. in Natural Disaster Reduction* (NDR96) 1996. Canberra: Institute of Engineers Australia.
[19] Australian Bureau of Meteorology. The heatwave threat. [Cited 11 November 2006]; Available from: http://www.bom.gov.au/weather/wa/sevwx/perth/heatwaves.shtml#trends
[20] Gaffen DJ and Ross RJ. Increased summertime heat stress in the US. *Nature*. 1998. 396: 529-30.
[21] Checkley W, Epstein L, Gilman R, et al. Effect of El Nino and ambient temperature on hospital admissions for diarrhoeal diseases in Peruvian children. *Lancet*. 2000. 355: 442-50.
[22] Wikipedia, Hurricane Katrina.[Cited 11June 2007]; Available from: http://en.wikipedia.org/wiki/Hurricane_Katrina
[23] Whitman S, Good G, Donoghue ER, et al. Mortality in Chicago attributed to the July 1995 heat wave. *American Journal of Public Health*. 1997. 87: 1515-8.
[24] Ando M. Risk assessment of global warming on human health. *Global Environmental Research*. 1998. 2: 69-78.
[25] Erda L. Agricultural vulnerability and adaptation to global warming in China. *Water, Air, and Soil Pollution*. 1996. 92: 63-73.
[26] Khasnis A and Nettleman M. Global warming and infectious diseases. *Archives of Medical Research*. 2005. 36: 689-96.
[27] Hales S, Weinstein P and Woodward A. Public health impacts of global climate change. *Reviews on Environmental Health*. 1997. 12: 191-9.
[28] Weinstein P. An ecological approach to public health intervention: Ross River virus in Australia. *Environmental Health Perspectives*. 1997. 105: 364-6.
[29] Bi P, Tong S, Donald K, Parton KA and Ni J. Climatic, reservoir and occupational variables and the transmission of haemorrhagic fever with renal syndrome in China. *International Journal of Epidemiology*. 2002. 31: 189-93.
[30] Tam C, Rodrigues L, O'Brien S and Hajat S. Temperature dependence of reported Campylobacter infection in England, 1989-99. *Epidemiology and Infection*. 2006. 134: 119-25.
[31] Kovats R, Edwards S, Hajat S, et al. The effect of temperature on food poisoning: a time-series analysis of salmonellosis in ten European countries. *Epidemiology and Infection*. 2004. 132: 443-53.
[32] Zhang Y, Bi P and Hiller J. Climate change and the transmission of vector/rodent-borne diseases. *Asia Pacific Journal of Public Health*. Article in press. Accepted March 2007.
[33] Isabel S, Casas CD and Carcavallo RU. Climate change and vector-borne diseases distribution. *Social Science Medicine*. 1995. 40: 1437-40.
[34] *World Health Organization. Roll back malaria.* [Cited 5 May 2004]; Available from: http://who.int/int-fs/en
[35] Kovats S, Ebi KL and Menne B. Methods of assessing human health vulnerability and public health adaptation to climate change. in *Health and global environmental change. Series* No. 1. 2003. Copenhagen: World Health Organization.
[36] Zhou G, Minakawa N, Githeko A and Yan G. Association between climate variability and malaria epidemics in the East African highlands. *Proceedings of National Academy of Sciences of the USA*. 2004. 101: 2375-80.

[37] Kuhn K, Campbell-Lendrum D, Armstrong B and Davies C. Malaria in Britain: past, present, and future. *Proceedings of National Academy of Sciences of the USA*. 2003. 100: 9997-10001.

[38] Abeku TA, van Oortmarssen GJ, Borsboom G, de Vlas SJ and Habbema JD. Spatial and temporal variations of malaria epidemic risk in Ethiopia: factors involved and implications. *Acta Tropica.* 2003. 87: 331-40.

[39] Tanser FC, Sharp B and Sueur Dl. Potential effect of climate change on malaria transmission in Africa. *Lancet*. 2003. 362: 1792-8.

[40] Bi P, Tong S, Donald K, Parton KA and Ni J. Climatic variables and transmission of malaria: a 12-year data analysis in Shuchen county, China. *Public Health Reports*. 2003. 118: 65-7.

[41] Small J, Goetz S and Hay S. Climatic suitability for malaria transmission in Africa, 1911-1995. *Proceedings of National Academy of Sciences of the USA*. 2003. 100: 15341-5.

[42] Singh N and Sharma V. Patterns of rainfall and malaria in Madhya Pradesh, central India. *Annals of Tropical Medicine and Parasitology*. 2002. 96: 349-59.

[43] Poveda G, Rojas W, Quinones M, et al. Coupling between annual and ENSO timescales in the malaria-climate association in Colombia. *Environmental Health Perspectives*. 2001. 109: 489-93.

[44] Kleinschmidt F, Sharp B, Clarke G, Curtis B and Fraser C. Use of generalized linear mixed models in the spatial analysis of small-area malaria incidence rates in Kwazulu Natal, South Africa. *American Journal of Epidemiology*. 2001. 153: 1213-21.

[45] Van der Hoek W, Konradsen F, Perera D, Amerasinghe P and Amerasinghe F. Correlation between rainfall and malaria in the dry zone of Sri Lanka. *Annals of Tropical Medicine and Parasitology*. 1997. 91: 945-9.

[46] Bouma M, Poveda G, Rojas W, et al. Predicting high-risk years for malaria in Colombia using parameters of El Nino Southern Oscillation. *Tropical Medicine and International Health*. 1997. 2: 1122-7.

[47] Bryan J, Foley D and Sutherst R. Malaria transmission and climate change in Australia. *Medical Journal of Australia*. 1996. 164: 345-7.

[48] Bouma M and van der Kaay H. The El Nino Southern Oscillation and the historic malaria epidemics on the Indian subcontinent and Sri Lanka: an early warning system for future epidemics? *Tropical Medicine and International Health*. 1996. 1: 86-96.

[49] Walker J. Malaria in a changing world: an Australian perspective. *International Journal for Parasitology*. 1998. 28: 947-53.

[50] Al-Mansoob M and Al-Mazzah M. The role of climate on malaria incidence rate in four governorates of Yemen. *Medical Journal of Malaysia*. 2005. 60: 349-57.

[51] World Health Organization. *Dengue prevention and control.* [Cited 6 May 2004]; Available from: http://www.who.int/gb

[52] World Health Organization. *Global defence against the infectious disease threat.* [Cited 6 May 2004]; Available from: http://www.who.int/gb/ebwha/pdf-files/eb109/eeb10916.pdf

[53] De Wet N, Ye W, Hales S, et al. Use of a computer model to identify potential hotspots for dengue fever in New Zealand. *New Zealand Medical Journal*. 2001. 114: 420-2.

[54] Woodruff R, McMichael T, Butler C and Hales S. Action on climate change: the health risks of procrastinating. *Australia and New Zealand Journal of Public Health*. 2006. 30: 567-71.

[55] Watts D, Burke D, Harrison B, Whitmire R and Nisalak A. Effect of temperature on the vector efficiency of Aedes aegypti for dengue 2 virus. *American Journal of Tropical Medicine and Hygiene*. 1987. 36: 143-52.

[56] Russell RC. Mosquito-borne arboviruses in Australia: the current scene and implications of climate change for human health. *International Journal for Parasitology*. 1998. 28: 955-69.

[57] Yi B, Zhang Z, Xu D and Xi Y. Relationship of dengue fever epidemic to Aedes density changed by climate factors in Guangdong Province. *Wei Sheng Yan Jiu*. 2003. 32: 152-4.

[58] Li CF, Lim TW, Han LL and Fang R. Rainfall, abundance of Aedes aegypti and dengue infection in Selangor, Malaysia. *Southeast Asian Journal of Tropical Medicine and Public Health*. 1985. 16: 560-8.

[59] Bi P, Donald K, Parton K and Hobbs K. Climate variability and the dengue outbreak in Townsville, Queensland, 1992-93. *Environmental Health*. 2001. 1: 54-60.

[60] Pontes RJ, Freeman J, Oliveira-Lima JW, Hodgson JC and Spielman A. Vector densities that potentiate dengue outbreaks in a Brazilian city. *American Journal of Tropical Medicine and Hygiene*. 2000. 62: 378-83.

[61] Zhang Y and Bi P. Vector-borne diseases in Australia. *Public Health Bulletin South Australia*. 2006. 4: 9-11.

[62] Harley D, Sleigh A and Ritchie S. Ross River virus transmission, infection and disease: a cross-disciplinary review. *Clinical Microbiology Reviews*. 2001. 14: 909-32.

[63] Tong S, Bi P, Donald K and McMichael AJ. Climate variability and Ross River virus transmission. *Journal of Epidemiology and Community Health*. 2002. 56: 617-21.

[64] Tong S and Hu W. Different responses of Ross River virus to climate variability between coastline and inland cities in Queensland, Australia. *Occupational and Environmental Medicine*. 2002. 59: 739-44.

[65] Tong S, Hu W and McMichael AJ. Climate variability and Ross River virus transmission in Townsville Region, Australia, 1985-1996. *Tropical Medicine & International Health*. 2004. 9: 298-304.

[66] Tong S and Hu W. Climate variation and incidence of Ross River virus in Cairns, Australia: a time-series analysis. *Environmental Health Perspectives*. 2001. 109: 1271-3.

[67] Lindsay M, Mackenzie J and Condon R. Ross River virus outbreaks in Western Australia: epidemiological aspects and the role of environmental factors. in Health in the greenhouse: the medical and environmental health effects of global climate change. Ewan CE, Calvert G, Bryant E and Garrick J, Editors. 1993. Canberra: AGPS Press.

[68] Mackenzie J, Lindsay M and Broom A. Climate change and vector-borne diseases: potential consequences for human health. in Health in the greenhouse: the medical and environmental health effects of global climate change. Ewan CE, Calvert G, Bryant E and Garrick J, Editors. 1993. Canberra: AGPS Press.

[69] WHO. Japanese Encephalitis in Uttar Pradesh, India. [Cited 22 September 2006]Available: http://www.searo.who.int/en/Section10/Section392_10243.htm

[70] Bi P, Tong S, Donald K, Parton KA, Ni J. Climate variability and the transmission of Japanese encephalitis in eastern China. *Vector Borne and Zoonotic Diseases* 2003;3:111-15.

[71] Khan S, Narain K, Handigue R, Dutta P, Mahanta J, Satyanarayana K, Srivastava V. Role of some environmental factors in modulating seasonal abundance of potential Japanese encephalitis vectors in Assam, India. *Southeast Asian J of Trop Med Public Health* 1996;27: 382-91.

[72] Sakai T, Takahashi K, Hisasue S, Horimoto M, Takizawa T. Meteorological factors involved in Japanese encephalitis virus infection in cattle. *Nippon Juigaku Zasshi* 1990;52: 121-27.

[73] Mogi M. Relationship between number of human Japanese encephalitis cases and summer meteorological conditions in Nagasaki, Japan. *Am J Trop Med Hyg* 1983; 32: 170-74.

[74] Bi P, Zhang Y and Kevin A Parton. Weather Variables and Japanese Encephalitis in the Metropolitan Area of Jinan city, China. *Journal of infection*. 2007. Article in press.

[75] China Ministry of Health. National infectious disease report 2003. [Cited 29 April 2004]; Available from: http://www.moh.gov.cn

[76] Bi P, Wu X, Zhang F, Parton KA and Tong S. Seasonal rainfall variability, the incidence of Hemorrhagic Fever with Renal Syndrome, and prediction of the disease in low-lying areas of China. *American Journal of Epidemiology*. 1998. 148: 276-81.

[77] Rose JB, Epstein PR, Lipp EK, et al. Climate variability and change in the United States: potential impacts on water- and food-borne diseases caused by microbiologic agents. *Environmental Health Perspectives*. 2001. 109: S211-21.

[78] Rayner M and Scarborough P. The burden of food related ill health in the UK. *Journal of Epidemiology and Community Health*. 2005. 59: 1054-7.

[79] Neville L and McAnulty J. Communicable enteric disease surveillance, NSW, 2000-2002. *New South Wales Public Health Bulletin*. 2004. 15: 18-23.

[80] Health Canada. Climate change and health: research report. 2004.

[81] Checkley W, Epstein LD, Gilman RH, et al. Effects of El Nino and ambient temperature on hospital admissions for diarrhoeal diseases in Peruvian children. *Lancet*. 2000. 355: 442-50.

[82] D'Souza R, Becker N, Hall G and Moodie K. Does ambient temperature affect foodborne disease? *Epidemiology*. 2004. 15: 86-92.

[83] Hu W, McMichael AJ and Tong S. El Nino Southern Oscillation and the transmission of hepatitis A virus in Australia. *Medical Journal of Australia*. 2004. 180: 487-8.

[84] Patrick M, Christiansen L, Wain M, et al. Effects of climate on incidence of Campylobacter spp. in humans and prevalence in broiler flocks in Denmark. *Applied and Environmental Microbiology*. 2004. 70: 7474-80.

[85] Zhang Y, Bi P, Wang G, Hiller J. El Nino Southern Oscillation and dysentery in Shandong province, China. *Environmental Research*. 2007. 103: 117-20.

[86] Zhang Y, Bi P, Hiller J, Sun Y, Ryan P. Climate variations and bacillary dysentery in northern and southern cities of China. Journal of Infection. Article in press. Available online January 2007

[87] Kovats RS, Edwards S, Charron DF, et al. Climate variability and campylobacter infection: an international study. *International Journal of Biometeorology*. 2005. 49: 207-14.

[88] Curriero F, Patz J, Rose J and Lele S. The association between extreme precipitation and waterborne disease outbreaks in the United States, 1948-1994. *American Journal of Public Health*. 2001. 91: 1194-9.

[89] Singh R, Hales S, Wet N, et al. The influence of climate variation and change on diarrheal disease in the Pacific Islands. *Environmental Health Perspectives*. 2001. 109: 155-9.

[90] Bentham G and Langford IH. Climate change and the incidence of food poisoning in England and Wales. *International Journal of Biometeorology*. 1995. 39: 81-6.

[91] Nath G, Singh S and Sanyal S. Childhood diarrhoea due to rotavirus in a community. *Indian Journal of Medical Research*. 1992. 95: 259-62.

[92] Pinfold J, Horan N and Mara D. Seasonal effects on the reported incidence of acute diarrhoeal disease in northeast Thailand. *International Journal of Epidemiology*. 1991. 20: 777-86.

[93] Konno T, Suzuki H, Katsushima N, et al. Influence of temperature and relative humidity on human rotavirus infection in Japan. *Journal of Infectious Diseases*. 1983. 147: 125-8.

[94] Louis V, Gillespie I, O'Brien S, et al. Temperature-driven Campylobacter seasonality in England and Wales. *Applied Environmental Microbiology*. 2005. 71: 85-92.

[95] Fleury M, Charron DF, Holt JD, Allen OB, Maarouf AR. A time series analysis of the relationship of ambient temperature and common bacterial enteric infections in two Canadian provinces. *International Journal of Biometeorology*. 2006. 50: 385-91.

[96] Murray C and Lopez A. The global burden of disease. 1996. Geneva: World Health Organization.

[97] Word Health Organization. Assessing the environmental burden of disease. [Cited 6 May 2004]; Available from: http://www.who.int/peh/burden/burdenindex.htm

[98] McMichael A, Campbell-Lendrum D and Kovats S. Annex 6.7: Comparative risk assessment of the health effects of climate change. in *Proceeding of the 12th annual meeting of the International Society for Environmental Epidemiology (ISEE)*. 2000.

[99] Zhang Y. The relationship between climate variation and selected infectious diseases: Australian and Chinese perspectives. PhD thesis. 2007.

[100] Kovats R, Campbell-Lendrum D and Matthies F. Climate change and human health: estimating avoidable deaths and disease. *Risk Analysis*. 2005. 25: 1409-18.

[101] Woodruff RE, Guest C, Garner MG, et al. Prediction of Ross River virus epidemics from regional weather data. *Epidemiology*. 2002. 13: 384-93.

[102] Hales S, de Wet N, Maindonald J and Woodward A. Potential effect of population and climate changes on global distribution of dengue fever: an empirical model. *Lancet*. 2002. 360: 830-4.

[103] Shumway RH and Stoffer DS. *Time series analysis and its applications*. 2000. New York: Springer.

[104] Zhang Y, Bi P, Hiller J. Climate variations and Salmonellosis transmission: A Comparison of Regression Models. *International Journal of Biometeorology*. Article in press. Accepted 7 May 2007

[105] Lindsay S, Joyce A. Climate Change and the Disappearance of Malaria from England. *Global Change & Human Health*. 2000. 1: 184-187.

[106] CSIRO. Australia State of the Environment Report 2001. *http://www.deh.gov.au/soe/2001/atmosphere/atmosphere04-6d.html. Accessed 25 August 2006.*

[107] Stieb, D. M., Judek, S. & Burnett, R. T. Meta-anaylsis of time-series studies of air pollution and mortality: Effects of gases and particles and the influence of cause of death, age and season. *Journal of the Air and Waste Management Association*. 2002. 52: 470-84.

[108] Smith, R. Invited commentary: Timescale-dependent Mortality Effects of Air Pollution. *American Journal of Epidemiology*. 2003.157; 1066-70.

[109] Ponce de Leon, A., Anderson, H. R., Bland, J. M., Strachan, D. P. & Bower, J. Effects of air pollution on daily hospital admissions for respiratory disease in London between 1987-88 and 1991-92. *Journal of Epidemiology and Community Health*. 1996.33; S63-S70.

[110] Medina-Ramon, M., Zanobetti, A. & Schwartz, J. The Effect of Ozone and PM_{10} on Hospital Admissions for Pneumonia and Chronic Obstructive Pulmonary Disease: A National Multicity Study. *American Journal of Epidemiology*. 2006.163; 579-88.

[111] Simpson, R., Denison, L., Petroeschevsky, A., Thalib, L. & Williams, G. Effects of ambient particle pollution on daily mortality in Melbourne, 1991-1996. Journal of Exposure *Analysis and Environmental Epidemiology*. 2000,10; 488-96.

[112] Zanobetti, A. & Schwartz, J. Air pollution and emergency admissions in Boston, MA. *Journal of Epidemiology and Community Health*. 2006. 60; 890-5.

[113] Barnett, A. G., Williams, G. M., Schwartz, J., Best, T. L., Neller, A. H., Petroeschevsky, A. L. & Simpson, R. W. The Effects of Air Pollution on Hospitalizations for Cardiovascular Disease in Elderly People in Australian and New Zealand cities. *Environmental Health Perspectives*. 2006.114; 1018-23.

[114] Hales, S., Salmond, C., Town, G. I., Kjellstrom, T. & Woodward, A. Daily mortality in relation to weather and air pollution in Christchurch, New Zealand. *Australian and New Zealand Journal of Public Health*. 2000.24; 89-91.

[115] Campbell-Lendrum D and Woodruff RE. Comparative risk assessment of the burden of disease from climate change. *Environmental Health Perspectives*. 2006. 114: 1935-41.

In: Climate Change Research Progress
Editor: Lawrence N. Peretz, pp. 269-289

ISBN: 978-1-60021-998-6

Chapter 8

GLOBAL WARMING, CROP SELECTION AND ADAPTATION OPTIONS IN KENYAN AGRICULTURE[1]

Jane Kabubo-Mariara*

School of Economics, University of Nairobi, P.O. Box 30197, 00100, Nairobi

Abstract

Available evidence suggests that the world is getting warmer and this is expected to have adverse impact on agriculture in the future. Global circulation models predicted that global warming will lead to increased temperatures of up to 8.7°C and cause variability of rainfall by up to 34% in Kenya by the year 2100. From these predictions, the two extreme climate events that may adversely impact on the agricultural sector are drought and flooding in both the arid and semi-arid lands (ASALs) and the higher potential areas. The major challenges to sustainable agricultural production in Kenya are low, poorly distributed and unreliable rainfall, climate and ecological extremes. The impacts of the extreme weather and climate events could have severe socio-economic impacts including water and energy shortages, and also shortage of other essential basic commodities and long term food insecurity. Previous studies show that the country will incur huge losses in farm revenues due to climate change. This makes it important to understand adaptive mechanisms by farmers to circumvent the potential welfare loss due to climate change in Kenya. To address this issue, this chapter examines the impact of climate change on crop selection and also assesses farmers' awareness of climate change. The chapter also examines the main adaptation measures employed by farmers to counter adverse impacts of climate change, upon perceiving climate change. The results suggest that the decision to farm a crop or a portfolio of crops responds to climate change and that the choice of most crops is more sensitive to temperature than to precipitation. Further, choice of most crops exhibit a significant hill shaped relationship with temperature and precipitation.

---

1 The author is grateful to the Global Environment Facility/World Bank for financing the project under which the data used in this chapter were obtained. Thanks to Robert Mendelsohn, Pradeep Kurukulasuriya and Rashid Hassan for project support. The author is however responsible for all errors and omissions in this chapter.
* E-mail address: jmariara@mail.uonbi.ac.ke. Tel: 254 20 318262, ext 28122. Fax: 254 20 245566

1.0. Introduction

There is increased evidence that it is getting hotter. There is also enough scientific evidence to show that any significant change in climate on a global scale will affect local agriculture and therefore the world's food supply (Houghton et al., 2001). The highest damages from climate change are predicted to be in the agricultural sector in sub-Saharan Africa because the region already endures high heat and low precipitation and also because of the semi-arid nature of a large portion of the continent, frequency of droughts and scarcity of water. In Kenya, global circulation models predict that global warming will lead to increased temperatures of up to 8.7°C and cause variability of rainfall by up to 34% by the year 2100 (Strzepek and McCluskey 2006). From these predictions, the two extreme climate events that may adversely affect the agricultural sector are drought (crop water stress leading to declining yields) and flooding (resulting in water logging) in both the arid and semi-arid lands (ASALs) and the high potential areas. The ASALs will however bear the largest impact of global warming. Water resources in the country are most vulnerable in the ASALs of the country where severity of drought and floods is expected to increase, while the already strained ground water resources in the coastal regions will be most vulnerable in the future (Kabubo-Mariara and Karanja, 2007). In addition to the predicted climate change scenarios, unpredicted climate events such as high frequency of flooding may still occur.

The major challenges to sustainable agricultural production in Kenya are low, poorly distributed and unreliable rainfall, especially in the ASALs. Kenya has climate and ecological extremes, with altitude varying from sea level to over 5000m in the highlands. The mean annual rainfall ranges from less than 250mm in the arid and semi-arid areas to 2000mm in high potential areas. The country can be divided into seven agro-climate zones using a moisture index based on annual rainfall expressed as a percentage of potential evaporation. Areas with an index greater than 50% have high potential for cropping and are designated zones I, II and III. These zones account for about 18% of Kenya's land area. The semi-humid to arid regions (zones IV, V, VI and VII) have indexes of less than 50% and a mean annual rainfall of less than 1100 mm. These zones are generally referred to as the Kenyan range-lands (the ASALs). Ninety percent of the arid and semi-arid areas lie below 1260m and mean annual temperatures range from 22°C to 40°C (Kabubo-Mariara and Karanja, 2007). In the medium to high potential areas, the major problems include soil degradation and unfavourable development domain dimensions (Kabubo-Mariara et al., 2006).

The vulnerability of agriculture to climate is a major issue warranting urgent attention because of declining productivity and contribution to economic growth; given the important role the sector plays (food security, national income, export earnings, rural employment and non-farm enterprises) in many African countries. The continent fares worst in terms of income losses and the drop in consumption of low income households because of its low substitution possibilities between imported and domestic foods (Maddison, 2006). The overall adverse impacts of the extreme weather and climate events could also have severe socio-economic impacts such as food, water and energy shortages as well as shortage of other essential basic commodities and long term food insecurity.

Though there is convincing evidence that climate significantly affects farm revenue, studies on the impact of climate change on agriculture suggest that farmers can mitigate the impact of global warming on farm revenue through several adaptation mechanisms

(Kurukulasuriya and Mendelsohn, 2006, Benhin, 2006, Kabubo-Mariara and Karanja 2006). Some studies also suggest that adaptation will reduce the damages and increase the benefits of warming (Mendelsohn et al., 1994). The major adaptation mechanisms in African agriculture include crop diversification, altering planting dates, increased use of irrigation, shading/sheltering, water and soil conservation (Maddison 2006). Some farmers adapt by switching from crops towards livestock management, while livestock producers adapt by diversifying species and also by switching to small ruminants (Seo and Mendelsohn, 2006, Kabubo-Mariara 2007).

Few empirical studies have directly addressed adaptation to climate change in Africa. Maddison (2006) is the single most comprehensive study on adaptation that we are aware of[2]. The main objective of his study was to determine the ability of farmers in Africa to detect climate change and to ascertain how they have adapted to the perceived climate change. Kurukulasuriya and Mendelsohn, (2006) examine whether choice of crops responds to climate change in Africa. Other studies have focused on livestock adaptation (Seo and Mendelsohn, 2006, Kabubo-Mariara 2007). This chapter contributes to the limited empirical evidence on adaptation to climate change by examining the impact of climate change on crop selection, given the estimated damage response of agriculture to climate change in Kenya. The chapter also examines general farmer awareness of climate change, mitigation strategies and barriers to adaptation. In the absence of adequate arable land to allow for extensive agriculture and the predicted global warming, it is important to understand crop choice and adaptive mechanisms by farmers to circumvent the potential welfare loss due to climate change in Kenya.

The rest of the chapter is organized as follows. The next section briefly presents the data types and sources; section 3 discusses climate change predictions and also presents an overview of the impact of climate change on Kenyan agriculture; section 4 discusses the empirical results. Section 5 concludes.

2. The Data

Household Data: The primary data on which this chapter is based was drawn from a sample of 724 households. The data were collected from 38 districts drawn from six out of eight provinces in Kenya between June and August 2004. The districts chosen captured variability in a wide range of agro-climatic conditions (rainfall, temperatures and soils), market characteristics (market accessibility, infrastructure, etc.) and agricultural diversity, among other factors. Each district was divided into agro-ecological zones and samples of three different farm types/sizes: large, medium and small chosen from each ecological zone. The data were collected using a common questionnaire designed jointly by the School of Forestry and Environmental Studies of Yale University and the Centre for Environmental Economics and Policy in Africa (CEEPA), University of Pretoria for 11 countries participating in the regional Global Environmental Facility/World Bank project. More details on sampling and key summary statistics are presented in Kabubo-Mariara and Karanja (2006).

2 The paper by Maddison is based on evidence gathered by participating country teams (including Kenya) on perceptions and adaptations to climate change in Africa as part of the GEF/World Bank project. To the best of my knowledge, no comprehensive analysis of adaptation mechanisms has been carried out for any individual country that participated in the project.

Secondary data: In addition to the household data, the study also made use of satellite and ARTES (Africa Rainfall and Temperature Evaluation System) climate data. The temperature data came from satellites which measure temperatures twice daily via a Special Sensor Microwave Imager mounted on US Defense Department satellites (Basist et al. 1998). The ARTES dataset was interpolated from weather stations by the National Oceanic and Atmospheric Administration based on ground station measurements of precipitation and minimum and maximum temperature (World Bank, 2003). The data were constructed from a base with data for each month of the survey year and for morning and evening. The monthly mean temperatures were estimated from approximately 14 years of data (1988–2003) and the mean monthly precipitation was estimated for 1961–1990 to reflect long-term climate change. The summary statistics for seasonal climate variables are presented in Table 1. Hydrological data (mean flow) were obtained from The International Water Management Institute (IWMI) and the University of Colorado and were based on monthly values from 1961–1990 time series. The final values were estimated using hydrological models for Africa (IWMI and UOC 2003). The mean flow for the country was estimated at about 301,000m^3. Soil data was obtained from the Food and Agricultural Organization (FAO 2003)-(see Kabubo-Mariara and Karanja, 2006 for a detailed description of the hydrological and soil data).

Table 1. Sample statistics for temperatures and precipitation by season

	Temperatures (°C)		*Precipitation (mm/mo)*	
Season	**Mean**	**Std dev.**	**Mean**	**Std dev.**
Fall (December–February)	19.29	2.67	88.80	41.45
Summer (March–May)	19.07	2.74	103.71	31.57
Winter (June–August)	18.50	2.36	62.40	40.82
Spring (September–November)	19.09	2.66	71.89	26.95
Annual average	18.99	2.58	84.53	18.60
Long rains (March–August)	19.33	2.73	90.90	34.97
Short rains (September–February)	18.65	2.46	81.27	23.71

3. Climate Change and Impact on Agriculture in Kenya

3.1. AOGCMS and Climate Change Scenarios

The IPCC predicts emission scenarios on the basis of complex climate models. A range of scenarios of future greenhouse gases and aerosols emissions have been developed based on certain assumptions of population and economic growth, land use, technological change and energy availability (Houghton et al. 2001). Though there are a wide range of Special Report on Emissions Scenarios (SRES), only A2 and B2 have been integrated by many Atmosphere-Ocean Global circulation models (AOGCMS) because of the assumptions on which each is based. These scenarios represent a range of equally plausible future climates (expressed as anomalies of the baseline 1961-1990 climate) with differences attributable to the different climate models used and to different emission scenarios that the world may follow. For

Kenya, 10 scenarios are derived by using five different models (CSIRO2, HADCM3, CGCM2, ECHAM and PCM)[3] in conjunction with two different emission scenarios (A2 and B2) (Strzepek and McCluskey 2006). Strzepek and McCluskey obtained climate change estimates for Kenya between 2000 and 2100 by using the two scenarios to modify real-time climate information. In each scenario, climate values at the Global Resources Information Database (GRID) cell level were summed to predict climate change.

Table 2. Climate predictions of AOGCMS and SRES for 2000-2100

			2000	2020	2050	2080	2100
SRES	**Model**	**Temp.(°C)**					
A2	CSIRO		21.81	23.59	25.24	27.93	29.92
	CGCM2		21.81	23.41	24.94	27.43	29.23
	EHAM		21.81	23.27	24.78	27.26	29.07
	HADCM		21.81	23.69	25.52	28.50	30.66
	PCM		21.81	22.98	24.10	25.91	27.19
B2	CSIRO		21.81	23.77	25.41	27.03	28.09
	CGCM2		21.81	23.31	24.56	25.76	26.55
	EHAM		21.81	23.32	24.63	25.91	26.75
	HADCM		21.81	23.80	25.48	27.08	28.13
	PCM		21.81	23.07	24.07	25.01	25.60
		Precip.(mm)					
A2	CSIRO		82.92	87.68	91.23	97.20	101.63
	CGCM2		82.92	84.70	85.76	87.66	89.16
	EHAM		82.92	88.01	92.44	99.74	105.04
	HADCM		82.92	87.73	90.85	95.93	99.61
	PCM		82.92	86.03	88.23	91.77	94.27
B2	CSIRO		82.92	86.06	87.78	89.48	90.61
	CGCM2		82.92	85.41	86.70	88.03	88.93
	EHAM		82.92	89.51	94.71	99.76	103.08
	HADCM		82.92	86.78	89.05	91.25	92.68

Computed from raw data provided by Strzepek and McCluskey, (2006).

The predicted temperature and precipitation for the period 2000 to 2100 are presented in Table 2. Table 3 shows the predicted decadal average changes in annual climate variables and stream flow changes for 2050 and 2100, relative to the year 2000. The predictions for precipitation can be positive or negative because some regions are expected to gain in terms of rainfall but others are expected to lose (Kabubo-Mariara and Karanja, 2006). For instance, the CGCM2-a2 scenario predicts that precipitation could increase in some regions by 6% but fall in others by 6%. The figures for temperatures are predicted increases in degree Celsius. The predicted figures for stream flows are percentage changes. The highest predicted global warming impacts are from HADCM3 and CSIRO scenarios, but the lowest are from PCM. For precipitation, the highest predicted changes are from the ECHAM but the lowest are from the CGCM2 and the PCM. For stream flow, the highest predicted changes are from the

3 CGCM2 is a Canadian Model, CSIRO2 is an Australian model, ECHAM is a German Model, HADCM3 is a British Model, PCM is a Parallel Climate Model

ECHAM model and the lowest from the CGCM2 model. From the predicted scenarios, one can observe that temperatures are predicted to rise by between 2.2 °C and 8.7°C, while precipitation is expected to vary by between 4% and 34% by the year 2100.

Table 3. Predicted decadal average changes in annual values of precipitation, temperature and flow: 2050-2100

	Precipitation (Percentage change)									
	CGCM2		**CSIRO2**		**ECHAM**		**HADCM3**		**PCM**	
Year	2050	2100	2050	2100	2050	2100	2050	2100	2050	2100
A2- Scenarios	106	116	109	123	113	134	110	124	106	115
B2- Scenarios	104	109	105	109	116	129	108	115	106	110
	Temperature (increases °C)									
	CGCM2		**CSIRO2**		**ECHAM**		**HADCM3**		**PCM**	
Year	2050	2100	2050	2100	2050	2100	2050	2100	2050	2100
A2- Scenarios	3.0	7.4	3.4	8.2	2.8	7.2	3.6	8.7	2.2	5.4
B2- Scenarios	2.7	4.7	3.6	6.3	2.8	4.9	3.6	6.3	2.3	3.8
	Stream flow (Percentage change)									
	CGCM2		**CSIRO2**		**ECHAM**		**HADCM3**		**PCM**	
Year	2050	2100	2050	2100	2050	2100	2050	2100	2050	2100
A2- Scenarios	80	67	97	100	107	128	95	98	95	97
B2- Scenarios	81	74	91	89	112	126	94	95	94	95

Source: Strzepek and McCluskey, (2006).

3.2. Impact of Climate Change on Agriculture

The predicted climate scenarios are likely to have devastating impacts on agriculture in Kenya. Empirical evidence shows that high summer temperatures are harmful to crop production while high winter temperatures are beneficial (Kabubo-Mariara and Karanja, 2006, 2007)[4]. This is because summer (March–May) is the planting period followed by formative crop growth, while winter (June–August) is the period for ripening and maturing of crops. High summer temperatures would therefore slow down or destroy crop growth, while higher winter temperatures are crucial for ripening and harvesting. In the Kenyan highlands, winters can be quite chilly and excessively low winter temperatures have been associated with crop damage from frost. Summer temperatures exhibit a U-shaped relationship with net

4 This sub-section borrows heavily from results of a Ricardian analysis of the impact of climate change on Kenyan crop agriculture, (Kabubo-Mariara and Karanja, 2006). The Ricardian model is a reduced form model that examines how a set of exogenous climate and other control variables affect farm value. The standard Ricardian model relies on a quadratic formulation of climate, which allows for testing of the expected non-linear shape of the response function between net revenues and climate (Mendelsohn et al. 1994). See Kabubo-Mariara and Karanja (2006) for full discussion of the Ricardian model and its application in Kenya.

crop revenue and winter temperatures a hill-shaped one. Precipitation was found to be positively correlated with net crop revenue and to exhibit a hill-shaped relationship with revenue. The results also showed that climate exhibits a non-linear relationship with net revenue. The estimated marginal impacts of climate change on crop revenue further showed that crop revenue is inelastic with respect to temperature but elastic with respect to precipitation (Table 4)[5]. The temperature elasticities suggested that global warming is harmful for agricultural productivity. Though precipitation elasticities were much higher than temperature elasticities, the magnitude of marginal impacts suggested that the temperature component of global warming may have more serious repercussions than rainfall.

Table 4. Marginal impacts of climate on net crop revenue (US$/ha)

Variable	Marginal impact	Std. error
Summer temperature	-94.77*	45.76
Winter temperature	84.87***	34.26
Overall temperature	-9.90	16.52
Temperature elasticity	-0.55	
Fall rainfall	7.11***	2.45
Summer rainfall	5.95**	2.58
Overall rainfall	13.06***	4.83
Precipitation elasticity	3.18	

*** significant at 1% level. Standard errors in parenthesis

Table 5. Predicted impacts of different climate scenarios by zone (loss in US$)*

Climate change scenario	Medium & low potential	High potential	All zones
3.5°C rise in temperature	97.01 (28)	-0.11 (-0.03)	57.91 (17)
+4°C rise in temperature	125.31 (37)	15.67 (4)	80.80 (23)
20% reduction in rainfall	111.33 (33)	47.67 (13)	80.92 (23)
+3.5°C temperature + 20% reduction in precipitation	208.34 (61)	47.56 (13)	138.71 (40)
+4°C temperature + 20% reduction in precipitation	236.63 (69)	63.34 (18)	161.69 (47)

*Percentage loss in brackets

Predicted impacts of uniform climate change scenarios show that with precipitation remaining the same, a 3.5°C increase in temperature in all districts would result in a 0.03% (US$0.11 per hectare) gain in high potential zones but a 28% (US$97 per hectare) loss in medium and low potential zones (Table 5). These results therefore show that medium and low potential zones will bear the brunt of global warming in Kenya. With a 4°C rise in temperature and a 20% fall in rainfall, Kenya is likely to experience losses of up to US$125

5 See Kabubo-Mariara and Karanja, (2006) for the regression models and results on which the marginal impacts and impacts of predicted climate scenarios are based.

per hectare by the year 2100 for medium and low potential zones compared to losses of only US$16 for high potential zones and US$81 for the whole country. These results suggest that if the maximum predictions of the AOGCMS and SRES (+ 8.7°C temperature and -34% rainfall) are to take effect, then the agricultural sector will literary collapse unless appropriate adaptation measures are undertaken, combined with a reduction of green house gases.

4. Empirical Results

4.1. Impact of Climate Change on Crop Selection

Descriptive statistics

The survey sought to find out the six main crops produced by households, but no attempt was made to gather information on mono-cropping, so it is quite difficult to make crop choice mutually exclusive. A detailed exploration of the data suggests that is it quite difficult to isolate the single most important crop chosen by any particular household. Due to these difficulties, our analysis focuses on individual crop choices within the household, though we acknowledge that the choice of one crop may not necessarily be independent of the choice of other crops. The main crops farmed by households in the sample are presented in Table 6. Maize is the most popular crop adopted by 93% of all households. Further analysis suggests that there are no major differences in crop choice between high potential (95%) and medium and low potential zones (91%). This is probably because there are a number of different varieties of maize which can be adapted to different climatic conditions (Kurukulasuriya and Mendelsohn, 2006, Karanja, 2006). Maize can be grown in climates ranging from temperate to tropical and tolerates high temperatures (15°C to 45°C) (Wahaj, Muñoz and Maraux, 2006).

Table 6. Crop choices in sampled households

Crop	No. of Households	% of households
Maize	670	93
Beans	517	71
Banana	183	25
Potatoes	198	27
Tomatoes	87	12
Cowpeas	89	12
Sorghum	84	12
Millet	68	9
Coffee	86	12
Tea	96	13
Sugarcane	55	8
All other crops	383	53
Maize-bean	684	94
Coffee- tea	158	22

Beans is the second most popular crop. The main diet of a majority of Kenyan households is a combination of maize and beans, which explains the popularity of the two crops as well as a maize-beans portfolio. Beans grow well with medium rainfall, but the crop is not suited to the humid, wet tropics. Its ideal temperatures range from 10°C to 27°C. All other crops are farmed by a relatively small proportion of households. Though other crops (wheat, vegetables, groundnuts e.t.c.) are produced by 41% of all households, the number producing each of these crops is quite small. The main cash crops in the sampled households included coffee, tea and sugarcane, but all were farmed by relatively few households. Coffee and tea were basically found in the same households on small holder farms.

Determinants of crop selection

To examine the determinants of crop choice[6], we divide the crops into three categories: major food crops; minor food crops and cash crops. We test the impact of climate related variables together with a set of household level covariates. All food crops are seasonal and in most cases there are two cropping season. For this reason, we use seasonal rather than annual climate variables as regressors. The cash crops are perennial and therefore seasonal climatic conditions may not have much impact. For this reason, we use annual climate variables as regressors. We include both linear and quadratic terms of the climate variables in-order to test for the non-linear relationship between climate change and crop choice (Kurukulasuriya and Mendelsohn, 2006). The results are presented in Tables 7, 8 and 9. The chow test results (Wald chi(2)) suggests that the null hypothesis that the non-intercept coefficients are jointly equal to zero is rejected at all conventional levels of testing for adoption of all crops. This implies that the underlying empirical probit models are stable and have a strong predictive power in explaining choice of particular crops.

The marginal impacts and significance of coefficients for temperature variables for most crops suggest that crop choice is sensitive to seasonal climate variations. The decision to farm all major food crops exhibit a hill shaped relationship with summer temperatures, but a U shaped relationship with winter temperatures. These results are consistent with results for crop choice in Africa (Kurukulasuriya and Mendelsohn, 2006). The response of choice of bananas to changes in seasonal temperatures is however insignificant. The results for choice of minor crops except sorghum are consistent with those of major food crops, but only the results for millet are significant. The insignificance of the coefficients and the low marginal effects suggest that choice of most minor food crops, such as cowpeas and sorghum are not very responsive to temperature changes. This is because most of these crops are drought resistant crops and can thrive with little rainfall (Wahaj, Muňoz and Maraux, 2006; Karanja, 2006). The marginal impacts for precipitation are much lower and less significant than for temperatures suggesting that crop choice is more sensitive to temperature than to precipitation changes. This is consistent with results for choice of some crop portfolios in Africa (Kurukulasuriya and Mendelsohn, 2006), and also for the response of crop revenue to climate change in Kenya (Kabubo-Mariara and Karanja 2006). The impact of precipitation on crop choice differs from one crop to another. Maize exhibits a U shaped relationship with both winter and summer precipitation. The marginal impacts are significant. Though this result may be interpreted to imply that maize does not need rainfall, it suggests that at low rainfall, an increase in rainfall may make farmers to opt for other crops (faster growing crops such as potatoes) especially in areas where maize is only produced in one season per year. But as rainfall increases, farmers switch to increased maize cropping and a reduction of other crops.

6 See appendix 2 for the analytical framework for the determinants of crop choice.

Excess rainfall will however destroy maize crop and therefore discourage choice of the crop. This most likely explains the hill shaped relationship between seasonal precipitation and choice of most other major food crops.

Table 7. Probit regression results for choice of major food crops

Variable	*Maize*	*Beans*	*Bananas*
Summer temperature	0.297	0.8297	0.2619
	[2.12]**	[2.66]***	[0.92]
Summer temperature squared	-0.0057	-0.0194	-0.011
	[1.81]*	[2.45]**	[1.50]
Winter temperature	-0.3502	-0.9511	-0.0612
	[2.01]**	[2.28]**	[0.15]
Winter temperature squared	0.0069	0.0237	0.0079
	[1.69]*	[2.20]**	[0.77]
Summer precipitation	-0.0067	-0.0035	0.0219
	[2.79]***	[0.40]	[2.41]**
Summer precipitation squared	0.00002	-0.00002	-0.00004
	[2.48]**	[0.79]	[1.37]
Winter precipitation	-0.046	0.0231	0.2601
	[2.48]**	[0.43]	[4.30]***
Winter precipitation squared	0.0002	-0.0001	-0.001
	[2.48]**	[0.36]	[4.22]***
Log mean flow	-3.5206	6.9711	14.7122
	[2.14]**	[2.19]**	[4.68]***
Log mean flow squared	0.3347	-0.6496	-1.3478
	[2.16]**	[2.20]**	[4.64]***
Soils (andosols)	0.0095	0.08	-0.2341
	[0.13]	[0.29]	[0.89]
Average years of education of household	0.0019	0.0176	0.0217
members	[0.86]	[2.73]***	[3.55]***
Log farm size	0.0529	-0.0051	-0.0614
	[3.03]***	[0.09]	[1.23]
Log farm size squared	-0.0143	-0.0126	0.0047
	[4.02]***	[1.02]	[0.42]
Log household size	0.0526	0.0128	0.1358
	[2.96]***	[0.23]	[2.45]**
Log male wage rates	0.0256	0.129	-0.1408
	[1.40]	[2.54]**	[2.78]***
Log child wage rates	-0.032	-0.1737	-0.0789
	[1.59]	[2.38]**	[1.14]
Observations	715	715	715
Wald chi2(17)	39.93***	43.87***	108.24***
Pseudo R2	0.156	0.0518	0.1748

Robust z statistics in brackets
* significant at 10%; ** significant at 5%; *** significant at 1%

Except for sorghum, other minor food crops also exhibit a hill shaped relationship with precipitation but the marginal impacts are quite small. The coefficients are only significant for cowpeas, sorghum and all other minor crops combined. Like for major food crops, choice of food crops are more sensitive to temperatures than to precipitation. Among these crops, choice of potatoes and millet respond significantly to temperature variations, while cowpeas, millet and other crops only respond significantly to changes in precipitation. This is because these crops are mostly produced in more dry areas. The finding of different response of individual crops to seasonal climate changes is not uncommon in the literature. In an Africa wide study, Kurukulasuriya and Mendelsohn (2006), found that the choice of different crops is sensitive to seasonal climate variables, but the effect varies by crop.

Table 8. Probit regression results for choice of minor food crops

	Potato	*Cowpea*	*Sorghum*	*Millet*	*Other crops*
Summer temperature	1.6453	0.0005	-0.0355	0.4952	-0.2985
	[3.90]***	[0.46]	[0.70]	[3.74]***	[0.86]
Summer temperature squared	-0.0461	-0.00001	0.0016	-0.0122	-0.0003
	[4.18]***	[0.61]	[1.16]	[3.56]***	[0.03]
Winter temperature	-1.3124	-0.0017	0.0702	-0.6926	0.5686
	[2.67]***	[1.08]	[1.01]	[4.10]***	[1.25]
Winter temperature squared	0.04	0.00004	-0.0027	0.0178	-0.0042
	[3.03]***	[1.18]	[1.40]	[4.03]***	[0.35]
Summer precipitation	0.0047	0.0003	-0.0023	0.0046	0.029
	[0.52]	[3.81]***	[1.04]	[0.81]	[2.64]***
Summer precipitation squared	-0.00002	-0.00001	0.0001	-.00002	-0.0001
	[0.81]	[3.81]***	[0.93]	[1.20]	[2.59]***
Winter precipitation	0.0298	0.001	-0.0271	-0.0175	0.1896
	[0.49]	[3.29]***	[2.34]**	[0.58]	[2.71]***
Winter precipitation squared	-0.0001	-0.00001	0.0001	0.0001	-0.0007
	[0.52]	[3.34]***	[2.36]**	[0.54]	[2.69]***
Log mean flow	6.2177	-0.043	-0.4333	0.9916	5.1062
	[1.82]*	[2.77]***	[0.69]	[0.64]	[1.38]
Log mean flow squared	-0.5711	0.004	0.0382	-0.095	-0.4525
	[1.79]*	[2.76]***	[0.66]	[0.67]	[1.32]
Soils (andosols)	0.2757	-0.0032	-0.0475	-0.1699	-0.2199
	[1.12]	[1.76]*	[0.48]	[1.01]	[0.71]
Average years of education of household members	-0.0082	0-.00004	-0.0023	0.001	0.0072
	[1.42]	[3.00]***	[2.62]***	[0.38]	[1.07]
Log farm size	-0.0117	0.0002	0.0258	0.0344	0.1255
	[0.21]	[1.07]	[1.76]*	[1.33]	[2.02]**
Log farm size squared	0.0049	-.00004	-0.0065	-0.0046	-0.0106
	[0.34]	[1.03]	[1.32]	[0.63]	[0.72]
Log household size	-0.0627	0.0001	0.0321	0.0214	0.1449
	[1.16]	[0.45]	[4.00]***	[0.91]	[2.22]**
Log male wage rates	0.0309	0.0001	-0.0609	0.0093	0.0455
	[0.61]	[0.36]	[4.48]***	[0.36]	[0.75]
Log child wage rates	0.2497	0.0013	-0.0624	-0.0414	0.1671
	[3.00]***	[3.05]***	[2.98]***	[1.21]	[1.97]**
Observations	715	715	715	715	715
Wald chi2(17)	146.66***	164.93***	130***	65.42***	79.94***
	0.2131	0.4449	0.3445	0.1283	0.0836

Robust z statistics in brackets
* significant at 10%; ** significant at 5%; *** significant at 1%

The results for choice of cash crops show that annual temperature and precipitation are both quite important to crop choice. This is consistent with results for food crops for the

African continent by Kurukulasuriya and Mendelsohn (2006). Coffee and sugarcane exhibit a hill shaped relationship with annual temperature, but tea and a coffee-tea portfolio exhibit a U shaped relationship. The results suggest that increased temperatures may encourage choice of coffee and sugarcane other factors held constant. Very low temperatures destroy coffee berries before they mature, but moderate temperatures are important for ripening of coffee berries so long as there is adequate rainfall. Excessive temperatures will lead to drying up of berries before they are harvested. For sugarcane, rising temperatures, coupled by adequate rainfall will promote growth of young and maturing cane, but excessive temperatures retard growth and maturity. The results are however insignificant, implying that sugarcane choice is not very responsive to temperature changes. This is contrary to findings by Deressa et al., (2005) who found sugarcane production to be quite sensitive to climate change. Though coffee and tea thrive mostly in the same zone (the humid zones), tea requires more moderate temperatures than coffee, and therefore high temperatures will discourage tea farming. This explains the inverted U-shaped relationship between temperatures and choice of tea crop on one hand and a coffee-tea portfolio on the other.

Choice of coffee, tea and a combined portfolio of the two exhibits a hill shaped relationship with annual precipitation, suggesting that the two crops require adequate rainfall. The marginal impacts of annual climate variables on tea are higher than for coffee, suggesting that tea is more responsive to climate variations than coffee. Sugarcane exhibits a U shaped significant relationship with annual precipitation, suggesting that at low levels of rainfall, farmers may be substituting sugarcane for other crops as an adaptation to variations in precipitation. However, at higher levels of rainfall, the same farmers switch to choice of cane.

The control variables (mean flow, soil and household characteristics) are introduced with the expectation that they shape farmers adaptation decisions in the face of climate change, (Maddison, 2006). The results show that the impact of stream flow on crop choice varies by crop, just like the climate variables. This is certainly due to the root system of particular crops. Some crops are better suited to hold the soil against stream flow and runoff (e.g. regumes and tubers) but others (e.g. maize) have less impact. Lower stream flow is associated with choice of maize, cowpeas and sorghum. The choice of these crops exhibit a U shaped relationship with flow. This result is consistent with findings by Kurukulasuriya and Mendelsohn (2006) for Africa. Increased flow is associated with choice of potatoes, beans and bananas and choice of these crops exhibit a hill shaped relationship with stream flow. Similar results are obtained for choice of coffee, tea and a combined portfolio. The results for choice of sugar cane are not significant. Holding other factors constant, soils do not seem to be important factors in choice of crops except for cowpeas. The negative significant impact suggests that farmers are less likely to grow cowpeas in fertile soils. Other control variables exhibit mixed results. Average level of education in a household suggests that more educated households are more likely to choose beans, bananas, cowpeas, coffee and tea. There is a non-linear relationship between cropland and choice of maize and very minor crops, but cropland does not seem to matter for choice of all other crops. House hold size is a positive and significant correlate of choice of maize, banana, sorghum, other minor crops and tea. This suggests the importance of availability of labour in choice of labor intensive crops. High adult wage rates discourage choice of bananas, coffee and sorghum, but child wage rates have mixed impacts, and are inversely correlated with choice of beans, sorghum and coffee. Choice of potatoes, cowpeas and other crops is positively correlated with child wage rates.

Table 9. Probit regression results for choice of cash crops

	Coffee	*Tea*	*Coffee-tea*	*Sugar cane*
Annual temperature	0.2997	-0.5101	-0.2482	0.1013
	[2.70]***	[4.91]***	[1.63]*	[1.30]
Annual temperature squared	-0.007	0.0119	0.0056	-0.0024
	[2.71]***	[4.81]***	[1.57]	[1.34]
Annual precipitation	0.0472	0.0609***	0.1119	-0.0228
	[4.38]***	[4.32]***	[5.55]***	[2.21]**
Annual precipitation squared	-0.0003	-0.0004	-0.0006	0.0001
	[4.13]***	[4.35]***	[5.49]***	[2.37]**
Log mean flow	4.1535	6.1756	10.0696	-0.5049
	[2.45]**	[3.39]***	[3.40]***	[0.44]
Log mean flow squared	-0.3871	-0.5764	-0.9388	0.046
	[2.45]**	[3.42]***	[3.41]***	[0.43]
Soils (andosols)	-0.1515	0.1265	-0.1313	0.1395
	[1.45]	[0.98]	[0.70]	[1.09]
Average years of education of household members	0.0083	0.0133	0.0258	-0.0037
	[2.79]***	[3.89]***	[4.72]***	[1.72]*
Log farm size	-0.0211	-0.0329	-0.0415	0.0586
	[0.76]	[1.07]	[0.88]	[1.59]
Log farm size squared	0.0027	0.006	0.0068	-0.014
	[0.47]	[0.78]	[0.60]	[1.23]
Log household size	-0.0102	-0.0482	-0.0444	0.0676
	[0.36]	[1.57]	[0.89]	[3.40]***
Log male wage rates	-0.0808	-0.0081	-0.0574	-0.0283
	[3.36]***	[0.30]	[1.39]	[1.42]
Log child wage rates	-0.0554	0.0117	-0.0583	0.0247
	[1.89]*	[0.34]	[1.10]	[0.95]
Observations	715	715	715	715
Wald chi2(13)	100.02	145.59***	108.86***	114.11***
Pseudo R2	0.2528	0.1692	0.1442	0.2277

Robust z statistics in brackets
* significant at 10%; ** significant at 5%; *** significant at 1%

4.2. Awareness of and Adaptations to Climate Change

Economic adaptation has been argued to significantly reduce vulnerability to anticipated future impacts of climate change. Previous studies have shown that the potential contribution of adaptation to reducing the negative impacts of global warming is large. The basic forms of adaptation identified in the literature including micro-level adaptations, market responses, institutional changes and technological developments (Kurukulasuriya and Rosenthal 2003; Darwin et al. 1995). In this chapter, we focus on micro-level adaptations which include farm production adjustments such as diversification and intensification of crop and livestock production, changing land use, irrigation and altering the timing of operations.

Perceptions of and adaptations to short-term climate variations

In this section, we analyze the perceptions and adaptation of farmers to short-term climate variations and also constraints on adaptation mechanisms. In the first instance, we analyze the adaptations to short-term climate variations shown in Table 10. The tabulations show that though households practice a range of adaptation measures, the most popular one is crop diversification or mixed cropping, adopted by 37% of all households, and tree planting, adopted by 16%. The results for the first measure support literature which argues that farmers are likely to adopt diversification of crop and livestock varieties, including the replacement of plant types, in order to increase productivity in the face of temperature and moisture stress (Kurukulasuriya and Rosenthal 2003). The relatively low proportion adopting adjustments to livestock management could be explained by land scarcity in more arable areas, which may hinder large scale livestock production. Nevertheless, such a measure is expected to reduce soil erosion and improve moisture and nutrient retention (Kurukulasuriya and Rosenthal 2003).

Table 10. Adaptations to short-term variations in weather

Variable	All regions	High potential	Med-low potential
Crop diversification / Mixed / Multi-cropping	37% (0.48)	44% (0.50)	33% (0.47)
Different planting dates	13% (0.34)	16% (0.37)	11% (0.31)
Adjustments to livestock management	6% (0.24)	4% (0.19)	7% (0.26)
Increased use of irrigation	14% (0.34)	9% (0.29)	16% (0.37)
Increased water conservation techniques	13% (0.33)	12% (0.32)	13% (0.34)
Soil conservation techniques	11% (0.31)	13% (0.34)	10% (0.30)
Shading and shelter / Tree planting	16% (0.37)	14% (0.34)	18% (0.38)
Other adaptation measures	12% (0.32)	9% (0.29)	13% (0.34)
No adaptation	13% (0.33)	17% (0.38)	10% (0.30)

Most other measures were adopted by between 11% and 14% of all households. Though the percentage of households involved is small, results for irrigation, water and soil conservation support the argument that a range of management practices can help reduce vulnerability through less runoff and erosion, and promoting nutrient restocking in soils, while other techniques may improve the soil structure and fertility (Kurukulasuriya and Rosenthal 2003). It was reported by 13% of households that they did not do anything to counter the impact of short-term variations in weather. Analysis by agro-ecological potential reveals that the use of mixed cropping, different planting dates and soil conservation techniques are more common in high potential zones, while livestock, irrigation, water conservation and shading/sheltering are more important in medium and low potential zones.

Table 11 shows the constraints on adaptations to climate change. About 60% of all households are hindered from adapting by lack of credit and savings (poverty). This supports findings in the literature that diversification is costly in terms of the income opportunities that farmers forgo (Kurukulasuriya and Rosenthal 2003). Another 19% fail to adopt any measure because of lack of knowledge about appropriate adaptations. This is consistent with Maddison

(2006), who found that education is the most important determinant of adaptations to climate change. The other constraints are reported by a relatively small proportion of households. Only 8% of households reported that there were no barriers to adaptation. Poverty and lack of knowledge seem to be more critical constraints in medium and low potential zones than in high potential zones.

Table 11. Constraints to short-term adaptations (% of households)

Constraint faced	All regions	High potential	Med-low potential
Lack of information about short-term climate variation	8% (0.27)	7% (0.26)	10% (0.29)
Lack of knowledge of appropriate adaptations	19% (0.39)	16% (0.36)	25% (0.43)
Lack of credit or savings	59% (0.49)	56% (0.50)	64% (0.48)
No access to water	8% (0.27)	12% (0.32)	3% (0.16)
Lack of appropriate Seed	5% (0.21)	4% (0.19)	6% (0.24)
Other constraints	13% (0.33)	12% (0.32)	14% (0.35)
No barriers to adaptation	8% (0.28)	9% (0.28)	8% (0.27)

Perceptions of and adaptations to long-term climate variations

Addressing long-term climate change should entail a comprehensive long-term response strategy at the national or local level and requires a dynamic approach (Kurukulasuriya and Rosenthal 2003). However, in the absence of directed policy responses, farmers choose their own adaptation measures depending on their farm (e.g. farm size) and household (e.g. education and experience) characteristics (Maddison, 2006). This adaptation could take several forms: changing crop types, such as introduction of drought tolerant varieties, and shifting from crop production to game ranching, among other measures.

Table 12. Households (%) observing long-term changes in temperatures

Observed variation	All regions	High potential	Med-low potential
Increased temperature	47% (0.50)	52% (0.50)	43% (0.50)
Decreased temperature	5% (0.22)	5% (0.22)	5% (0.22)
Altered climate range	18% (0.38)	19% (0.39)	17% (0.37)
No change	28% (0.45)	4% (0.20)	2% (0.13)
Don't know	3% (0.16)	4% (0.20)	2% (0.13)

Table 12 shows the percentage of households that reported having observed long-term changes in temperatures: 47% of all households noticed long-term increases in mean temperatures, while only 5% noticed decreased temperatures; 18% noticed climate variations but did not indicate the direction of change; and 28% reported that they had not noticed any change. Farmers in high potential zones were more aware of long-term climate change than their counterparts in medium and low potential zones. In Table 13, we present the percentage of households that reported having noticed long-term changes in precipitation: 56% of all households reported that precipitation had decreased over the years, compared to only 5%

who reported increased precipitation, while 16% reported that they had not noticed any change. There were no clear patterns in the differences in perceptions of high potential zone farmers and medium and low potential zone farmers, though a higher percentage of the latter (59%) reported having observed shorter rainy seasons. The pattern of perceptions to climate change seems to be consistent across the 11 African countries (Maddison, 2006)[7].

Table 13. Households (%) observing long-term changes in precipitation

Observed variation	All regions	High potential	Med-low potential
Increased precipitation / extended rainy season	5% (0.22)	10% (0.30)	2% (0.14)
Decreased precipitation / shorter rainy season	56% (0.50)	52% (0.50)	59% (0.49)
Change in timing of rains / earlier / later	18% (0.39)	21% (0.41)	17% (0.37)
Change in the frequency of droughts	7% (0.26)	7% (0.25)	7% (0.26)
No change	16% (0.37)	13% (0.34)	19% (0.39)

From Tables 12 and 13, it is clear that farmers in Kenya have perceived increased global warming though a few households are not clear on the direction of change. Next we consider the range of adjustments that the farmers make to counter the impact of long-term temperature and precipitation changes, conditional on perceiving global warming. The results are presented in Tables 14 and 15. The last row of these Tables show that only 60% of all households made any effort to counter long-term temperature changes (40% reported no adaptations), compared to 78% in the case of precipitation. For long-term temperature changes, two main adaptations emerge: crop diversification and shading/sheltering or planting of trees. As in the case of short-term climate change, crop diversification is more common in high potential zones, while shading/sheltering is more common in medium and low potential zones. In the case of crop diversification, farmers in Kenya are known to switch to more adaptive crop varieties such as fast growing and hybrid varieties. This supports the literature which shows that long-term adaptation measures mainly take the form of changing crop types, such as drought tolerant varieties. However, the low percentages adopting this measure could be due to lack of knowledge, skills and finances (Kurukulasuriya and Rosenthal 2003). Shading/sheltering/tree planting is important because in addition to countering climate change it is also a form of soil conservation. The adaptation options chosen by Kenyan farmers are consistent with the continent wide options (see Maddison, 2006:37).

Improved water management is one of the most important long-term adaptation options that countries must pursue. Though few households reported water management measures for long-term changes in temperature, this contrasts with adjustments for long-term precipitation changes, which included increased water conservation, crop diversification and increased use of irrigation. The low overall adoption rates could be attributed to scarcity of resources, including water for irrigation, and lack of knowledge about the importance of these options. Only 10% of the full sample of households reported having used any irrigation at all.

7 Maddison however questions the reliability of data (Maddison, 2006:18). It would however be unlikely that random errors in data collection would have occurred consistently in 11 African countries unless the design of the questionnaire and suggested codes were faulty, which I believe was not the case. The key issue as

Table 14. Adaptation to long-term temperature changes (% of households)

Variable	All regions	High potential	Med-low potential
Crop diversification / mixed / multi-cropping	25% (0.43)	38% (0.49)	17% (0.38)
Different planting dates	6% (0.23)	7% (0.25)	5% (0.21)
Adjustments to livestock management	4% (0.19)	4% (0.20)	4% (0.19)
Increased irrigation / groundwater / watering	6% (0.23)	6% (0.24)	5% (0.22)
Increased water conservation techniques	7% (0.25)	6% (0.24)	7% (0.25)
Decreased water conservation techniques	6% (0.24)	10% (0.29)	4% (0.19)
Shading and shelter / tree planting	22% (0.41)	12% (0.32)	28% (0.45)
No adaptation	40% (0.49)	37% (0.48)	41% (0.49)

Table 15. Adaptation to long-term precipitation changes (% of households)

Variable	All regions	High potential	Med-low potential
Crop diversification / mixed / multi-cropping	34% (0.47)	35% (0.48)	34% (0.47)
Different planting dates	15% (0.36)	19% (0.40)	13% (0.33)
Adjustments to livestock management	6% (0.23)	3% (0.16)	7% (0.26)
Increased irrigation / groundwater / watering	16% (0.37)	15% (0.35)	18% (0.38)
Increased water conservation techniques	21% (0.41)	17% (0.37)	23% (0.42)
Decreased water conservation techniques	13% (0.34)	16% (0.37)	12% (0.33)
Shading and shelter / tree planting	9% (0.28)	4% (0.20)	12% (0.32)
No adaptation	22% (0.41)	24% (0.43)	21% (0.41)

The above analysis shows that farming households in Kenya are aware of both short- and long-term climate changes. Further, more than 80% of households have implemented various adaptation mechanisms to counter short-term climate variations, compared to 60% and 78% that have implemented various mechanisms to counter long-term temperature and precipitation changes respectively. These percentages are rather high, but in no way reflects appropriateness or success of the adaptation measures. Maddison (2006) argues that some adaptation options are adopted autonomously rather than as a result of awareness of climate change or other control variables that are can be influenced by policy.

5. Conclusion

This chapter examines crop selection and adaptation options taken by farmers in the face of climate change in Kenya. The analysis is based on primary data collected from a sample of 724 households from 38 districts in 2004. The primary data is enriched with secondary climate data, which reflect long term climate change in Kenya. The probit regression method is used to analyse the impact of climate change on crop choice, while descriptive analysis is used to assess awareness of climate change, the key adaptation options and barriers to

Maddison acknowledges may be that not very comprehensive data was collected on adaptations because this was a relatively minor objective of the project heavily geared towards Ricardian analysis.

adaptation. The crop choice analysis is based on three groups of crops: major food crops (maize, beans and bananas), minor food crops (potatoes, cowpeas, sorghum, and millet) and cash crops (coffee, tea and sugarcane).

The results suggest that the decision to farm a crop or a portfolio of crops responds to climate change. The results further show that choice of most crops is more sensitive to temperature than to precipitation. For the major crops, choice of maize is found to be responsive to both temperature and precipitation, but choice of beans and bananas are only responsive to temperatures and rainfall respectively. The decision to farm all major food crops exhibit a hill shaped relationship with summer temperatures, but a U shaped relationship with winter temperatures. For minor food crops, choice of potatoes and millet are found to be responsive to temperatures only, but choice of cowpeas, sorghum and all other minor crops are found to respond significantly to precipitation. Choice of potatoes and millet exhibit a hill shaped relationship with temperatures. Except for sorghum, choice of other minor food crops exhibit a hill shaped relationship with precipitation but the marginal impacts are quite modest.

Choice of tea and coffee respond significantly to both temperature and precipitation, but choice of sugarcane only responds to precipitation. The responses support the expectation that farmers choose crops which are best adapted to the local conditions (Kurukulasuriya and Mendelsohn, 2006). Choice of coffee and sugarcane exhibit a hill shaped relationship with annual temperature, but tea and a coffee-tea portfolio exhibit a U shaped relationship. Choice of coffee and tea exhibit a hill shaped relationship with annual precipitation, but the marginal impacts on tea selection are higher than for coffee, suggesting that tea is more responsive to climate variations than coffee. Sugarcane exhibits a U shaped significant relationship with annual precipitation.

The analysis of awareness and adaptation of farmers to climate change show that farming households in Kenya are aware of both short- and long-term climate change and some have implemented various adaptation mechanisms. The analysis also shows differences between the perceptions and adaptations of medium/low potential zone farmers and their counterparts in high potential zones. Crop diversification is the most common adaptation measure, particularly in high potential zones, while water conservation, irrigation and shading/sheltering of crops are the main adaptation measures in drier regions.

The results suggest that farmers in Kenya adapt crop selection to climate change and also adopt other mechanisms to counter the adverse impact of climate change on agriculture. Though the survey did not investigate the role of government in promoting farm level adaptations to climate change, it can play a critical role in promoting the level and diversity of adaptations. For instance, improving knowledge through agricultural extension could encourage both short- and long-term adaptations to climate change. Farmers should be provided with information on expected long-term climate change and the appropriate adaptations for each scenario. Availability of credit and policies that improve household welfare would also be important, given that most of the households reported that adaptation is constrained by lack of credit and income.

Appendix: Analysis of Crop Choice: Theory and Methods

To analysis crop choice, we assume that a farmer chooses the crop that is expected to maximize profits. The farmer must first determine whether or not it is profitable to farm a particular crop and also the most profitable portfolio of crops, given land scarcity and competing alternatives. The profits of farmer i from selecting crop j can be defined as in Kurukulasuriya and Mendelsohn, (2006):

$$\pi_{ij} = V(K_j, S_j) + \varepsilon(K_j, S_j) \tag{1}$$

Where K is a vector of exogenous characteristics of the farm, including climate, mean flow and soils and S is a vector of characteristics of farmer i. The profit function is composed of two components: the observable component V and an error term, ε. The disturbance, ε, is unknown to the researcher but may be known to the farmer (the farmer is more likely to choose a crop that is most profitable, but the cumulative distribution function (CDF) is a function of ε that is known up to a finite parameter vector (Kurukulasuriya and Mendelsohn, 2006; Seo and Mendelsohn, 2006). If we let Z= (K, S), the farmer will choose crop *j* over all other crops *k* if and only if:

$$\pi^*(Z_{ji}) > \pi^*(Z_{ki}) \text{ for } \forall k \neq j.[or \text{ if } \varepsilon(\mathrm{Z_{ki}}) - \varepsilon(\mathrm{Z_{ji}}) < V(\mathrm{Z_{ji}}) - V(\mathrm{Z_{ki}}) \text{ for } k \neq j] \tag{2}$$

Specifically, the farmer's problem is

$$\arg\max\left[\pi^*(Z_{1i}), \pi^*(Z_{2i}), ..., \pi^*(Z_{ji})\right] \tag{3}$$

Following Kurukulasuriya and Mendelsohn, (2006), the probability P_{ji} that the jth crop will be chosen can be specified as:

$$P_{ji} = \Pr[\varepsilon(\mathrm{Z_{ki}}) - \varepsilon(\mathrm{Z_{ji}}) < V_{\mathrm{j}} - V_{\mathrm{k}}]\forall k \neq j \text{ where } \mathrm{V_j} = V(Z_{ji}) \tag{4}$$

Assuming ε is $IN(0, \delta^2)$ and $V_k = Z_{ki}\gamma_k + \alpha_k$, the probability that farmer i will choose crop j can be expressed as

$$P_{ji} = \frac{\exp(Z_{ji}\gamma_j)}{\sum_{k=1}^{j} \exp(Z_{ki}\gamma_k)} \tag{5}$$

The probability that a farmer chooses crop j or a portfolio of crops is estimated using a probit model[8] (Maddala, 1995).

[8] The data on which this chapter is based was collected for the purpose of Ricardian analysis of the impact of climate change on net farm revenue. For this reason, adequate effort was not made to capture the primary

References

Basist A, Grody N.C, Peterson T.C & Williams C.N, 1998. Using the Special Sensor Microwave Imager to monitor land surface temperatures, wetness, and snow cover. *Journal of Applied Meteorology* **37**(8): 888–911.

Benhin J., 2006. Climate change and South African agriculture: impacts and adaptation options. *CEEPA Discussion Paper* No. **21**, Centre for Environmental Economics and Policy in Africa, University of Pretoria.

Darwin R, Tigra M, Lewandrowski J & Raneses A, 1995. World agriculture and climate change: *Economic adaptations*. *AER*-**703**, US Department of Agriculture, Economic Research Service, Washington DC.

Deressa T., Hassan R. & Poonyth, D. 2005. Measuring the economic impact of climate change on South Africa's sugarcane growing regions. *Agrekon* **44**(4): 524–542.

FAO (Food and Agriculture Organization), 2003. *The digital soil map of the world*: Version 3.6 (January), Rome, Italy.

Houghton J.T., Ding Y., Griggs D.J., Noguer M., van der Linden P.J., Dai X., Maskell K. & Johnson C.A, 2001. Climate change: The scientific basis. Contribution of working group I to the third assessment report of the Intergovernmental Panel on Climate Change. New York: Cambridge University Press.

IWMI (International Water Management Institute) & The University of Colorado. 2003. Hydroclimatic Data for the GEF/CEEPA/World Bank project on climate, water and agriculture: Impacts and adaptation of agro-ecological systems in Africa.

Kabubo-Mariara J. and Karanja F., 2006. The Economic impact of climate on Kenyan crop agriculture: A Ricardian approach. *CEEPA Discussion Paper* No. **12**, Centre for Environmental Economics and Policy in Africa, University of Pretoria.

Kabubo-Mariara J. & Karanja F., 2007. The economic impact of climate change on Kenyan crop agriculture: A Ricardian approach. *Global and Planetary Change*. **57**: 319–330.

Kabubo-Mariara J., 2007. *Climate change adaptation and livestock activity choices in Kenya: An economic analysis*. Mimeo. University of Nairobi. Nairobi.

Kabubo-Mariara J., Linderholf V., Kruseman G., Atieno R. & Mwabu G., 2006. Housheold welfare, Investment in soil and water conservation and tenure security: Evidence from Kenya. *PREM working paper* **06-06.** Poverty reduction and environmental management. Vrije Universiteit.

Karanja F., 2006. Cropwat model analysis of crop water use in six districts in Kenya. *CEEPA Discussion Paper* No. **35**, Centre for Environmental Economics and Policy in Africa, University of Pretoria.

Kurukulasuriya P & Rosenthal S, 2003. Climate change and agriculture: A review of impacts and adaptations. *Climate Change Series* **91**, Agriculture and Rural Development Department and Environment Department, the World Bank, Washington, DC.

Kurukulasuriya P. & Mendelsohn R., 2006. Crop selection: Adapting to Climate Change in Africa. *CEEPA Discussion Paper* No. **26**, Centre for Environmental Economics and Policy in Africa, University of Pretoria.

---

crops produced by farmers and given the intensity of farming on small holder farms, it is quite difficult to make the crops mutually exclusive so as to use multinomial logit methods of analysis, given that crop choices may not be independent.

Maddala, G.S. (1995), *Limited Dependent and Qualitative Variables in Econometrics*, New York: Cambridge University Press.

Maddison D., 2006. The perceptions of and adaptation to climate change in Africa. *CEEPA Discussion Paper* No. **10**, Centre for Environmental Economics and Policy in Africa, University of Pretoria.

Mendelsohn R, Nordhaus W & Shaw D, 1994. The impact of global warming on agriculture: A Ricardian analysis. *American Economic Review* **84**: 753–771.

Seo S. N & Mendelsohn R., 2006. Climate change adaptation in Africa: A Microeconomic analysis of livestock choice. *CEEPA Discussion Paper* No. **19**, Centre for Environmental Economics and Policy in Africa, University of Pretoria.

Strzepek K. & McCluskey A., 2006. District level hydroclimatic time series and scenario analysis to assess the impacts of climate change on regional water resources and agriculture in Africa. *CEEPA Discussion Paper* No. **13**, Centre for Environmental Economics and Policy in Africa, University of Pretoria.

Wahaj R, Muňoz G. & Maraux F., 2006. Actual crop water use in project countries: A synthesis at the regional level. *CEEPA Discussion Paper* No. **38**, Centre for Environmental Economics and Policy in Africa, University of Pretoria.

World Bank, 2003. Africa rainfall and temperature evaluation system (ARTES). World Bank, Washington DC.

In: Climate Change Research Progress
Editor: Lawrence N. Peretz, pp. 291-312

ISBN: 978-1-60021-998-6

Chapter 9

THE POTENTIAL FOR INCREASING THE CONTRIBUTION OF RENEWABLE ENERGIES TOWARDS REPLACING FOSSIL FUELS IS TOO SMALL TO JUSTIFY THE EXPENDITURE OF GOVERNMENT RESEARCH FUNDS

H. Douglas Lightfoot *
Global Environmental Climate Change Centre, McGill University
805 Sherbrooke Street West, Montreal, Quebec, Canada, H3A 2K6

Abstract

To ensure future world energy supplies and to reduce carbon emissions an urgency is developing to find replacements for fossil fuels. Renewable energies have serious limitations that prevent a significant increase in their contribution to replace fossil fuels. This paper examines the technical reasons why the renewable energies, i.e., solar, biomass, wind, and hydro, currently supply only about 9% of current world energy demand. This percentage dropped from almost 100% three hundred years ago and is still falling. The limitations of renewable energies are such that they are useful in only relatively small niche applications. Hydro is no exception even though it provides electricity when required, is about 75% of total renewable energy, and currently contributes about 7% of world primary energy demand. Research funds must be withdrawn from renewable energies and directed towards areas with potential to replace significant quantities of fossil fuels.

1. Introduction

Renewable energies such as wood, wind and solar have been valuable sources of energy since the beginning of man. They have been used for cooking, heating, making tools and a host of other activities. As populations grew, so did the pressure on energy sources. In England in

* E-mail address: dlightfo@aei.ca. Tel: 514 457 5637. 8 Watterson, Baie d'Urfe, Quebec, Canada H9X 3C2

1700 there were 600,000 people living in London and there was a shortage of wood for heating and cooking. That was when England turned to coal as the fuel of choice [1]. Since then, renewable energies plummeted from virtually 100% of world energy demand to about 9% today for very good technical and resource limitation reasons [2]. These physical limitations are going to continue because no amount of research can change them.

The world began turning away from renewable energies centuries ago because the supply was inadequate, and that was when energy consumption was very much less than today. For example, total U.S. energy consumption today is more than thirty-five times that in 1850.

The purpose of this paper is to show that renewable energies have specific technical limitations that cannot be overcome by research. Research funds are always in short supply and can be safely removed from renewable energies. These funds are better spent in more productive areas.

2. What Are Renewable Energies?

The sun is the source of all renewable energies except for volcanic geothermal, which is the result of radioactive decay inside the earth. Renewable really means inexhaustible.

Some of the sun's radiation is converted to heat at the surface of the earth, thereby causing winds. Other radiation energy evaporates water that falls as rain or snow at higher elevations and provides hydro power. Visible light drives photosynthesis for growing plants, which are at the bottom of the food chain. On average, only 0.1-0.5 % of sunlight falling on plants is converted to plant material [3].

Solar, or sunlight, biomass, wind, and hydro are the renewable energies, i.e., those available in real-time, season after season. They accounted for about 39.7 exajoules (EJ) [1] [4] of world energy in 2005, or about 9% of the 457 EJ [5]consumed by the world in that year. In 2005, world fossil fuel energy consumption was 388 EJ. Therefore, whatever replaces fossil fuels must supply 388 EJ of energy annually. If world energy demand increases as in IPCC IS92a, the "business as usual" scenario, it will be more than three times current consumption in 2100.

There are good reasons why the contribution of renewable energies to world energy demand fell from virtually 100% a couple of centuries ago to about 9% today [6]. The reasons are still the same today and make it difficult to maintain, let alone increase, the current 9% of renewable energies as world energy demand grows.

Each renewable energy will be discussed in turn, except volcanic geothermal because it is small and is limited to a few specific geographic locations [7].

3. Solar Energy

Although there is enough solar radiation falling on the earth's surface each day in one hour to power the world for a year, the problem is to capture the energy in sufficient quantity and at a useful time. Sunlight is dilute and does not arrive at a constant rate. It is cyclical daily, intermittent daily and cyclical seasonally.

[1] Energy Information Administration scale for measuring primary energy [5]. 1 EJ = 10^{18} Joules

The variation is shown in Figure 1 [8] for Madison, Wisconsin, which is at 43° North Latitude [9]. These figures are for a horizontal flat plate and would approximate the situation for fixed panel PV arrays on clear days. As the equator is approached, the difference between summer and winter becomes somewhat smaller, but is always there. The solar radiation on the left vertical axis is given in watts per square meter. These are instantaneous rates, and are different from the 24 hour averages given in solar insolation tables[2]. For nine months of the year in Madison more than one half of the solar radiation arrives between 10 AM and 2 PM.

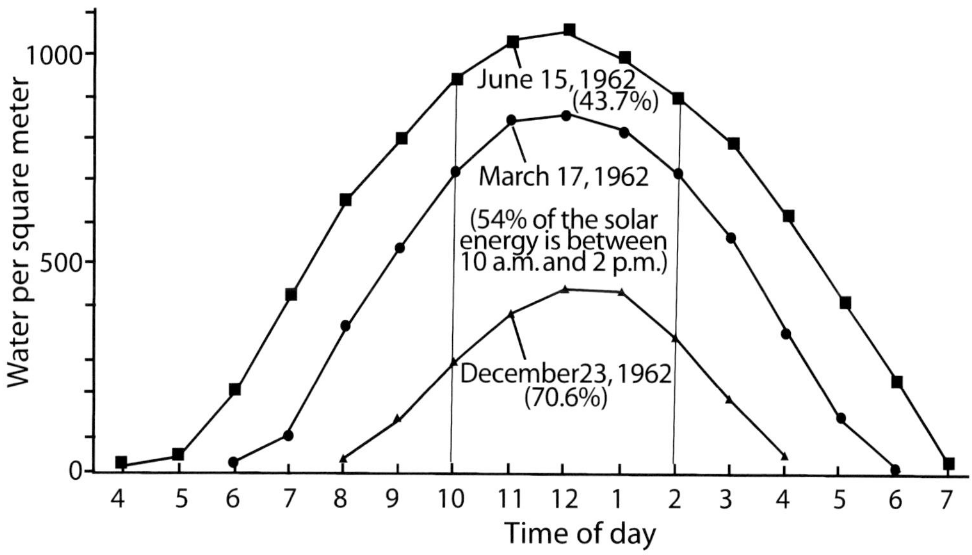

Figure 1. Typical solar radiation on clear days, Madison Wisconsin, 43° N.

Except for hot water heating and biomass liquids, all renewable energy enters the world's energy system as electricity.

Sunlight is converted to electricity either directly by photovoltaic cells or indirectly, such as with concentrating mirrors that heat oil to boil water to make steam for a steam/electric power plant. The Solar Energy Conversion System (SEGS) concept installation at Kramer Junction in California's Mojave Desert is an example of the latter.

4. SEGS Hot Oil System

4.1. Land Area

Configuration of the land surface area is critical for any large solar collector field. An aerial view of the SEGS installation reveals one aspect of the problems associated with any solar electricity generation facility [10] [11]. They require the most desirable land, i.e., horizontal or slightly sloping to the south in the northern hemisphere.

[2] Insolation refers to exposure to the sun's rays.

Statements about the land area required to produce a given electricity output are often vague, such as, "... extensive studies by the U.S. Department of Energy have identified four Southwest states (Arizona, California, Nevada and New Mexico) with deserts large enough to generate 4000 GW, more than twice the output needed for our plan." [12] This is a general statement and does not have the detail required to make it useful. For example, is 4000 GW the gross amount of sunlight energy? Or, is it something else? The method of calculating the 4000 GW has to be understood before valid conclusions can be drawn.

Configuration of the land surface must be taken into account. The large blocks of land required for 4000 GW are not contiguous flat plains. Large blocks of land not currently allocated to other uses are not a common commodity among the mountains, valleys, lakes, streams, towns, highways, rail lines, homes, farms, nature reserves, etc., that characterize most land areas, even in the desert. The side of mountains sloping away from the sun and areas with too much shading by mountains are not desirable locations for solar collector fields.

The current SEGS facilities are a small niche operation and land area is not a serious problem. The way the small facility fits into the specific location makes it easy to overlook the difficulties of scaling up by several orders of magnitude required to provide a significant contribution to energy supply.

Once a group of sites for solar collectors is selected, the basic size of the facility must be increased to allow for the shape of the individual parcels of land; accommodation for people to operate the facility; the energy necessary to operate and maintain the system; and the transmission losses to move the electricity to markets up to thousands of kilometers away.

To be seriously considered, any proposal for constructing very large solar fields must show exactly how many arrays are required, exactly where each is physically located as well as factoring in all of the previously mentioned items.

Fortunately, Google Earth is now available. It brings satellite views of the land under consideration to your computer screen. The potential problems in selecting sites for large solar collector fields can now be more readily identified.

4.2. Solar Collector Land Area: IPCC Estimates

It is worthwhile at this point to digress and discuss the estimates of potential solar photovoltaic electricity made by Working Group III (WG III) of the Intergovernmental Panel on Climate Change (IPCC) available in Climate Change 2001: Mitigation [13]. The method used by the authors of Section 3.8.4.3.4, Solar Energy, was simply to use 1% of unused land for the minimum value and 10% for the maximum value. The surface configuration of the land does not appear to have been taken into account.

Using the same data base of renewable energies as in Climate Change 2001: Mitigation, Eliasson [14], and Lightfoot and Green [15] , it can be calculated that to generate 1 EJ/yr of electricity from sunlight requires an average of 1,905 km^2 to 3,333 km^2 of land.

IPCC WG III suggests that 1% of unused land, or 393,000 km^2, would be available to produce solar electricity from photovoltaic cells. Almost 60% of this solar land is in Africa and the Middle East, which is very far from where most electricity in the world is consumed. Most of the remaining land is in areas of relatively poor solar insolation. Increasing solar land area from 1% to 10% of unused land does little, if anything, to ameliorate either of these problems.

Because of the intermittent nature of wind and solar electricity, installed capacity is limited to up to 20% of fossil fuel installed electricity generating capacity in large integrated systems, or less than 10% of annual fossil fuel generated electricity supply [13]. See Section 7 for additional information.

Thus, for solar electricity to be available on a scale larger than 10%, it must be stored in some manner. The question arises as to whether or not solar electricity can be stored on a large scale by converting it to hydrogen. This question is answered later in this paper under Section 5. Solar Hydrogen.

4.3. Helpers for SEGS

The SEGS system has two helpers to get around the variability of solar insolation. The first is a hot oil storage system large enough to cover short periods of little or no sunlight, such as overnight. The second helper is a gas turbine fuelled by natural gas to provide electricity for longer periods of cloud cover or changes in annual sunlight available and can provide up to 25% of annual electricity output.

4.4. Land Required per Unit of Electricity

The quantity of land required per unit of electricity depends on the efficiency of the solar collector for converting sunlight into electricity and the distance required between collectors. These factors are important because only a fixed amount of sunlight falls on a square meter of land area. If twice the solar energy is required, then twice the land area is needed. Therefore, solar systems must be judged on the portion of the actual sunlight available that they utilize.

Silicon PV cells are about 15% efficient in converting sunlight into electricity. The average spacing for fixed panel PV cells is about two areas of land for one area of PV surface [16] [2]. Therefore, (15%/2) = 7.5% of the sunlight falling on a given area of land is converted to electricity.

Two-axis tracking panels collect about twice the amount of electricity per day, but the spacing between panels is much larger to prevent one panel from shading another at the winter solstice. Two actual examples, which have since been dismantled, are the two axis tracking panels at the Hesperia, California (34.5° North latitude) solar power plant in 1982 where land to collector ratio was about 8:1, and for Carrisa Plains, California (35.3° North latitude) in 1983 where the ratio was about 11:1. At lower latitudes the land to collector ratio might be as low as 5:1 for 2-axis tracking panels. In this case, 15% x (2/5) = 6% of the potential sunlight could be collected.

For a SEGS unit, the peak efficiency from sunlight to electricity is 21% [17], whereas the average is 15.2% and the land to solar collector area is 3.5:1 [18]. On average (15.2/3.5) = 4.3% of sunlight falling on a given land area is converted to electricity.

4.5. Cooling Water Supply

The efficiency of any steam/electric system, such as the SEGS system, depends on the difference between the steam temperature and that of the cooling water for the exhaust steam condenser. The steam condenser can be cooled by air, but the efficiency is lower than with water as the cooling medium.

To minimize water consumption, water is recycled through cooling towers to cool it for re-use. Assuming steam/electric efficiency is 33%, then, for every 1 kWh of electricity produced, 2 kWh of heat in the condenser must to be dissipated in the cooling tower through the evaporation of 3.2 kg of water, or 3.2 liters.

Unfortunately, there is little water available in deserts.

4.6. Annual Variation of Solar Insolation

Annual solar variation must be taken into account if the required electricity flow is to be delivered in a timely and reliable manner. Figure 1 shows the maximum and minimum electricity production with an annual solar insolation ratio of 1.8:1[3] [12] around an average monthly electricity production rate of 100%. Starting on March 21, the amount of hot oil produced is higher than the average and must be stored for future use. Starting on September 21, hot oil production rate is lower than the average and must be made up from storage. The quantity of hot oil that must be stored for an average of three months over March to September is much too large to be practical.

The SEGS system depends on natural gas to provide the necessary reliability and to compensate for annual solar insolation variability. When fossil fuels, such as natural gas, become scarce, it will be difficult to wean SEGS off them.

Annual solar variation could also be compensated for by installing 40% more collectors than required on December 21, i.e., (100 – 71.4)/71.4 = 40%. The "extra" collectors would be turned off when not required. The land area required would also increase by 40%.

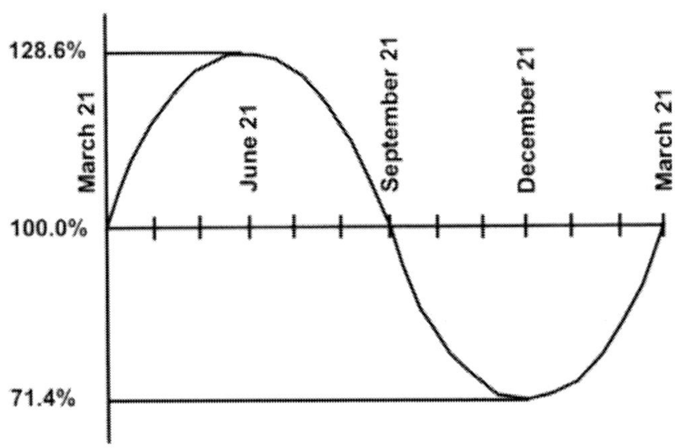

Figure 2. Annual maximum and minimum electricity output from the SEGS array at Kramer Junction, California.

[3] For single axis tracking parabolic reflectors.

4.7. Cleaning Mirrors

In the August 1991 issue of National Geographic, there is a beautiful time-lapse photograph on pages 84 and 85 showing the route of the cleaning truck during a mirror cleaning operation. The caption to the photo says: "Lights from a maintenance truck thread through the world's largest solar-thermal complex, Luz International, in California's Mojave Desert. Working at night, the 20-person crew takes a week and a half to wash the company's 1.5 million mirrors, which produce 95 percent of U.S. solar electricity."

Howard Hayden, author of The Solar Fraud [19], describes solar collector cleaning at the SEGS facilities as follows:

> "….. The optical efficiency varies from 71% (units I and II) to 80% (units VIII and IX). That is, between 71% and 80% of the sunlight that strikes the mirrors is actually reflected to the pipes containing the Therminol. They achieve this high efficiency by washing the mirrors every five or so days, and with a high pressure wash every ten-to-twenty days. Let's repeat that: they wash the several million square meters of mirror – much more than the 2.3 million m^2 of aperture – about 25 times a year!"

This statement and the caption from National Geographic are consistent.

From a PVUSA test in Sacramento, "a simple side-by side test of two identical modules installed on a fixed-tilt roof top. One is cleaned three times a week and the other is left to the forces of nature. Annualized soiling losses can be expected to exceed 7% during a normal year, 4% during a wet year and during drought years may exceed 10%" [20].

The parabolic mirror shape and orientation of the SEGS collectors is likely to collect dust at a significantly higher rate than the fixed slope PV cell described. Cleaning the SEGS mirrors regularly is an essential part of maintaining efficiency and is not a trivial operation.

From the photograph in the National Geographic, the land used by the SEGS system does not appear to be environmentally friendly. It looks as if it has been rendered completely barren—nothing appears to live there—or would be allowed to live there.

4.8. Summary RE SEGS

The average efficiency of collecting solar energy and transferring it from the collector to the boiler is well documented over the 16 years since the SEGS facilities were completed in 1991.

Efficiency of converting sunlight to electricity and spacing of collectors control the number of collectors required and the land area for a given output of electricity. Even if reflector efficiency could be increased to 100%, the physical limitations previously described are the critical limiting factors for a SEGS unit.

Consequently, there is no incentive to invest research funds to improve solar collector efficiency for systems such as SEGS, or similar ones.

5. Solar Hydrogen

One of the major problems with solar energy and wind energy is that they are intermittent and erratic. The benefits of both of these renewable energies might be realized if the electricity could be stored on a large scale and released to the grid at the necessary rates. Large scale storage of electricity is desirable, but is not possible now and there are no possibilities in sight. Currently, relatively small amounts of electricity can be stored by such methods as chemical storage batteries and "pumped storage" in hydroelectric systems.

It is often suggested that the manufacture of hydrogen by electrolysis of water (H_2O) into its components, hydrogen (H_2) and oxygen (O_2), could be used as a means of storing large quantities of electricity from intermittent sources.

To put hydrogen production from any source into perspective, it is necessary to examine the quantity of hydrogen required to power a specific energy application. For example, 1 EJ of liquid hydrogen per year would supply about one week of world transportation fuel in 2002.

The main factors associated with the production of 1 EJ/yr of hydrogen by solar electricity are: (1) the supply and transmission of solar electricity; (2) the supply and treatment of the water required by the electrochemical plant; (3) the electrochemical plant; (4) liquefaction, storage and transportation of liquid hydrogen, and (5) disposal of oxygen. Although oxygen is non-flammable, it must be treated carefully because it vigorously supports combustion of materials which will not usually burn in air. Concentrations of 25% oxygen in air, versus the normal 20.8%, will significantly increase the rate of combustion for some materials [21].

5.1. A Solar Hydrogen Production Facility

For purposes of this analysis, we assume the solar field and electrochemical plant are located near Luxor, Egypt, 26° North latitude, about 170 km north of Aswan. The solar insolation at 290 W/m^2 is close to the highest in the world [22] and the Nile River is a plentiful source of water.

The overall concept is shown in Figure 3, a simplified process flow sheet, and described below.

5.2. The Area of Solar Collector Field

Photovoltaic cells of 15% efficiency are mounted on fixed panels, and the land to collector ratio is 1.65 for 26° North latitude to prevent shading of panels between 10 AM and 2 PM at the winter solstice. With 10% added to allow for roads, fences, ground conditions, etc., the land occupied by the photovoltaic cells is 1,343 km^2. Further increases are necessary to compensate for the 90% efficiency of the electrolysers [23], 10% electricity losses in the transmission system to the electrolysis plant, and 25% for the energy to liquefy hydrogen [24]. A 5% allowance is provided for auxiliary power and maintenance for when the sun is not shining, and another 5% to compensate for evaporation losses of hydrogen during storage and shipping. The electricity output of the solar field and its area would have to be increased by a combined factor of: (1/0.9) x 1.10 x 1.25 x 1.05 x 1.05 = 1.68, and the land area of the solar field would be 1.68 x 1,343 = 2,250 km^2, which is 71% of the area of Rhode Island.

5.3. The Electricity Supply System

The voltage that can be applied to a photovoltaic cell is limited to about 600 volts. To minimize transmission losses, the few hundred volts of direct current from the photovoltaic cells is converted to alternating current by an inverter and then increased by a transformer to around 115,000 volts. When the electricity reaches the electrochemical plant, it is transformed to a lower voltage and converted back to direct current by a rectifier. The flow of electricity, i.e., flow of electrons through the electrolysers, would typically be 5,150 amperes. It is the total number of electrons transferred that is important because two electrons must be transferred for each molecule of hydrogen formed. Annual electricity output of the solar collector field of 1.68 EJ/yr is about 3.6% of world electricity production in 2002.

5.4. Water Requirement and Treatment

The water requirement just to make the hydrogen is 217,000,000 liters per day (l/d), on average. This was increased by 10% to 239,000,000 l/d to provide additional hydrogen for auxiliary power and for evaporation losses from liquid hydrogen storage. The water in the Nile River water is relatively turbid, has a pH of about 8.0, total alkalinity of about 220 mg/l, total hardness of near 190 mg/l and calcium hardness of about 120 mg/l [25]. Thus, more than 40 tonnes/day of suspended and dissolved solids must be removed and disposed of from the 239,000,000 liters per day of make-up water. The water must be pure to 1-2 megohm/cm, i.e., distilled water quality, to prevent scale build-up in the electrolyser and deterioration of the electrolyte properties.

The electrochemical plant would have about 50,000 electrolytic cell units operating at an average of 10.6 hours per day. The building for just the electrolytic cells would have an area of 2.5 km^2 to 3 km^2.

5.5. Other Considerations

In estimating the size of the electrochemical plant needed to electrolyze water into 1 EJ/yr of hydrogen, it was assumed that the plant would be operating at full capacity whenever the sun was on the collector field. This may well not be true, as the time to start up and shut down such a large electrochemical plant may cut significantly into the time available for the production of hydrogen.

The design of the solar field is based on fixed panel PV arrays. If two-axis tracking panels were used instead, the daily amount of electricity generated from each panel would double. The peak electricity output would last longer and might make operation of the electrolysers somewhat easier. However, two and a half times more land would be required

Storing and transporting liquid hydrogen is also a major effort. One liter of liquid hydrogen has one quarter the energy density of one liter of gasoline and requires four times the volume to transport the same quantity of energy. Further, hydrogen is a liquid at -253°C, and is continually venting. Care would have to be taken to minimize evaporation losses. It would require about 420 round trips by super tankers with the capacity of the Exxon Valdez to move 1 EJ/yr of hydrogen to the necessary terminals versus 105 of the same tankers filled with oil.

Liquid hydrogen is stored in vacuum insulated vessels to reduce the rate of heat transfer from the surrounding air into the hydrogen and, hence, the loss by evaporation. The rate of loss is 0.3% per day from a 28,000 USG rail car with an insulation layer consisting of a vacuum jacket and multilayer radiation shielding [24].

The parabolic mirrors of the SEGS in the Mojave Desert in California require cleaning 25 times a year to maintain efficiency. It is likely that photovoltaic surfaces subject to the dust of the Sahara Desert would require cleaning more often [19].

5.6. Summary RE Solar Hydrogen

The efficiency of PV cells is critical to determining the quantity of land required. This analysis is based on 15% efficient silicon cells. Even if the efficiency could double to 30%, which is close to the theoretical limit, it would make little difference to the viability of this project.

It is important to note that the cost of the silicon wafer is a very small part of the cost of the entire facility, about 25% of the module cost [26]. Even if the silicon wafers were free, the effect on the viability of this project is virtually negligible. Thus, research into cost reduction of the active PV surface is not justified for a large scale project such as this one.

5.7. Small Niche Applications for Solar

5.7.1. Heating Water for Domestic and Commercial Use

The best use for solar energy is for heating water for domestic use. The solar heated water can be stored until needed. If the water temperature is too low because of too little sun for a given period, the water can be heated by an auxiliary source of energy and the temperature raised to the desired level. There are some commercial water heating installations where sunlight is sufficient, or subsidies are large enough.

5.7.2. Solar Power Towers

A large field of two-axis tracking mirrors is focused at the top of a tower. The temperature at the focal point can be in the range of 550°C to 1500°C and is used to make steam to power a turbine for generating electricity [27]. Some systems use the heat to melt a salt mixture to store the energy for future use. The molten salt is used to make steam when electricity is required [28].

Solar Power Towers work, but the output is inherently small because sunlight is dilute. The land utilization in terms of collecting sunlight and converting it to electricity is likely towards the low end of the range for solar systems.

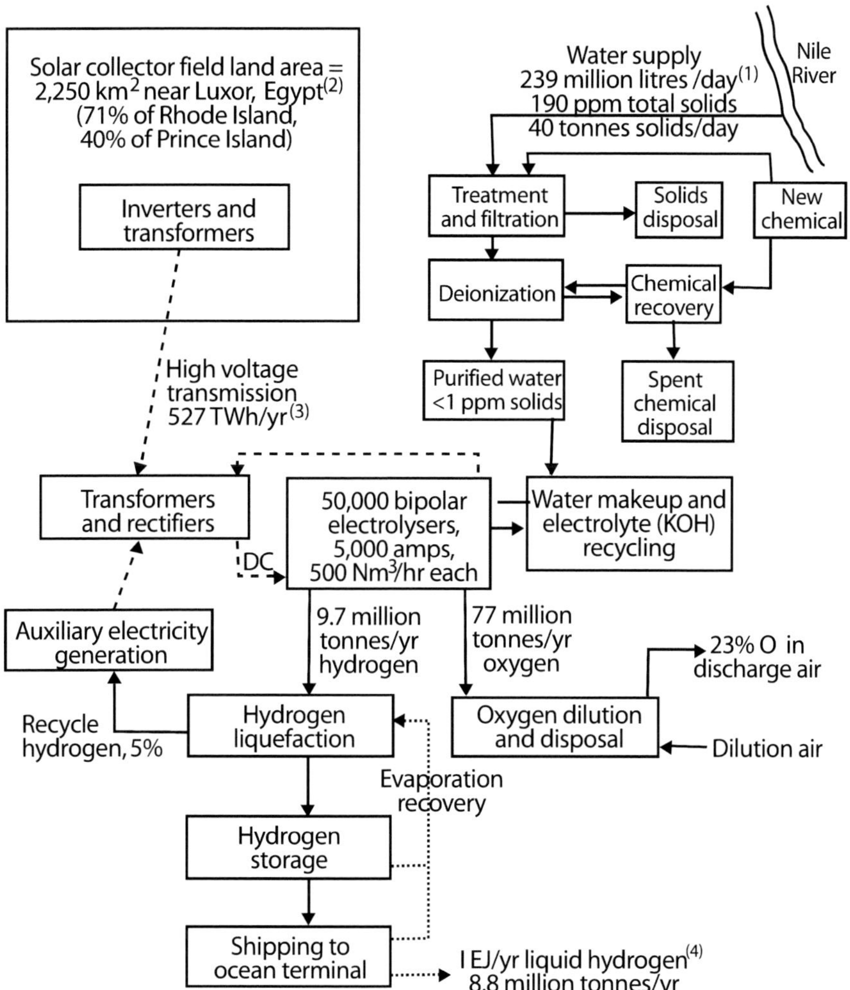

Notes:

1) Enough water for a city of 0.5 million people

2) Average 10.6 hours of sunshine daily

3) 3.6%of 2002 world hydroelectricity production

4) One week/yr of fuel for the worlds cars, trucks, airplanes, and ships, etc.,2% of world transportation energy in 2002

Figure 3. Simplified process flow sheet for solar hydrogen production at Luxor, Egypt.

5.7.3. Solar Updraft Towers

A solar tower is a chimney that can be hundreds of meters tall with turbines near the base of the chimney to generate electricity from the resulting updraft. Increasing the height of the chimney increases the air flow, provided the diameter is adjusted accordingly to compensate for pressure drop up the chimney. Part of the air flow is caused by the height of the chimney, which can be thought of as a pipe carrying air from the ground to a higher elevation where the air pressure is lower. The difference in air pressure between the bottom and top of the 195 meter high by 10 meter diameter chimney that operated for eight years at Manzanares in

Spain was 22 centimeters of water. To increase the pressure drop available to the turbines, a solar collector greenhouse of 46,000 m^2 heated the incoming air to lower its density – hot air rises. Maximum power output was about 50 MW [29].

Solar Updraft Towers work, but output is inherently small because the pressure difference across the turbines is small. There may be small niche applications for this technology.

5.7.4. Heating of Building Ventilation Air

In areas of the far north where fuel is expensive because it must be flown in, solar heating of ventilation air is a viable project. It is accomplished by exposing the metal outer skin of a vertical wall to sunlight. Enough heat can be collected from the top of the wall to reduce the quantity of fossil fuel required.

5.7.5. Battery Storage of Solar Electricity

PV cells are an excellent way of charging batteries for applications that require small amounts of electricity, such as calculators, communication equipment and some lighting. Applications such as these are particularly useful in areas remote from an electricity grid.

5.7.6. Net Metering

In some areas, home owners can install a PV array on the roof of their house and use the electricity in place of electricity from the grid. They can draw electricity from the grid when needed and sell electricity back to the grid when there is a surplus. This arrangement is termed "net metering". This system will work up to the point where about 20% of installed electricity is supplied by surplus electricity from PV cells. At this point, the amount of intermittent solar electricity is large enough to make the grid unstable.

6. Biomass

Biomass grows relatively slowly and requires much more land per unit of energy than any other renewable energy [18]. It has proved to be inadequate several times throughout history simply because it did not grow nearly fast enough. England began to use coal as fuel about 1700 [1] and the US in about 1850 because the forests were depleted and firewood was in short supply. Trees are the preferred form of biomass fuel because they can accumulate and store energy for several years until needed. For this reason, trees are a more concentrated source of biomass than crops, such as corn or sugar cane, which must be harvested every year.

People in undeveloped countries are forced to use biomass as wood, scrounged from wherever it is available, and animal dung for cooking and heating. Excessive harvesting of wood is common and devastates the local environment. Burning of dung hurts the environment both by eliminating it as a fertilizer for the soil, and by releasing black carbon to the atmosphere as it burns, thereby contributing to global warming [30].

It is impossible to return to firewood and other biomass fuels as a significant source of energy. Biomass was inadequate in the U.S. in 1850, and current energy consumption is more than 35 times higher than in 1850.

Currently, biomass consumption for fuel is about 7.5 EJ per year; mostly waste from lumbering, pulp and paper, and agricultural waste, such as sugar cane stalks. During the paper making process, wood is chemically treated to recover half the wood as cellulose for paper. The remainder, such as lignin, is burned to provide heat and electricity for the manufacturing plant. Wood also supplies building materials for most countries in the world.

It is often proposed that solid biomass be converted to liquid fuels for use as transportation fuel, such as converting corn to ethanol. However, almost all of the energy in the ethanol comes from fossil fuels, and the energy gain is less than 10% [31]. The approximately 60% surplus energy is low grade energy – it cannot be used to drive cars or to make more ethanol. The fossil fuel energy input for ethanol, from planting to distillation, comes 70% from coal, 15% from natural gas and 15% from oil. This leads to the rationale that ethanol displaces imported oil on a seven to one basis.

Solid biomass is inadequate to power the world, and liquid biomass is even more so because of the losses in conversion and the large quantity of fossil fuel required. A more efficient and environmentally friendly way of using corn as fuel is to burn it directly to displace coal. Carbon dioxide emissions are two and a half times less than fermenting corn to ethanol and displacing gasoline.

Ultimately, corn for ethanol production competes with food supply. Such competition will cause severe disruptions in the food supply for everyone, especially the poor of the world, who often already have trouble feeding themselves. Burning food fit for human consumption as fuel appears to be an unsound practice.

Thus, as in the past, commercial biomass will always be a small niche contributor to world energy supply.

The IPCC estimates the annual consumption of non-commercial biomass at about 45 EJ/yr [32]. Non-commercial biomass, often called traditional biomass, is material such as wood, animal dung, etc. and is used mainly in developing countries in the residential sector for heating and cooking.

The difference between what is possible and the scale on which it can be achieved is always important. Manufacture of ethanol from sugar cane, corn or switchgrass is possible, but not on a scale that can make a significant contribution to world transportation energy demand, or an impact on carbon emissions.

6.1. Summary RE Biomass

There is no research that can significantly increase the contribution of biomass fuels to satisfying world energy demand. Therefore, programs to do so must be discontinued.

Funds for biomass research are much better directed towards preserving diversity and developing varieties of plants and food crops better suited to specific growing conditions.

7. Wind Power

Some of the energies required to manufacture electricity, such as fossil fuels and hydro, can be stored and used to make electricity as needed. However, wind cannot be stored for future use.

Because large-scale electricity storage is not possible, production and consumption is balanced on a "knife-edge" of supply and demand. At any given moment, only enough electricity is produced to exactly equal the amount being consumed. Wind cannot deliver power when needed because the wind blows intermittently. This is a serious limitation, because it does not interface well with human energy use patterns, which require electricity to be available when needed.

Figure 3 is a record of wind variability in the control area of E.ON Netz, a large wind farm operator in Europe [33] [34]. The record of wind electricity input to the grid in 2004 shows that it varied continually from 0.2% to as high as 38% of daily peak grid load.

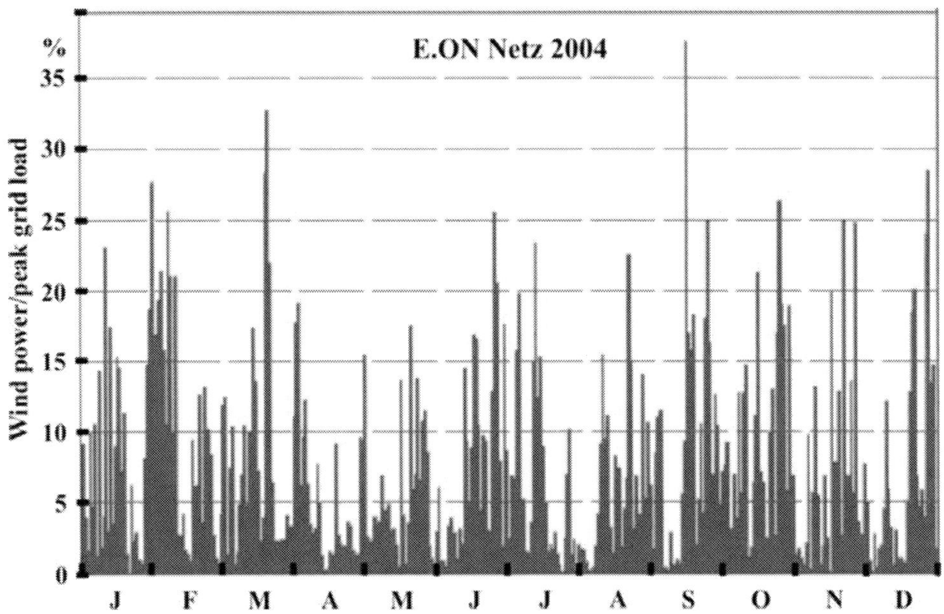

Figure 4. Wind is intermittent and erratic: it always needs a reliable helper.

The variability and erratic nature of wind electricity shown in Figure 3 explains why wind always requires a reliable helper. Sometimes wind essentially provides no electricity. The significance of this fact is that the helper must be large enough to supply the peak electricity load all by itself when the wind speed drops below 4 meters per second (m/s)[4], exceeds 25 m/s [34], or the temperature is too low [35], and the turbines stop turning. In other words, when a wind farm is built, a helper must also be built. The helper can be about one-third the installed capacity of wind because wind is only available one-third of the time, whereas the helper must be available 100% of the time.

Enmax Corp., the Alberta, Canada, power utility has more than 4% of its power coming from wind farms. It announced in early May 2007 that it was imposing a ban on additional wind farms until a new 1200 MW natural gas fuelled station was completed and operating to reduce the vulnerability of disrupting the system when the wind stopped blowing. The typical wind farm in Southern Alberta can harvest wind only 35% of the time [36].

[4] The power in the wind varies as the cube of the wind speed, i.e., v^3. The lower cut-off is at 4 m/s because there is little power in the wind. At 25 m/s, there is 244 times more power in the wind, and the turbine stops to protect it from damage by excessive stresses.

Wind electricity is limited to 20% of installed capacity, or less than 10% of annual electricity production, otherwise the erratic supply causes the grid to become unstable and collapse. The 20% limit of installed capacity is based on practical experience and was described in a paper by Grubb and Meyer in 1993 [37] [33].

Consequently, fuel saving is less than 10% and carbon emission reduction is less than 10%. Hydro as a helper has the same 10% limitation as fossil fuels. Continually changing wind electricity flow results in continually changing river flows [34]. On the positive side, wind can save up to 10% of water behind a dam.

Figure 3 shows that the supply of wind electricity can, and does, stop suddenly. In order to be ready for when the turbines stop, the helper, such as a fossil fuel plant, must operate in "spinning reserve mode" to be ready to pick up the load the instant the flow of electricity stops or changes sharply. The boilers are burning fuel, keeping the steam pressure maintained, and the generators are turning in synchronization with the electrical frequency on the grid. "Spinning reserve mode" is something like having your foot on the brake of your car waiting at a red light. Your car doesn't move, but the engine is still running. And the engine must run, because when the light turns green, you need the power immediately. Unfortunately, just like your car at a stoplight, spinning reserve mode consumes energy.

With a limit of 10%, how can Denmark supply 20% of its electricity by wind power? Because Denmark is a small part of the Nordel grid which consists of Denmark, Sweden, Norway and Finland. Denmark's wind electricity is 2% of the Nordel system output [38], and is well below the 10% limit.

It is often said that wind can supply the electricity for say, 100,000 houses. This is a misleading statement because wind cannot supply the electricity for even one house without its reliable helper.

In the past, a good application in rural areas was for pumping water with the pump physically connected to the wind mill shaft. A correctly sized water tank could be filled in times of good wind, and it would last through calm periods to the next good wind. There are good small niche applications for wind power in remote areas with very high diesel fuel costs, good wind resource and small enough for battery storage as the helper [39].

Another disadvantage of wind is that most of the good wind areas are remote. The best winds are usually far from where most people live and where electricity is needed. For example, in North America, that is in the far north, over the oceans and in the mountains. Wind power requires lots of land—20,000 sq. km/EJ[5] of electricity produced. Turbines must be spaced apart by ten times their rotor diameter, otherwise they interfere with the air flow to adjacent turbines.

A Canada Wind Map [40] and a US wind map [41] are available.

7.1. Wind Power Is Heavily Subsidized

Wind electricity is inherently difficult and would be rarely used were it not for generous subsidies. In the United States, "two-thirds of the value" of a wind energy project comes from federal tax breaks [42]. The situation in Canada is similar. For example, Hydro Quebec is studying proposals to buy wind electricity at 6.5 ¢/kWh [43] plus 1.3 ¢/kWh for integration

---

[5] 1 terawatt (TW) year = 1000 MW years = 31.5 EJ per year.

and connection plus 0.9 ¢/kWh for balancing for a total of 8.7 ¢/kWh delivered to a substation. The Hydro Quebec cost for hydro electricity is 2.8 ¢kWh, or about one-third that of wind electricity.

In Germany in 2004, E.ON Netz, a large wind farm operator in Europe, was paid 9 Euro cents/kWh [33], which is equivalent to 13 cents[6] Canadian/kWh.

7.2. Summary RE Wind Power

There is little potential gain in wind electricity generation efficiency through additional research funding. Wind turbines are highly developed. The best turbines are close to 44% efficient, which is 75% of the maximum of 59%, ie., the Betz limit. The turbine blades are so aerodynamic that insects smeared on the surface can reduce efficiency by up to one-half [44], somewhat like the fact that an airplane cannot fly with frost on its wings. The tip speeds of what might appear to be a slowly turning blade can approach 300km/hr – insects can smear over the surface of the blades just like on the bumpers and windshield of your car.

The key point of this analysis is that wind power is a small energy source now for very good technical reasons, and always will be.

Research funds can be safely diverted away from wind power.

8. Hydroelectric Power

The potential energy of rain and snow falling at high elevations is concentrated by rivers and stored behind dams for use as required to generate electricity.

Hydro power is almost the perfect form of energy. The potential energy stored as water behind a dam is forced by gravity through turbines to generate electricity. The water flow can be changed almost instantaneously to meet electricity demand. Dams are also useful for providing irrigation water to nearby farmland and for controlling floods.

Today, world hydropower displaces about 30 EJ[1] of fossil fuels. Hydropower plants supply about 20% of the world's electricity for more than one billion people [45]. If the full potential of world hydro could be exploited, it would displace about another 30 EJ for a total of about 60 EJ of fossil fuels.

About one half of the suitable world hydro sites have been developed. Hydropower potential is already 65% developed in Western Europe and 76% in the United States, where there are over two thousand hydro power plants.

8.1. Summary RE Hydro Power

Hydro-electric generating efficiency has been 85% efficient for more than half a century. Thus, there is little opportunity for research funds to improve efficiency.

[6] In 2006, the average was $1 Can = $0.882 US.

9. Renewable Energy Footprints

Renewable energies are usually considered to be desirable sources of energy because they are natural and inexhaustible. Although there are positive and valuable aspects to renewable energies, each one has a land footprint and a footprint on the environment.

Table 1 gives the land footprint area required to produce 1 Megawatt (MW) of electricity from various primary energy sources [46].

Table 1. Estimated footprint in km^2 of the energy source to produce 1 MW of electricity per year

	Energy Source	Upper end of range, km^2/MW	Lower end of range, km^2/MW
1	Biomass	5.2	-
2	Wind	0.79	-
3	Ocean thermal	0.28	-
4	Photovoltaic	0.12	-
5	Solar thermal	0.08	-
6	Geothermal	0.003	-
7	Hydro	0.37	0.07
8	Coal	0.04	0.001
9	Nuclear	0.01	0.001

Biomass fuel, such as wood and liquid fuels from corn, switchgrass, etc. require large quantities of land for the amount of energy produced compared to other energy sources.

Although wind power is more land efficient, it exerts a strong ecological footprint. A range of groups oppose new installations of wind turbines, and press for removal of others. The winds that attract wind power installations are often migratory routes for birds, which are sometimes killed by turbine blades. Vibrations transmitted through the ground and noise as blades pass the tower annoy people living nearby. Objections to wind turbines in scenic locations are becoming more common.

Ocean tidal energy is remote, intermittent and small and will remain small.

Large scale solar energy PV and solar-thermal energy are limited as discussed earlier.

Volcanic geothermal energy requires very little land per unit of output, however it is geographically limited to specific hot spots in the earth's crust, such as Iceland. Corrosion, scale build-up and pollution are often problems.

The range of land area required by hydro to produce electricity varies greatly. This is because of the vast differences in flood plains between hydroelectric projects. In mountainous countries, such as Switzerland, Norway and Canada, flood plains are small and contained by mountain valleys. In relatively low-lying countries, such as Thailand and Brazil flood plains are often much larger.

Hydro power exerts a physical and ecological footprint. Objections to building more hydro dams are raised because flooding of land forces people to move from their homes and archeological sites are destroyed. Downstream, the nature of the river changes and fish and other species become endangered or extinct as traditional spawning grounds are cut-off [47].

Older dams are often removed rather than maintained [48], thereby restoring the health of the river and opening it again to fish and other water creatures.

The footprint of coal is shown for comparison.

Nuclear energy has the smallest footprint except for volcanic geothermal.

10. Solar Hydrogen, Footprints and Capacity Factors

The solar hydrogen production facilities in Figure 1 are based on the quantity of solar energy available in Luxor, Egypt as the source of electricity.

Luxor, Egypt was selected for this example because it is in the area with the highest solar insolation on earth, and it is near a large source of water. If the facilities were located anywhere else, solar insolation would be less and land area and equipment would be larger. Table 2 shows the relative increase in the size of these components if the hydrogen production facilities were located in Spain or in the southwestern portion of the United States.

Table 2. Relative size of hydrogen production facility components compared with solar at Luxor, Egypt

Location or electricity supply	Landarea	Electricitys ystem size	Number of electrolyser cells	Liquefaction system size
Luxor (290 W/m^2)	1.0	1.0	1.0	1.0
US southwest (250 W/m^2)	1.2	1.2	1.2	1.2
Spain (167 W/m^2)	1.7	1.7	1.7	1.7
Wind	7[7]	1.4[8]	1.4	1.4
Nuclear	0.01-0.1[7]	0.5[9]	0.5	0.5

For additional comparison, wind and nuclear electricity have been added to Table 2. Nuclear, the only continuous source of electricity, reduces land area by a factor of 10 to 100 and equipment by a factor of two.

Therefore, it is reasonable to conclude that for hydrogen manufactured by electrolysis of water to become a significant part of world energy supply, the electricity supply is much more likely to be from nuclear than from solar or wind.

11. Conclusion

To ensure future world energy supplies and to reduce carbon emissions there is a developing urgency to find replacements for fossil fuels. Renewable energies have serious limitations that

[7] From Table 1.
[8] Wind capacity factor of 0.35.
[9] Nuclear capacity factor of 0.95.

prevent a significant increase in their contribution to replace fossil fuels. There is little that research funds can do to move any of the renewable energies beyond this position.

Today, renewable energies contribute about 9% of world energy demand. This share has continually dropped since coal became the fuel of choice about 1700, and is likely to continue to diminish as world energy consumption increases. Nevertheless, we must use renewable energies where it makes sense, and we can do so without the expenditure of government research funds.

Research funds must be withdrawn from renewable energies and directed towards areas with potential to replace significant quantities of fossil fuels.

References

[1] Roberts P, The end of oil, On the Edge of a Perilous New World, Houghton Mifflin, New York, 2004.

[2] Lightfoot H D, Green C, The dominance of fossil fuels: technical and resource limitations to alternative energy sources, Centre for Climate and Global Change Research Report 92-6, May 1992. Available from: URL: *http://www.mcgill. ca/gec3/*

[3] Hayden, H C, The Solar Fraud, Why Solar Energy Won't Run the World, Second Edition, 2004, Vales Lake Publishing, LLC, page 116.

[4] International Energy Outlook (IEO) 2003, Energy Information Administration (EIA), Report DOE/EIA-0484(2003), March 2003, page 183, Table A2. Total World Other under 2005 is 37.6 Quads = 39.7 EJ/yr of total renewable energies. Available from: URL: *www.eia.doe.gov/oiaf/ieo/index.html.*

[5] Lightfoot H D, Understand the three different scales for measuring primary energy and avoid errors, Energy 32 (2007) 1478-1483, doi:10.1016/j.energy.2006.10.009

[6] Lightfoot H D, A Strategy for Adequate Future World Energy Supply and Carbon Emission Control, EIC Climate Change Technology, 2006 IEEE May 2006 Page(s):1 – 10, DOI: 10.1109/EICCCC.2006.277266. Available from: URL: *http://ieeexplore.ieee.org/Xplore/guesthome.jsp*

[7] Moreira, J. R., Technological and Biological Mitigation Potentials and Opportunities, CLA Chapter 3, WGIII, Slide 12, July 18, 2001. Available from: URL: *http://www.ipcc.ch/present/presentations.htm*

[8] Lightfoot H D, Green C, The dominance of fossil fuels: technical and resource limitations to alternative energy sources, Centre for Climate and Global Change Research Report 92-6, May 1992, pp. 47. Available from: URL: *http://www.mcgill.ca/gec3/*

[9] Kreith F, Kreider J F, Principles of Solar Engineering, McGraw-Hill, 1978, page 19, Figure 1.8.

[10] Flagsol SEGS Technology, Aerial view of the five 30MW parabolic trough plants at Kramer Junction in the Mojave desert of California built between 1985 and 1991. Available from: URL: *http://www.flagsol.com/SEGS_tech.htm*

[11] TREC-UK Pictures of Clean Solar Energy Plants. Available from: URL: *http://www.trec-uk.org.uk/pictures.htm*

[12] Shinnar R, Citro F, A Road Map to U.S. Decarbonization, Science, 1 September 2006, Vol. 313. no. 5791, pp. 1243-1244

[13] Metz, B., Davidson, O., Swart, R., Pan, J., editors, Climate Change 2001: Mitigation, Contribution of Working Group III to the Third Assessment Report of the Intergovernmental Panel on Climate Change, Cambridge University Press, Cambridge UK, 2001, pages 247-249. Available from: URL: *http://www.grida.no/climate/ipcc_tar/wg3/131.htm*

[14] Eliasson, Baldur, Renewable Energy, Status and Prospects, 1998, Energy and Global Change, ABB Corporate Research Ltd, CH-5405 Baden, Switzerland

[15] Lightfoot H D, Green C, An assessment of IPCC Working Group III findings in Climate Change 2001: Mitigation of the potential contribution of renewable energies to atmospheric carbon dioxide stabilization, Centre for Climate and Global Change Research Report No. 2002-5, November 2002. Pages 5-9, 22-23, 44-54. Available from: URL: *http://www.mcgill.ca/gec3/*

[16] Eliasson, Baldur, Renewable Energy, Status and Prospects, 1998, Energy and Global Change, ABB Corporate Research Ltd, CH-5405 Baden, Switzerland, Table 3.3, page 42.

[17] Philibert C, The present and future use of solar thermal energy as a primary source of energy, International Energy Agency, Paris, France, 2005, page 7. Available from: URL: *http://www.iea.org/textbase/papers/2005/solarthermal.pdf*

[18] Eliasson B, Renewable Energy, Status and Prospects, 1998, Energy and Global Change, ABB Corporate Research Ltd, CH-5405 Baden, Switzerland, Table 3.8, page 54.

[19] Hayden H C, The Solar Fraud – Why Solar Energy Won't Run The World, Vales Lake Publishing, LLC, Second Edition, 2004, pp 189

[20] Solar Electric Power Association: PV Basics: Photovoltaics Q&A. Available from: URL: *http://www.solarelectricpower.org/index.php?page=basics&subpage=pv&display=qa#06*

[21] Material Safety Data Sheet, oxygen, Air Liquide Australia, July 2004. Available at: URL: *http://www.msdsonline.com.au/airliquide/msds/msdsview.asp?SynonymCode=MOU500&msds_format=00&in_langcode=&uselogo=TRUE*

[22] Hayden H, The Solar Fraud, Why Solar Energy Won't Run The World, Vales Lake Publishing, P.O. Box 7595, Pueblo West, Colorado, 81007-0595, 2004, Table 5,page 98.

[23] Bellona Foundation, Oslo, Norway, Hydrogen, Report 6-2002, pp 20: Some of Norsk Hydro's elecrolysers have an efficiency of over 80% (high heating value).Available from: URL: *http://www.bellona.no/data/dump/0/04/37/3.html*

[24] Kirk-Othmer, Encyclopedia of Chemical Technology, Third Edition, Vol. 12, Hydrogen, page 971

[25] Central Laboratory for Aquaculture Research (CLAR), Abbassa, Abou Hammad, Sharkia, Egypt. The water analysis is for water that flows from the Nile River in Cairo through the Ismalia canal to the El Wadi El Quadim supply canal to the CLAR sedimentation pond. Available from: URL: *http://pdacrsp.orst.edu/ pubs/egypt.pdf*

[26] Potential Cost Reduction in PV systems, Contractor Cambridge Consultants Limited, prepared by Arthur D. Little, 2001, DTI/Pub URN 01/759, pp 27-28.

[27] Shukman D, Science correspondent, BBC News, Seville, Power station harnesses Sun's rays, 2 May 2007. Available from: URL: *http://news.bbc.co.uk/2/hi/science/nature/6616651.stm*

[28] Department of Communications Engineering at the FHTE (Fachhochschule für Technik Esslingen Germany). Available from: URL: *http://rhlx01.rz.fht-esslingen.de/projects/ alt_energy/sol_thermal/powertower.html#work*

[29] Solar Updraft Tower, Manzanares, Ciudad Real, Spain. Available from: URL: *http://en.wikipedia.org/wiki/Solar_updraft_tower*

[30] Jacobson, M. Z., Strong radiative heating due to the mixing state of black carbon in atmospheric aerosols, *Nature,* 409, 695-697, 2001. Available from: URL: *http://www.nature.com/nature/*

[31] Shapouri, H., Duffield, J., Wang, M., The 2001 Net Energy Balance of Corn-Ethanol (Preliminary). A report published by The U.S. Department of Agriculture and U.S. Department of Energy, 2004. Available from: URL: *http://www.ethanol-gec.org/otherstud.htm.*

[32] Metz, B., Davidson, O., Swart, R., Pan, J., editors, Climate Change 2001: Mitigation, Contribution of Working Group III to the Third Assessment Report of the Intergovernmental Panel on Climate Change, Cambridge University Press, Cambridge UK, 2001, Table 3.31. Available from: URL: *http://www.grida.no/climate/ipcc_tar/ wg3/131.htm*

[33] E.ON Netz, Wind Power 2005, Report: ENE_Windreport2005_e_eng.pdf. Available from: URL: *http://www.eon-netz.com/EONNETZ_eng.jsp*

[34] Bélanger C, Gagnon L, Adding wind energy to hydropower, Energy Policy, Vol. 30 No. 14, November 2002, pp. 279-284.

[35] Clarification of Wind Turbine Cold Weather Considerations: Manitoba Hydro Summary, Exhibit MH-NCN-1031 Section a, 4/7/2004. Available from: URL: *http://www.hydro.mb.ca/regulatory_affairs/wuskwatim/presentations/exhibits_1031.pdf*

[36] Cattaneo C, National Post (Canada), April 20th, 2007

[37] Grubb, M.J., Meyer, N.I., Wind Energy Resources, Systems and Regional Strategies. In Renewable Energy Sources for Fuels and Electricity, Johansson T., Kelly H., Reddy A., Williams R., (eds.), Island press, Washington, D.C. 1993

[38] Nordel Annual Report 2003, Available from: URL: *http://www.nordel.org/ Content/Default.asp?PageID=127*

[39] Kuntz, M., Dawe, J., Renewable. Rechargeable. Remarkable., Mechanical Engineering, October 2005, pp. 35-39.

[40] Canadian Wind Energy Atlas (CWEA), 2005. Available from: URL: *http://www.windatlas.ca/en/maps.php?field=EU&height=50*

[41] Map 2-1 United States annual average wind power. Available from: URL: *http://rredc.nrel.gov/wind/pubs/atlas/maps/chap2/2-01m.html*

[42] Taylor J, Wind Farms Costly for Kansans, New Study Finds, *Environment News,* The Heartland Institute, May 1, 2005. "On December 15, 2004, an official from the firm of Milbank, Tweed, Hadley & McCloy, LLP, pointed out to the American Bar Association's Renewable Energy Committee that 2/3 of the value of a wind energy project comes from two federal tax breaks."

[43] Hydro Quebec (HQ) Press Release, Gaspé, October 4, 2004, Call for tenders for electricity from wind power Hydro-Quebec selects eight bids for a total of 990 MW. Available from: URL: *http://www.hydroquebec.com/4d_includes/headlines/ PcAn2004-161.htm*

[44] Corten, G P, Veldkamp H F, Insects can halve wind-turbine power, Nature, Vol. 412, 5 July 2001, pages 41-42.
[45] How Hydropower plants work. Available from: URL: *www.howstuffworks.com*
[46] Eliasson B, Renewable Energy, Status and Prospects, 1998, Energy and Global Change, ABB Corporate Research Ltd, CH-5405 Baden, Switzerland, Figure 7.1, page 95.
[47] Grand Canyon Institute (GCI), 2005. Available from: URL: *http://www.glencanyon.org/ library/grandcanyon.php*
[48] American Whitewater (AW), 2005. Available from: URL: *http://www.american whitewater.org/archive/article/650/*

In: Climate Change Research Progress
Editor: Lawrence N. Peretz, pp. 313-329

ISBN: 978-1-60021-998-6

Chapter 10

THE ABILITY OF CLIMATE MODELS TO GENERATE TROPICAL CYCLONES: IMPLICATIONS FOR PREDICTION

Kevin Walsh
School of Earth Sciences, University of Melbourne

Abstract

Climate models are the main tools used for predicting the effect of global warming. Recently, these models have developed an increasingly sophisticated ability to simulate tropical cyclones, a very demanding task due to the relatively small size and the intense convection of these storms. This paper gives an evaluation of the state of the art of tropical cyclone climate modeling, including an examination of our current ability to simulate the correct numbers, structure and intensities of cyclones. The shortcomings of present models are outlined as well as specific recommendations for ways in which they can be improved. Additionally, a summary of current model-based predictions of the effects of global warming on tropical cyclones is given, including an assessment of how the uncertainty of these predictions might be reduced through improved simulation.

Introduction

Climate models of the general circulation (GCMs) are the main tools used to simulate and predict the global climate. They consist of numerical representations of the processes that govern the atmosphere and ocean, and as such are similar to the numerical weather prediction models that now serve as the main source of guidance for weather forecasting over periods of several days. Recent texts detailing their components are those of McGuffie and Henderson-Sellers [2005] and Washington and Parkinson [2005]. An assessment of their overall current performance can be found in Randall et al. [2007].

As little as fifteen years ago, the ability of climate models to generate tropical cyclones was exceedingly limited. It had been recognized for some time that climate models were able to generate tropical lows; since tropical cyclones are a feature of the observed climate system, it was also expected that they would be produced in realistic numbers in a good climate model. The simulation of tropical cyclones, however, poses several problems even for a state-of-the-art climate model. The horizontal resolution of typical coupled ocean-atmosphere global climate models, even after years of improvements in computing capabilities, is only of the order of 100-300 km [Randall et al. 2007]. This is exceedingly coarse compared to the scale of the physical processes occurring in tropical cyclones: typical eye diameters of intense storms are 10-100 km [Anthes 1982], and a good representation of processes in the eye and the surrounding eye wall, the region of most intense convection in the storm, is crucial for realistic simulation.

This is important for two main reasons. One of the main arguments employed to justify the use of climate models for simulations of climate change is that it is more likely that they will simulate climate change accurately if they are able to simulate the current climate well. In addition, experiments have shown that improvements in horizontal resolution alone can change the response of a climate model to increased greenhouse gases, including changes in the magnitude of the response of tropical cyclones [Bengtsson et al. 2007b]. Thus in order to obtain a more believable prediction of the effects of climate change on tropical cyclones, it is important to obtain the most realistic simulation possible of their characteristics in the current climate.

Improvements in resolution can be obtained by the use of regional climate models [Dickinson et al. 1989], in which only a small region of the global is simulated, thus increasing the available resolution for the same computing power. Regional climate model simulations have a number of advantages over global climate models for the simulation of tropical cyclones. Firstly, tropical cyclones only occur over certain areas of the globe, so there is no need to simulate all portions of the planet as a global model does. As a result, selected regions can be simulated at fine resolution. Another advantage is that regional climate models can be forced at their boundaries by reanalyses, thus giving an estimate of the best possible performance of the limited area climate model if it were forced with a representation of observations. A disadvantage of regional models for climate change simulations is that they have to be run nested within a global model, so a global climate change simulation already has to be performed before a regional model simulation can proceed. Another disadvantage is that most regional models now employ a large-scale forcing that keeps them from drifting too far away from the large-scale solution of the global model over the selected area [e.g. McGregor 2005; Knutson et al. 2007]. When implemented in the tropics, regional climate models needed to be carefully designed and tested to ensure that the large-scale climate remains realistic while at the same time ensuring that small scale features are generated in sufficient numbers to produce a realistic rate of tropical cyclone formation.

Improvements in resolution can also be obtained by using exceptionally fast computers and specifying the sea surface temperatures (SSTs) rather than calculating them with an ocean general circulation model (GCM) [e.g. Oouchi et al. 2006]. Horizontal resolutions of such global modelling studies are in the range of 10-50 km, which is still coarse compared to the structure of a typical tropical cyclone. Thus even the best climate model simulations presently have inadequate resolution, an issue that affects their simulations of cyclones across a whole range of phenomena.

In addition to the purely dynamical problem of having insufficient grid points to capture the observed sharp wind gradients in tropical cyclones, the lack of resolution causes numerous compromises to be made in the model representation of cloud and precipitation processes. Individual clouds are too small to be resolved by climate models, and so approximations of their effects need to be made in the form of parameterizations [e.g. McGuffie and Henderson-Sellers 2005]. Since convection processes are exceptionally important in determining the structure and intensity of tropical cyclones, the inability of climate models to resolve clouds limits their ability to represent these processes.

This chapter reviews the current state-of-the-art of the simulation of tropical cyclones by climate models. A review is given of early developments, followed by an assessment of the current capability of simulating the observed tropical cyclone formation rate and tracks, intensities and storm structure. A concluding section gives some recommendations for future work.

Early Work on Tropical Cyclone Simulation in Global Models

The study of Manabe et al. [1970] was the first to demonstrate the generation of tropical cyclone-like vortices in a global atmospheric model, using an approximate horizontal resolution of 200 km. Although the formation regions of the model-generated storms were unrealistic, the model did simulate the upper-level warm core maximum that is a feature of such storms. Subsequent work by Bengtsson et al. [1982] showed that operational ECMWF analyses were generating vortices with some of the characteristics of tropical cyclones, although the authors of this study did not attempt to determine the model climatology of these systems. The first model climatology of tropical cyclones was produced by Manabe and Broccoli [1990], using low-resolution (R15, about 500 km) and higher resolution (R30, about 300 km) simulations. A reasonable global climatology of tropical storm formation was produced, although there were substantial regional differences from observed numbers. The study also performed an experiment under enhanced greenhouse conditions, although the results from experiments using different representations of clouds were different. Similar studies by Wu and Lau [1992] and Haarsma et al [1993] further demonstrated the potential of this approach. The higher resolution work of Bengtsson et al. [1995] emploted a GCM with an approximate grid spacing of 125 km to generate an improved climatology of TCs. A companion study [Bengtsson et al. 1996] predicting a substantial decrease in TC numbers in a warmer world.

Note that all of these and subsequent studies employ varying means of detecting tropical lows in their simulations. Usually, a threshold value of storm maximum wind speed or vorticity is specified, along with other criteria designed to detected tropical cyclone-like lows and exclude other kinds of systems such as mid-latitude lows. The issue of what criteria to use is a significant one, in that the numbers of detected systems, and thus the model tropical cyclone climatology, depends upon the exact magnitude of these detection criteria. More will be said on this subject later in this chapter.

Development of the Climate Modelling of Tropical Cyclones

The obvious potential of high-resolution climate modelling to make projections regarding future climates led to a series of further studies on tropical cyclone simulation. These studies have been of several different types: studies using direct simulation of tropical cyclones using global models [e.g. Tsutsui and Kasahara 1996; Krishnamurti et al. 1998; Yoshimura et al. 1999; Sugi et al. 2002; Tsutsui 2002; Camargo et al 2005; McDonald et al 2005; Yoshimura et al. 2006; Bengtsson et al 2007a]; studies using a similar methodology but employing regional or variable-resolution climate models [Walsh and Watterson 1997; Walsh and Katzfey 2000; Walsh et al. 2004; Knutson et al. 2007; Leslie et al. 2007]; idealized studies whereby vortices are introduced into a modelling system, rather than being generated by it [Knutson et al 1998; Knutson and Tuleya 1999; Walsh and Ryan 2000; Knutson et al 2001; Knutson et al. 2004]; and studies that employ diagnostic parameters to deduce the effect of global warming on tropical cyclone characteristics from large-scale fields generated by models [Emanuel 1987; Royer et al. 1998; Chauvin et al. 2006; Camargo et al. 2007]. In addition, a statistical-dynamical model [Emanuel 2006] has been developed that uses synthetic storm tracks combined with a simple dynamical tropical cyclone model to generate a tropical cyclone climatology.

This work focuses mainly on the first two of these types of studies, as the purpose of this review is to assess the native ability of current models to generate and simulate tropical cyclones. This constitutes a severe test of the quality of a climate model, given that tropical cyclone simulations need fine resolution and a realistic representation of convective processes. An evaluation of current modelling ability involves an assessment of tropical cyclone numbers, tracks, intensity and structure.

Evaluation of Current Climate Modelling Capability

Numbers and tracks

It is difficult to make a general statement at this time regarding the performance of climate models in producing a realistic climatology of tropical cyclone formation.

Superficially, many past and current models have appeared to produce relatively good climatologies of tropical cyclone numbers. Recent model simulations have generated tropical lows in similar locations to the regions in which observed tropical cyclones form. Figure 1 shows results from Oouchi et al. [2006] that indicate the ability of a state-of-the-art simulation performed at a horizontal resolution of 20 km. Tropical cyclone formation locations occur in approximately the same regions as observed, with some exceptions: for instance, simulated formation is too far south in the South Pacific, too far north in the northwest Pacific, and is predicted to occur reasonably frequently in the South Atlantic, where present-day tropical cyclone formation is rare [Pezza and Simmonds 2005]. There is little simulated formation near the equator, as observed, and there is less formation in the central Pacific than in the east and west, also as observed. Other global model simulations have shown similar abilities to generate tropical cyclones in roughly the observed locations [e.g. McDonald et al 2005; Camargo et al. 2005].

But as stated earlier, the number of storms produced in these models depends very much on the magnitude of the detection criteria employed. Each study that has been performed evaluates their output by employing criteria that are designed to detect tropical cyclones in the model output, combined with a tracking scheme to determine the path of the detected lows. Almost every climate model study that has been performed on tropical cyclones has used different detection schemes [see also Walsh et al 2007], not all of which have been tested against observations. Typically, threshold values of low-level wind speed or vorticity are used, but often the selected values are almost arbitrary, or chosen to give a good agreement between the simulated and observed climatological formation of tropical cyclones. In contrast, the threshold value for observed tropical storms is quite specific: a 10-metre wind speed of 17.5 ms^{-1}, although both 1-minute and 10-minute wind averages are used, depending upon the formation region.

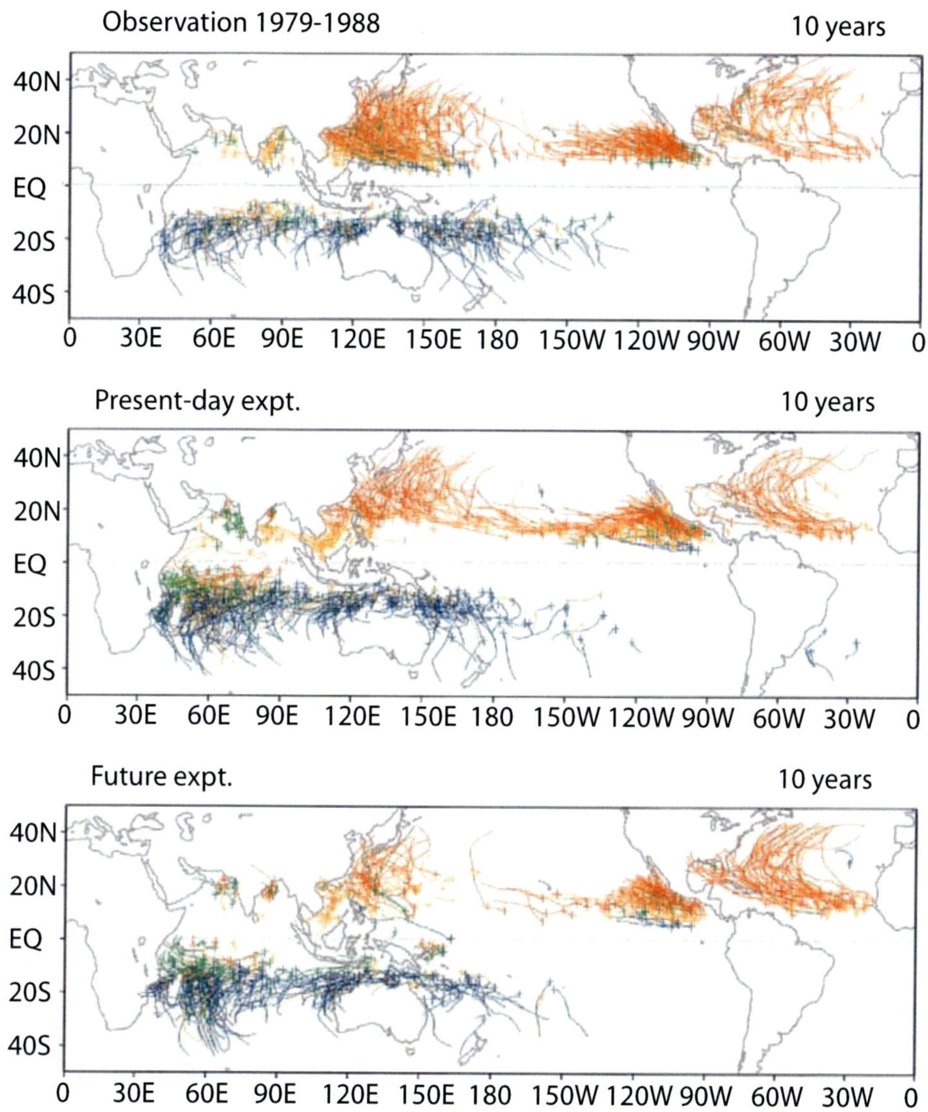

Figure 1. Comparison between observed tropical cyclone tracks (top); current climate simulation (middle) and enhanced greenhouse climate simulation (bottom) from the results of Oouchi et al. [2006]. Tropical cyclone formation locations are denoted by crosses.

For effective evaluation and comparison of simulations, Walsh et al. [2007] suggest a resolution-dependent 10-m wind speed detection threshold, calculated from observations averaged to that resolution (Figure 2). Thus this criterion represents reality as it would appear if it were averaged to the resolution of a climate model. In this way, the real ability of a climate model to generate tropical cyclones, compared with the real world, can be assessed.

The lack of a universal threshold detection criterion limits the assessment of model performance, as indicated in Table 1. Each of the studies shown uses well-validated climate models and performs comprehensive analysis of their results. But in general the detection criteria used are too vague to make conclusions about the models' ability to generate TCs. For example, in the impressive work of Oouchi et al. [2006], a global model is run at the exceptionally fine resolution of 20 km. Oouchi et al. [2006] use a threshold of 15 ms^{-1} at 850 hPa, whereas Walsh et al. [2007] suggest that for this resolution a threshold of 22 ms^{-1} at 850 hPa would be more appropriate. Thus this simulation may be generating too few tropical cyclones. The vorticity-based detection criterion used in Bengtsson et al. [2007a] was validated against reanalyses and the threshold tuned to obtain the best detection rate for tropical cyclones. However, the false alarm rate, the storms detected at this threshold that were not tropical cyclones, was not assessed in that study, and this is an issue in even relatively fine-resolution reanalyses [Walsh 1997]. The result is that this detection threshold may not be the best choice, which in this case would be a value that optimizes the combination of the false alarm rate and the detection rate. Thus again it is difficult to compare these state-of-the-art results to observed tropical cyclone formation rates.

Table 1. A selection of recent climate model simulations of tropical cyclone formation.

Study	Approximate resolution (km)	Threshold used	Apparent numbers generated	Recommended threshold	Implied real climatology
Camargo et al. (2005)	250	Basin-dependent	Too few	Wind speed threshold	?
McDonald et al. (2005)	100	Vorticity based	Too many	Wind speed threshold	?
Yoshimura et al. (2006)	125	15 ms^{-1} at 850 hPa	Too many	18 ms^{-1} at 850 hPa	Maybe good?
Chauvin et al. (2006)	55	15 ms^{-1} at 850 hPa	Too many	20 ms^{-1} at 850 hPa	Maybe good?
Oouchi et al. (2006)	20	15 ms^{-1} at 850 hPa	Good	22 ms^{-1} at 850 hPa	Too few
Leslie et al. (2007a)	55	15 ms^{-1} at 850 hPa	Good	20 ms^{-1} at 850	Too few
Bengtsson et al. (2007)	80	Vorticity based	Too many	Wind speed threshold	?

Regional climate model studies have also employed varying detection criteria. A recent example of this approach is that of Knutson et al [2007], who used an 18-km resolution model nested within reanalyses to simulate the observed increase in tropical cyclone formation that has occurred over the Atlantic basin in the past decade or so. They achieved very good results in simulating the observed year-to-year variability in tropical cyclone numbers, suggesting that this model would be a useful tool in explaining the causes of the observed increase. Nevertheless, they used a vorticity-based threshold, making comparison of their formation rates to reality more difficult.

While the direct detection of simulated tropical cyclone formation is the main technique for assessing climate model performance, indirect techniques also have some potential. The construction of tropical cyclone genesis parameters, derived from large-scale variables known to be conducive to tropical cyclone formation, has been attempted for some years. Gray [1975] originally defined the concept, although his version of the genesis parameter was shown to be likely too sensitive to sea surface temperature changes to be useful in climate change simulations [Ryan et al. 1992]. A modified version of Gray's parameter [Royer et al. 1998] has been shown to have some potential for climate change simulations, in that it is able to give a reasonable estimate of the directly detected cyclone numbers in climate model simulations under current and enhanced greenhouse conditions [McDonald et al. 2005; Chauvin et al., 2006]. Again, the validity of this comparison depends somewhat on whether there is complete confidence in the cyclone detection rate in the simulation. Nevertheless, this technique has considerable further potential in that it could enable a quick assessment of the cyclone-generating potential of a particular climate simulation, as well as determining those factors that could be improved in the model. Note, however, that a model with high horizontal diffusivity may give a good simulation of the mean climate but a poor simulation of transient, small-scale features in the tropics [McGregor, personal communication, 2005], leading to a mismatch between the prediction of a genesis parameter and the actual model formation rate of tropical cyclones.

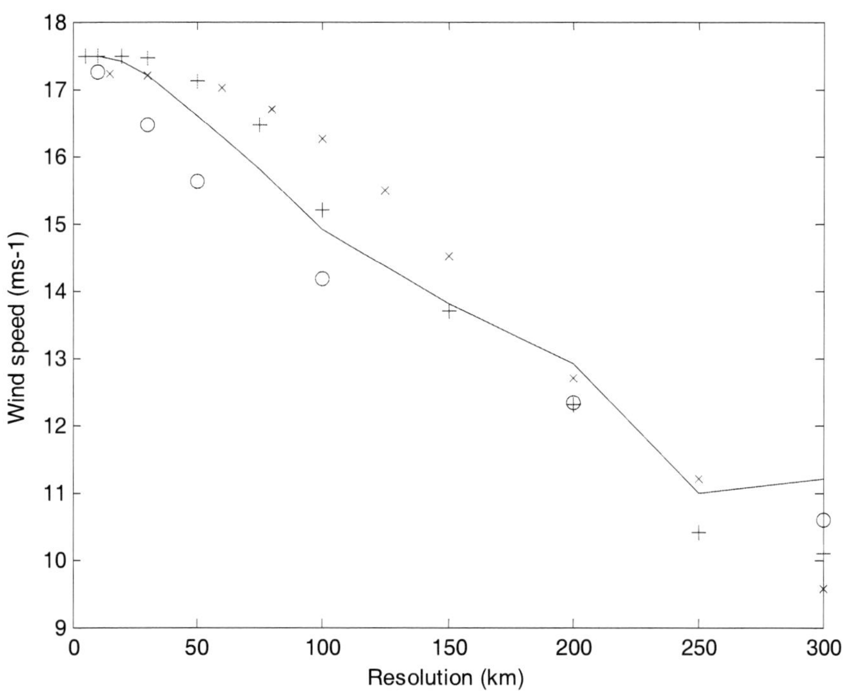

Figure 2. Variation of threshold detection with resolution, as derived from analytical curve fitting using the method of Holland [1980], for (solid line) grid box averages and (plus signs) grid point calculations. Also indicated (circles) are values derived from a selection of NOAA Hurricane Research Division wind analyses and values (x symbols) derived using the model of Hubbert et al. [1991]. From Walsh et al. [2007].

In addition to comparisons of formation numbers, comparisons of occurrence rates at different latitudes are needed, as tropical cyclones can often travel far poleward of their regions of origin and yet retain tropical characteristics. Many studies lack a systematic assessment of occurrence statistics, tending to concentrate on formation rate and intensity distribution. Most model studies include a cyclone track plot, giving an idea of poleward extent and track density in poleward regions. Camargo et al. [2005] compare observed and simulated track density for three different climate models, finding that there were both similarities and differences between observed and simulated tracks. One systematic result was that simulated tropical cyclone lifetimes tended to be considerably greater than observed, although as they point out this could be partly a result of their choice of detection criteria. In contrast, McDonald et al. [2005] found that simulated storms in their model tended to have shorter lifetimes than observed; again they speculate that this could be a result of their detection scheme. Chauvin et al. [2006] show some agreement between their model simulation and observations of days of tropical storm occurrence in the non-tropical Atlantic. For Oouchi et al. [2006], Fig. 1 shows what appears to be reasonably agreement between observed and simulated track density over the non-tropical North Atlantic and north Pacific, although there is a tendency for simulated track density to be too high in the latitudes near Japan and in the south Indian Ocean. Leslie et al. [2007a] give a good simulation of observed tropical cyclone lifetimes, while Bengtsson et al [2007a] show considerable track density in the eastern north Atlantic, which is not observed.

In summary, the evaluation of tropical cyclone formation rates in models requires the use of detection criteria that are directly comparable with observations. Of course, a very severe test of any climate model would be simply to try to reproduce the observed climatological formation rate using only the observed threshold intensity definition. The difficulty with this approach is that simulated tropical cyclone intensity increases as resolution increases, as Bengtsson et al. [2007b] and many others have shown. Thus as resolution increases, one would expect more storms to exceed the detection threshold simply as a result of increased resolution, which makes then the actual ability of lower-resolution models difficult to assess without a criterion that is adjusted for resolution.

Intensity

Some guidance on the resolution required to simulate reliably the intensity of tropical cyclones can be obtained from high-resolution numerical weather prediction models. These indicate that direct simulations would require horizontal resolutions finer than 5 km, with some work suggesting that 1 km would be preferable [Chen et al. 2007]. Without significant advances in techniques, it will be some time before climate models are run at this resolution. The finest-resolution global climate simulation that has been analysed for tropical cyclone formation and published is that of Oouchi et al. [2006], at 20 km resolution. Thus it is not surprising that climate model simulations universally simulate tropical cyclones that are weaker than observed, particularly for very intense storms.

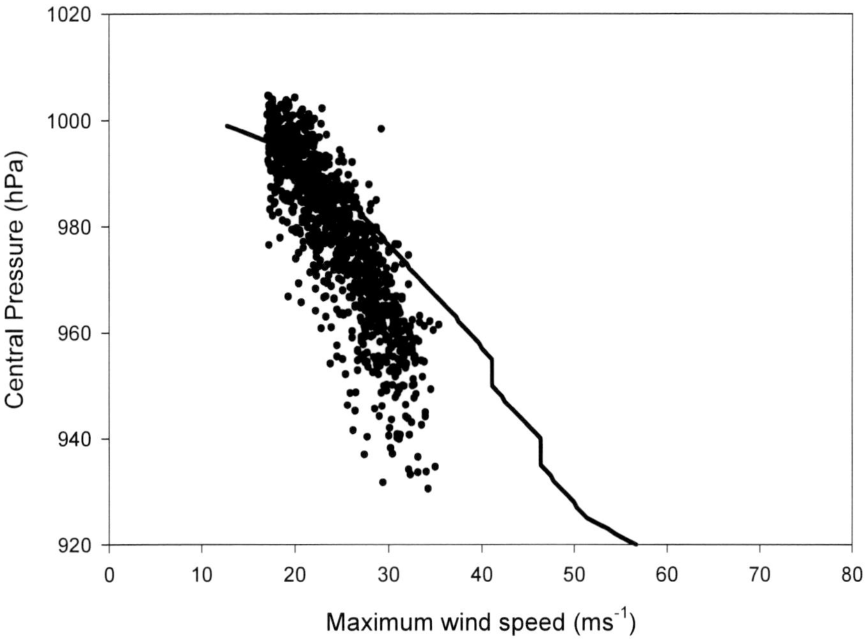

Figure 3. Observed relationship between tropical cyclone maximum wind speed and central pressure (solid line) compared with simulated TC maximum wind speeds and central pressures (dots). Observed relationship interpolated from data given in Bureau of Meteorology [1998]. From Walsh et al. [2004].

As an example, Fig. 3 shows results from the eastern Australian regional climate model simulations of Walsh et al. [2004], at a resolution of 30 km. This compares the relationship between central pressure and maximum wind speed for the simulations and the observations. For a given central pressure, the maximum wind speed in the observations is higher than that simulated, particularly for more intense storms. This implies that the pressure gradients in the simulation are insufficient to generate the high wind speeds observed, even though the central pressures of the simulated storms are reaching quite intense levels. For instance, in the 30-year high resolution simulation of Walsh et al. [2004], there were no storms with maximum 10m wind speeds above 40 ms^{-1}, which is barely hurricane force; in 30 years of observations in this region, one would expect at least 30 such storms. Similarly, in Oouchi et al. [2006], there were no storms with simulated maximum wind speeds greater than 55 ms^{-1}, although their agreement with the observed relationship between central storm pressure and maximum wind speed is better than that shown in Fig. 3. The 18-km regional simulations of Knutson et al. [2007] did not simulate storms more intense than 50 ms^{-1}.

Thus considerable improvements in techniques are needed. One possibility is the use of downscaling techniques from an existing regional climate model simulation. Leslie et al. [2007a] give results from a 50 km variable-resolution model over the region to the east of Australia. This version of the model gives a reasonably good simulation of the observed intensity distribution, although in common with other model simulations it does not simulate the peak intensities of the most intense storms. More recent results [Leslie et al 2007b] employ a new technique to improve the simulation of intensity. Results from a 20-km resolution simulation are "downscaled" using the Holland [1980] vortex model to estimate the peak winds. The resulting intensity distributions are very similar to reality and suggest that this downscaling technique can be employed to gain a better representation of intensity distributions in coarser-resolution models.

Another possibility is the use of multiply-nested grids, as employed in numerical weather prediction models of tropical cyclones [Kurihara et al 1998]. Tropical cyclones are rare events and thus the region over which they might develop does not need to be simulated at very fine resolution all of the time. Since lower-resolution climate models can simulate precursor vortices for tropical cyclones, a grid system could then identify the formation of these vortices and then simulate only the storm region at very fine resolution for the rest of its lifetime. This would combine the advantages of climate modelling with very fine resolution simulations, which would need to be finer than 5 km to capture the structural aspects of tropical cyclones that are important for a realistic simulation of intensity and other processes.

Structure

Like formation and intensity, the simulated structure of tropical cyclones in climate models resembles reality: the maximum wind speeds are at lower levels of the atmosphere, typical of warm core systems and quite different from the structure of mid-latitude cyclones, and the maximum central warm core anomaly is in the upper troposphere [see Frank 1977]. There remain certain deficiences, however. These are illustrated in Fig. 4, which compares the typical radially-average tangential wind structure for the same storm simulated at 125 and 30 km resolution. A number of differences are apparent. The intensity of the storm is greater at the finer resolution, and the maximum wind speed region is closer to the center of the storm: at a resolution 125 km, it is about 250 km, whereas at a resolution of 30 km, it is about 70 km. Both of these are considerably greater than the observed radius of maximum surface winds, which is usually 20-60 km in intense storms [Kimball and Mulekar 2004].

Similar structural differences have been found in recent simulations [e.g. Chauvin et al. 2006; Bengtsson et al. 2007a]. As a result, it is premature to assess the ability of climate models to simulate some of the finer-scale features that have been investigated using high-resolution numerical weather prediction models of tropical cyclones [e.g. Reasor et al 2005]. These might include eye wall replacement cycles, mesoscale vortex mergers, and so on.

Rainfall amounts

One of the most damaging aspects of tropical cyclones is the torrential rainfall that can occur from them, even from storms that are weak or dissipating. Predictions of a general increase in tropical precipitation in warmer world have been made for some time, and the latest consensus predictions of the Intergovernmental Panel on Climate Change (IPCC) also suggest that most tropical regions will experience higher rainfall [Meehl et al. 2007]. Several predictions have also now been made that tropical cyclone rainfall may increase substantially as a result of increased moisture availability [Knutson and Tuleya 1999; Walsh et al. 2004; Yoshimura et al. 2006; Bengtsson et al. 2007b]. Thus an assessment of the current ability of climate models to simulate tropical cyclone rainfall is needed.

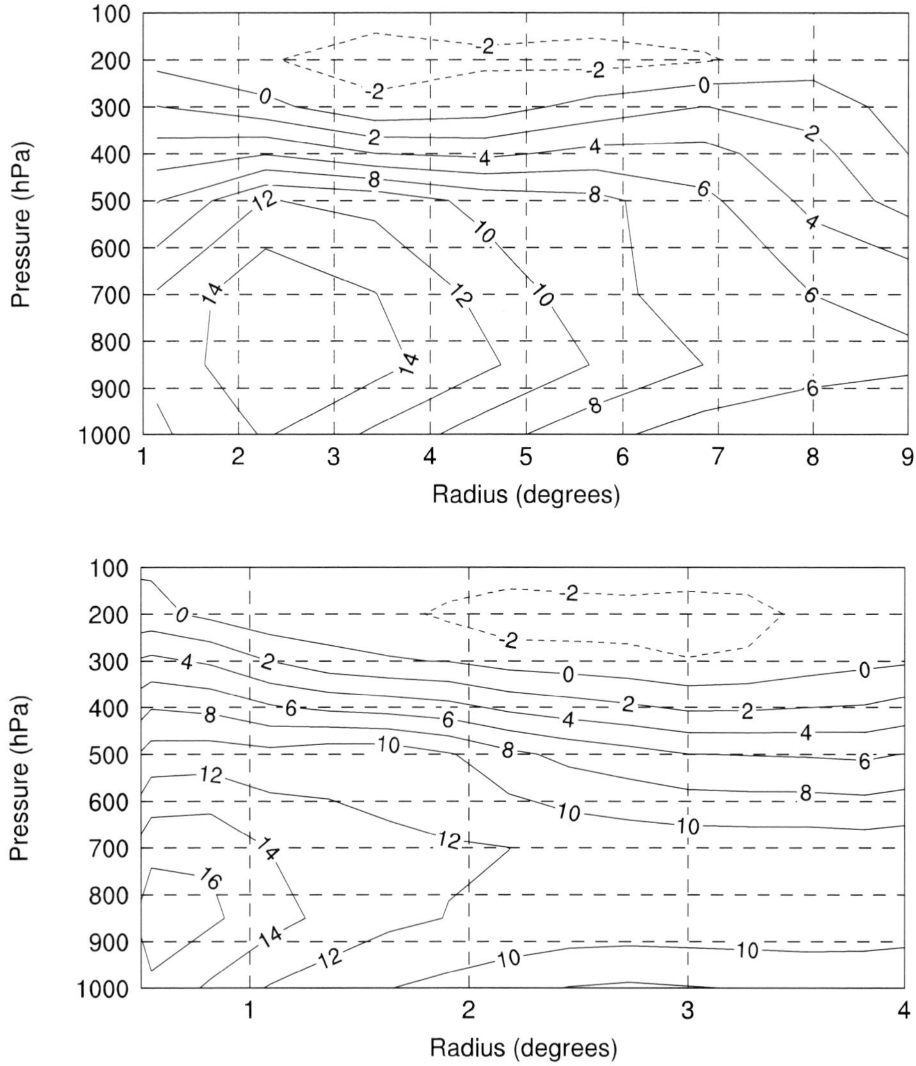

Figure 4. Radially-averaged tangential velocity from a selected storm simulation in a regional climate model, at a resolution of 125 km (top) and 30 km (bottom). After Walsh and Watterson [1997].

The major stumbling block in such an assessment has been the difficulty in obtaining reliable observations of tropical cyclone rainfall, particularly over sparsely-observed tropical ocean regions. The use of satellite estimates of rainfall amounts over these regions supplies a means of observing tropical cyclone rainfall. Lonfat et al. [2004] estimate mean tropical cyclone rainfall from estimates from the Tropical Rainfall Measuring Mission (TRMM) data. These data appear to estimate intense tropical rainfall amounts reasonably well [Huffman et al 2007], although with a considerable amount of scatter, and so can be compared to the simulations of tropical cyclone rainfall in climate models. These comparisons are summarized in Table 2, although the results of the individual studies required some interpretation due to different rainfall metrics being employed in each study. For comparison, the idealized tropical cyclone experiment of Knutson and Tuleya [2004] is also included. The results show that simulated tropical rainfall amounts are generally within the same order of magnitude as the available observations. An exception is the result of Knutson and Tuleya [2004], which is high even given that their storms were specified to be of hurricane intensity. This may be related to the high rainfall bias in the GFDL hurricane model used for this study [Tuleya et al 2007]. The analysis in Table 2 provides an estimate only; a comprehensive study of this issue would require detailed analysis both of original model output and of comparable observations.

Table 2. Estimates of tropical cyclone rainfall amounts simulated by climate models and idealized tropical cyclone experiments. Values are given for approximate typical average rainfall amounts within 100 km of storm center.

Study	Rainfall (mm h^{-1})
Lonfat et al. (2004)	5
Knutson and Tuleya (2004)	30 (hurricane intensity)
Chauvin et al (2006)	3
Yoshimura et al (2006)	3-6
Bengtsson et al. (2007b)	<10

Implications for Prediction of the Effects of Climate Change

The quality of climate model simulations of tropical cyclones has implications for the prediction of their future behavior, as well as for explanations of any current trends detected in observations.

As stated in the introduction, a prerequisite for building confidence in the predictions of climate models is that the model has a good simulation of the variable in question in the current climate. At present, it is difficult to determine exactly how good tropical cyclone simulations of formation are because of the differing detection methods that are used to declare the detected lows as "tropical cyclones". While it is clear that climate models are generating lows that have many of the characteristics of tropical cyclones, including regions of formation and tracks, it is hard to say whether they are being generated at the correct rate in these simulations. This is an important limitation in the validation of model performance and makes predictions of future numbers less confident as a result.

A similar issue arises when climate models are used to simulate climate change that has already occurred, to determine the possible causes of detected changes in phenomena, a process known as detection and attribution [Hegerl et al. 2007]. For example, in the promising results of Knutson et al. [2007], a large portion of the interannual variability of observed tropical cyclone formation in the Atlantic was able to be simulated by a regional climate climate nested within reanalyses. Even so, storms in these simulations were declared to be tropical cyclones by a vorticity-based threshold whose relationship to the actual vorticity of observed tropical cyclones is not clear. It would be useful to make this relationship more quantitative in order to build confidence in such simulations. This is particularly important in the Atlantic basin, where large trends in tropical cyclone numbers and measures of integrated storm intensity have been observed in recent years [Emanuel 2005; Kossin et al 2007]. As the Knutson et al. [2007] simulations are able to capture the increase in storm numbers in the basin over this period, similar simulations could be used as important tools to determine the physical factors causing these observed trends, including the possible contribution of anthropogenic climate change, particularly to increased tropical cyclone intensity.

The quality of intensity predictions is limited by the fact that no climate modelling system, with the recent exception of the downscaled results of Leslie et al. [2007b], has a good simulation of the observed intensity distribution of tropical cyclones. As Bengtsson et al. [2007b] show, finer resolution models with better simulations of intensity can simulate a different magnitude of response to increased global temperatures. Most simulations of future tropical cyclone intensity, though, show increased intensities for the most intense storms, so in conjunction with theoretical predictions [e.g. Emanuel 1987], an improved simulation of tropical cyclone intensities would presumably only lead to a better estimate of the size of the increase, rather than a change of the sign. Similar arguments can be made for the simulation of observed tropical cyclone structure, which affects the quality of the intensity simulations.

Like numbers, the quality of the rainfall simulations of tropical cyclones are a little uncertain, as authors have used different measures of rainfall amount and have in general not compared them systematically with available observations. Nevertheless, given that observed tropical cyclone rainfall amounts clearly increase with increasing intensities [Lonfat et al. 2004], there are good reasons to believe that increased tropical cyclone intensities in a warmer world will be accompanied by increased rainfall amounts.

Model Physics

Current climate models of tropical cyclones run at resolutions where parameterization of convective processes is needed. Nevertheless, with increased computing power, these models may soon be run at resolutions where explicit simulation of cloud processes will become important. Similarly, non-hydrostatic processes will become important as horizontal resolutions reach 5 km. For example, Knutson and Tuleya [2004], in their 9 km resolution idealized experiments with the Geophysical Fluid Dynamics Laboratory (GFDL) hurricane model [Kurihara et al 1998], ran simulations with four different convective parameterizations. They also ran a simulation without convective parameterization, even though, as they point out, 9 km is relatively coarse for a resolved convection simulation. They found that the resolved convection case gave a reasonable simulation of a hurricane, although the multiply-nested grid caused some discontinuities in the simulated precipitation field.

Regional numerical weather prediction model simulations of tropical cyclones have also addressed these issues. Rogers et al [2003] performed multiply-nested simulations down to a resolution of 1.67 km using the nonhydrostatic fifth-generation Pennsylvania State University–NCAR Mesoscale Model [MM5; Grell et al 1994]. No convective parameterization was employed in the innermost meshes. This simulation used an automated moveable fine mesh to follow the cyclone along its track; as indicated above, this technique could be adapted to climate model simulations. Subsequent simulations with this modelling system [Chen et al 2007] have shown that the inner-core structure of a hurricane could be well simulated at a resolution of 1.67 km but not at 5 km. This suggests that genuinely believable climate simulations of hurricane intensity distribution will only be achieved at very fine resolutions. As resolution becomes fine, the importance increases of the realism of the detailed cloud physics as represented in the model.

Few climate simulations of tropical cyclones have been performed with an ocean model that interacts with the tropical cyclone winds, despite the considerable importance of this two-way interaction for tropical cyclone development [e.g. Bender and Ginis 2000]. Some simulations have been performed that have included this effect [Scoccimarro et al 2007; Leslie et al 2007a], and this is certainly needed in future work, as intensities for some simulated storms may be overestimated in a model with fixed SSTs, which may change the intensity response somewhat in a warmer world [Knutson et al 2001; Hasegawa and Emori 2007].

Climate Model Predictions of Global Warming Effects

A detailed review of this topic will not be provided here, but rather the recent review of Walsh [2004] will be briefly updated. The main conclusions of Walsh [2004] were as follows:

- there was no consensus regarding possible changes in tropical cyclone numbers or tracks, although there might be substantial changes in regional formation rates;
- there was considerable evidence that peak tropical cyclone intensities would increase by 5-10% by around 2050, although there was a range of estimates;
- these intensity increases would likely be accompanied by increases in tropical cyclone precipitation; and
- there were likely to be no significant changes in regions of formation.

Since then, a number of simulations [Oouchi et al 2006; Yoshimura et al 2006; Bengtsson et al 2007b; Scoccimarro et al. 2007] have produced decreases in tropical cyclone formation rates in a warmer world, a consequence mostly of increases in tropical atmospheric stability. Even simulations where numbers decrease, though, have usually simulated increases in the peak intensities of cyclones, thus reinforcing this conclusion. Similarly, studies have also generally shown increases in peak tropical cyclone rainfall rates. Thus it appears unlikely that these two predictions will change in the future.

Conclusion

Climate models are demonstrating a constantly improving ability to generate the climatological characteristics of observed tropical cyclones, including numbers, tracks and intensities. The evaluation of their ability to simulate the correct rate of tropical cyclone formation is limited by the lack of genuinely objective intercomparison metrics. A method is proposed that addresses this issue through the development of a resolution-dependent detection criterion based upon low-level wind speed. The use of diagnostic formation parameters is showing increasing potential as a means of making quick estimates of the tropical cyclone-generating potential of a climate model.

Considerable further improvements in resolution will be needed before genuinely good climate simulations are obtained of tropical cyclone structure and wind speed distribution. Multiple-nesting techniques will likely be essential in the short term for achieving this goal, as well as the use of an interactive ocean.

There are now a number of groups working on this problem. It may be that there are enough results to begin the process of detailed intercomparison of climate model simulation of tropical cyclones. This process has led to improved understanding of the large-scale circulation of climate models [e.g. Covey et al 2003] and may do the same for the simulation of tropical cyclones by climate models.

Acknowledgments

The author would like to thank the Meteorological Society of Japan, the American Meteorological Society and Springer Publishing for permission to reproduce the figures used here. The author would also like to thank the University of Melbourne for providing support for this work.

References

Anthes, R.A. *Tropical Cyclones: Their Evolution, Structure and Effects*. American Meteorological Society, Boston, MA, 1982.
Bender, M. A.; Ginis, I. *Mon. Wea. Rev*., 2000, 128, 917-946.
Bengtsson, L.; Böttger, H.; Kanamitsu, M. *Tellus*. 1982, 34, 440-457.
Bengtsson, L.; Botzet, M.; Esch, M. *Tellus*. 1995, 47A, 175-196.
Bengtsson L.; Botzet, M.; Esch, M. *Tellus*. 1996, 48A, 57-73.
Bengtsson, L.; Hodges, K.I.; Roeckner, E. *J. Climate*. 2006, 19, 3518-3543.
Bengtsson, L.; Hodges, K.I.; Esch, M. *Tellus*. 2007a, in press.
Bengtsson, L.; Hodges, K.I.; Esch, M.; Keenlyside, N.; Kornbleuh, L.; Luo, J.-J.; Yamagata, T. submitted, 2007b.
Bureau of Meteorology. *Tropical Cyclone Warning Directive*. Bureau of Meteorology, Brisbane, Australia, 1998.
Camargo, S.J; Barnston, A.G.; Emanuel, K.A. *Tellus*, 2007, in press.
Camargo, S.J; Barnston, A.G.; Zebiak, S.E. *Tellus*. 2005, 57A, 589-604.
Chauvin, F.; Royer, J.-F.; Déqué, M. *Clim. Dyn*. 2006, 27, 377-399.
Chen, S.; Price, J.F.; Zhao, W.; Donelan, M.A.; Walsh, E.J. *Bull. Amer. Meteorol. Soc*. 2007, 88, 311-317.
Covey, C.; AchutaRao, K.M.; Cubasch, U.; Jones, P.; Lambert, S.J.; Mann, M.; Phillips, T.J.; Taylor, K.E. *Glob. Planet. Change*. 2003, 37, 103-133.
Dickinson, R.E.; Errico, R.M.; Giorgi, F.; Bates, G.T. *Clim. Change*, 1989, 14, 383-422.
Emanuel, K.A. *Nature*. 1987, 326, 483-485.
Emanuel, K.A. *Nature*. 2005, 436, 686-688.
Emanuel, K.A. *J. Climate*. 2006, 19, 4797-4802.
Frank, W.M. *Mon. Wea. Rev*. 1977, 105, 1119-1135.
Gray, W.M. *Tropical Cyclone Genesis*. Dept. of Atm. Sci. Paper No. 234, Colorado State Univ., Ft. Collins, CO, 1975.
Grell, G.A.; Dudhia, J.; Stauffer, D.R. 1994, NCAR Tech. Note NCAR/TN-398 1 STR.
Haarsma, R.J.; Mitchell, J.F.B.; Senior, C.A. *Clim. Dyn*. 1993, 8, 247-527.
Hasegawa, A.; Emori, S. 2007, submitted to *Geophys. Res. Letters*.

Hegerl, G.C., Zwiers, F.W.; Braconnot, P.; Gillett, N.P.; Luo, Y.; Marengo Orsini, J.A.; Nicholls, N.; Penner, J.E.; Stott, P.A. In *Climate Change* 2007*: The Physical Science Basis*; Solomon, S., D.; Qin, D.; Manning, M.; Chen, Z.; Marquis, M.; Averyt, K.B.; Tignor, M.; Miller, H.L.; Eds.. Cambridge University Press, Cambridge, United Kingdom and New York, NY, USA, 2007.

Holland, G. *Mon. Wea. Rev.* 1980, 108, 1212-1218.

Hubbert, G.D.; Holland, G.J.; Leslie, L.M.; Manton, M.J. *Wea. Forecast.* 1991, 6, 86-97.

Huffman, G.J.; Adler, R.F.; Bolvin, D.T.; Gu, G.; Nelkin, E.J.; Bowman, K.P.; Hong, Y.; Stocker, E.F.; Wolff, D.B. *J. Hydrometeor.* 2007, 8, 38-55.

Kimball, S.K.; Mulekar, M.S. *J. Climate*. 2004, 17, 3555-3575.

Knutson, T.R.; Sirutis, J.J.; Garner, S.T.; Held, I.M.; Tuleya, R.E. *Bull. Amer. Meteorol. Soc.* 2007, in press.

Knutson, T.R.; Tuleya, R.E. *Clim. Dyn.* 1999, 15, 503-519.

Knutson, T.R.; Tuleya, R.E.; Kurihara, Y. *Science*. 1998, 279, 1018-1020.

Knutson, T.R.; Tuleya, R.E.; Shen, W.; Ginis, I. *J. Climate*. 2001, 14, 2458-2468.

Knutson, T. R.; Tuleya, R.E.; *J. Climate*. 2004, 17, 3477-3495.

Kossin, J.P.; Knapp, K.R.; Vimont, D.J.; Murnane, R.J.; Harper, B.A. *Geophys. Res. Letters.* 2007, 34, L04815, doi:10.1029/2006GL028836.

Krishnamurti, T.N.; Correa-Torres, R.; Latif, M.; Daughenbaugh, G. *Tellus*. 1998, 50A, 186-210.

Kurihara, Y.; Tuleya, R.E.; Bender, M.A. *Mon. Wea. Rev*. 1998, 126, 1306-1322.

Leslie, L.M.; Buckley, B.W.; Karoly, D. *Meteorol. Atm. Phys*. 2007a, in press.

Leslie, L.M.; Leplastrier, M.; Buckley, B.W. 2007b, presented at 1st International Summit on Hurricanes and Climate Change. Crete, May 27-June 1, 2007.

Lonfat, M.; Marks, F.D. Jr.; S.S. Chen. *Mon. Wea. Rev*. 2004, 132, 1645-1660.

McDonald, R.E.; Bleaken, D.G.; Cresswell, D.R.; Pope, V.D.; Senior, C.A. *Clim. Dyn.* 2005, 25, 19-36.

McGregor, J.L. *C-CAM: Geometric Aspects and Dynamical Formulation. CSIRO Atmospheric Research Technical Paper* **70**, Aspendale, Australia, 2005.

McGuffie, K.; Henderson-Sellers, A. A Climate Modelling Primer, 3rd ed. John Wiley and Sons: 2005.

Manabe, S.; Holloway, J.L. Jr.; Stone, H.M. *J. Atmos. Sci.* 1970, 27, 580-613.

Manabe, S.; Broccoli, A.J. *Geophys. Res. Letters*. 1990, 17, 1979-1920.

Meehl, G.A.; Stocker, T.F.; Collins, W.D.; Friedlingstein, P.; Gaye, A.T.; Gregory, J.M.; Kitoh, A.; Knutti, R.; Murphy, J.M.; Noda, A.; Raper, R.C.B.; Watterson, I.G.; Weaver, A.J.; Zhao, Z.-C. In *Climate Change* 2007*: The Physical Science Basis*; Solomon, S., D.; Qin, D.; Manning, M.; Chen, Z.; Marquis, M.; Averyt, K.B.; Tignor, M.; Miller, H.L.; Eds.. Cambridge University Press, Cambridge, United Kingdom and New York, NY, USA, 2007.

Oouchi, K.; Yoshimura, J.; Yoshimura, H,; Mizuta, R. *J. Meteorol. Soc. Jpn.* 2006, 84, 259-276.

Pezza, A.B..; Simmonds, I. G*eophys. Res. Letters.* 2005, 32, Art. No. L15712.

Randall, D.A.; Wood, R.A.; Bony, S.; Colman, R.; Fichefet, T.; Fyfe, F.; Kattsov, V.; Pitman, A.; Shukla, J.; Srinivasan, J.; Stouffer, R.J.; Sumi, A.; Taylor, K.E. In *Climate Change* 2007*: The Physical Science Basis*; Solomon, S., D.; Qin, D.; Manning, M.; Chen, Z.;

Marquis, M.; Averyt, K.B.; Tignor, M.; Miller, H.L.; Eds.. Cambridge University Press, Cambridge, United Kingdom and New York, NY, USA, 2007.

Reasor, P.D.; Montgomery, M.T.; Bosart, L.F. *J. Atmos. Sci.* 2005, 62, 3151-3171.

Rogers, R.; Chen, S.; Tenerelli, J.; Willoughby, H. *Mon. Wea. Rev.* 2003, 131, 1577-1599.

Royer, J.-F.; Chauvin, F.; Timbal, B.; Araspin, P.; Grimal, D. *Clim. Change.* 1998, 38, 307-343.

Ryan, B.F.; Watterson, I.G.; Evans, J.L. *Geophys. Res. Letters*, 1992, 19, 1831-1834.

Scoccimarro, E; Gualdi, S.; Navarra, A. 2007, presented at 1st International Summit on Hurricanes and Climate Change. Crete, May 27-June 1, 2007.

Sugi, M.; Noda, A.; Sato, N. *J. Meteorol. Soc. Jpn.* 2002, 80, 249-272.

Tuleya, R.E.; DeMaria, M.; Kuligowski, R.J. *Wea. Forecast.* 2007, 22, 56-70.

Tsutsui, J.-I. *J. Meteorol. Soc. Jpn.* 2002, 80, 45-65.

Tsutsui, J.-I., Kasahara, A. *J. Geophys. Res.* 1996, 101, 15013-15032.

Walsh, K. *Clim. Res.* 2004, 27, 77-83.

Walsh, K.J.E.; Katzfey, J.J. *J. Climate.* 2000, 13, 1116-1132.

Walsh, K.J.E.; Ryan, B.F. *J. Climate*, 2000, 13, 3029-3036.

Walsh, K.J.; Watterson, I.G. *J. Climate.* 1997, 10, 2240-2259.

Walsh, K.J.E.; Nguyen, K.-C.; McGregor, J.L. *Clim. Dyn.* 2004, 22, 47-56.

Walsh, K.; Fiorino, M.; Landsea, C.; McInnes, K. *J. Climate.* 2007, 20, 2307-2314.

Washington, W.M.; Parkinson, C.L.. *An Introduction to Three-Dimensional Climate Modeling*, 2nd ed. Palgrave Macmillan, 2005.

Wu, G.; Lau, N.-C. *Mon. Wea. Rev.* 1992, 120, 958-977.

Yoshimura, J.; Sugi, M.; Noda, A. Proceedings of the 23rd Conference on Hurricanes and tropical meteorology, 10-15 January 1999, Dallas, American Meteorological Society, pp 1081-1084, 1999.

Yoshimura, J.; Sugi, M.; Noda, A. *J. Meteorol. Soc. Jpn.* 2006, 84, 405-428.

INDEX

A

B

C

D

E

H

I

M

S

T

U

V